The Biology of Schwann Cells

Schwann cells are a diverse group of cells formed from neural crest cells. They are essential components of the peripheral nerves of both vertebrate and invertebrate nervous systems. The diversity of Schwann cell subsets and function is seen in those Schwann cells that form myelin – that uniquely specialised part of the plasma membrane that spirals around axonal lengths to myelinate the peripheral nerves.
The Biology of Schwann Cells concentrates on Schwann cells of mammals and in particular humans. It covers the distinction between compact and non-compact myelin in depth, along with the perisynaptic cells which form the partnership between nerve terminals and muscle fibre. Developmental aspects are discussed alongside differentiation, together with the genetics of Schwann cells in health and disease. With chapters from world-renowned experts, this book is aimed at postgraduates and researchers in neuroscience and neurology, and anyone involved in the study of peripheral nerves.

PATRICIA J. ARMATI is an Associate Professor and Co-Director of the Nerve Research Foundation, Department of Medicine at the University of Sydney, Australia, with a long-standing research interest in the biology of Schwann cells.

The Biology of Schwann Cells

Development, Differentiation and Immunomodulation

Edited by

PATRICIA ARMATI
The University of Sydney, Australia

CAMBRIDGE UNIVERSITY PRESS
Cambridge, New York, Melbourne, Madrid, Cape Town, Singapore, São Paulo

Cambridge University Press
The Edinburgh Building, Cambridge CB2 2RU, UK

Published in the United States of America by
Cambridge University Press, New York

www.cambridge.org
Information on this title: www.cambridge.org/9780521850209

© Cambridge University Press 2007

This publication is in copyright. Subject to statutory exception
and to the provisions of relevant collective licensing agreements,
no reproduction of any part may take place without
the written permission of Cambridge University Press.

First published 2007

Printed in the United Kingdom at the University Press, Cambridge

A catalogue record for this publication is available from the British Library

Library of Congress Cataloging in Publication data

The biology of Schwann cells : development, differentiation, and
immunomodulation / edited by Patricia Armati.
    p. ; cm.
Includes bibliographical references.
ISBN-13: 978-0-521-85020-9 (hardback)
ISBN-10: 0-521-85050-7 (hardback)
 1. Neuroglia. I. Armati Patricia J.
[DNLM: 1. Schwann Cells—physiology. WL 102 B6164 2007] II. Title.

QP363.2.B563 2007
611'.0188—dc22

                                               2006037782

ISBN-13 978-0-521-85020-9   hardback
ISBN-10 0-521-85020-7   hardback

Cambridge University Press has no responsibility for the persistence or
accuracy of URLs for external or third-party internet websites referred to in
this publication, and does not guarantee that any content on such websites is,
or will remain, accurate or appropriate.

Dedication

This book is dedicated to all the 'Friends of the Schwann cell': Theodore Schwann who first named the cell; Richard and Mary Bunge and the unsung Patrick Wood; and also John Pollard my husband, whose ever-enquiring mind, excellence in research and dedication to all those with peripheral nerve diseases inspired my interest in the extraordinary tale of the Schwann cell.

Contents

	Preface	*page* ix
	Contributors	xi
1	Introduction to the Schwann cell	1
	EMILY MATHEY AND PATRICIA J. ARMATI	
2	Early events in Schwann cell development	13
	RHONA MIRSKY AND KRISTJÁN R. JESSEN	
3	The molecular organisation of myelinating Schwann cells	37
	EDGARDO J. ARROYO AND STEVEN S. SCHERER	
4	The role of the extracellular matrix in Schwann cell development and myelination	55
	MARIA LAURA FELTRI AND LAWRENCE WRABETZ	
5	The biology of perisynaptic (terminal) Schwann cells	72
	CHIEN-PING KO, YOSHIE SUGIURA AND ZHIHUA FENG	
6	Cytokine and chemokine interactions with Schwann cells: the neuroimmunology of Schwann cells	100
	ROBERT P. LISAK AND JOYCE A. BENJAMINS	
7	Schwann cells as immunomodulatory cells	118
	BERND C. KIESEIER, WEI HU AND HANS-PETER HARTUNG	
8	Mutations in Schwann cell genes causing inherited neuropathies	126
	MICHAEL E. SHY, JOHN KAMHOLZ AND JUN LI	
9	Guillain–Barré syndrome and the Schwann cell	158
	RICHARD A. C. HUGHES	

| 10 | Chronic idiopathic demyelinating polyneuropathy and Schwann cells | 171 |

JOHN D. POLLARD

| References | 185 |
| Index | 247 |

Colour plate section appears between pages 50 and 51

Preface

It is now over 200 years since Theodore Schwann first described the cell which bears his name. Such early descriptions of nervous system components were done without the powerful microscopes we have today, yet Schwann and Ramon Y. Cajal made foundation observations which still stand. Cajal's papers, especially, show the power of careful observation, an essential element of good science.

The Schwann cell has been historically underrated and poorly understood. In particular, the myelin-forming Schwann cells or their myelin are still often referred to as a simple 'sheath' for the neuron. However, Schwann cells in all their complexity form essential partnerships with neurons, and muscles. This is of particular relevance in the case of the myelin-forming Schwann cell, an enormous cell that expresses unique molecules and complex relationships related to maintenance of the compact and non-compact myelin regions of its plasma membrane. Schwann cells have other complex interactions, not least of which are found where nerve terminals and muscle fibres form the tripartite synapse in association with the perisynaptic Schwann cells. There are also the poorly understood satellite cells that surround the dorsal root ganglion nerve cell bodies, and of course the complexity of non-myelinated Schwann cells and their axonal associations.

It may be that the histopathological prominence of abnormalities of compact myelin has focussed research on this region of the Schwann cell. This is shown by the historical concentration on disturbance of compact myelin in diseases of the peripheral nervous system such as Guillain–Barré syndrome, chronic inflammatory demyelinating polyneuropathy (CIDP) and Charcot–Marie–Tooth disease. The study of the basic biology of this cell is therefore increasingly recognised as an essential element in understanding the development, function and potential for repair of the nervous system, including the central nervous system.

With the development of electron microscopy, molecular biology and genetic techniques, proteomics and other technologies pushing the boundaries of our knowledge, the unravelling of the astounding complexity of all cells including the Schwann cell is well underway. It is therefore timely that the current understanding of this cell be gathered into a book such as *The Biology of Schwann Cells*.

I would like to thank Dr Ariel Arthur, University of Sydney, Dr Martin Griffiths, Cambridge University Press, and 'my son' Mr Damien Pembroke, for their editorial assistance, encouragement and expertise.

Patricia Armati
October 2006

Contributors

Patricia Armati Ph.D.
Associate Professor, Nerve Research Foundation, Blackburn Building D06,
The University of Sydney, NSW 2006, Australia

Edgardo J. Arroyo Ph.D.
Research Associate, University of Pennsylvania, Department of Neurology, Rm 460 Stemmler Hall, 3600 Hamilton Walk, Philadelphia, PA 19104, USA

Joyce A. Benjamins Ph.D.
Professor and Associate Chair for Research, Department of Neurology, Wayne State University School of Medicine, Detroit, MI 48201, USA

Maria Laura Feltri M.D.
San Raffaele Scientific Institute, DIBIT, Via Olgettina 58, 20132 Milano, Italy

Zhihua Feng B.S.
Graduate Student, Neurobiology Section, Department of Biological Sciences, USC, Neuroscience Graduate Program, University of Southern California, 3641 Watt Way, Los Angeles, CA 90089-2520, USA

Hans-Peter Hartung M.D.
Professor, Department of Neurology, Heinrich-Heine-Universitat, Moorenstrasse 5, Dusseldorf, D40225, Germany

Wei Hu M.D.
Department of Neurology, Heinrich-Heine University, Moorenstrasse 5, Dusseldorf, D40225, Germany

Contributors

Richard A.C. Hughes M.D.
Professor of Neurology, Department of Clinical Neuroscience, King's College, London, SE1 9UL, UK

Kristján R. Jessen, Ph.D.
Professor of Developmental Neurobiology, Department of Anatomy and Developmental Biology, UCL, Gower Street, London WC1E 6BT, UK

Bernd C. Kieseier M.D.
Professor of Neurology, Heinrich-Heine-Universitat, Moorenstrasse 5, Dusseldorf, D40225, Germany

Chien-Ping Ko Ph.D.
Professor, Neurobiology Section, Department of Biological Sciences, University of Southern California, 3641 Watt Way, Los Angeles, CA 90089-2520, USA

Jun Li M.D., Ph.D.
Associate Professor, Division of Neuromuscular Disease, Department of Neurology, Wayne State University School of Medicine, 4201 St Antoine, UHC-8D, Detroit, MI 48201, USA

Robert P. Lisak, M.D.
Parker Webber Chair in Neurology, Professor and Chair of Neurology, Professor of Immunology and Microbiology, Wayne State University School of Medicine, 4201 St Antoine, UHC-8D, Detroit, MI 48201, USA

Emily Mathey, Ph.D.
Research Fellow, Department of Medicine and Therapeutics, Institute of Medical Sciences, University of Aberdeen, Foresterhill, Aberdeen AB25 2ZD, UK

Rhona Mirsky, Ph.D.
Professor of Developmental Neurobiology, Department of Anatomy and Developmental Biology, UCL, Gower Street, London WC1E 6BT, UK

John D. Pollard, M.B. B.S., Ph.D.
Bushell Professor of Medicine, Blackburn Building D06, The University of Sydney, NSW 2006, Australia

Steven Scherer M.D., Ph.D.
William N. Kelley Professor of Neurology, The University of Pennsylvania School of Medicine, Room 460 Stemmler Hall, 36th Street and Hamilton Walk, Philadelphia, PA 19104-6077, USA

Michael Shy, M.D.
Professor of Neurology, Molecular Medicine and Genetics Department, and Department of Neurology, Wayne State University School of Medicine, Detroit, MI 48201, USA

Yoshie Sugiura Ph.D.
Research Assistant, Department of Biological Sciences, Neuroscience Graduate Program, University of Southern California, 3641 Watt Way, Los Angeles, CA 90089-2520, USA

Lawrence Wrabetz M.D.
San Raffaele Scientific Institute, DIBIT, Via Olgettina 58, 20132 Milano, Italy

1

Introduction to the Schwann cell

EMILY MATHEY AND PATRICIA J. ARMATI

THEODOR SCHWANN 1810−1882

The Schwann cell is named in honour of the German physiologist Theodor Schwann (1810−1882, Figure 1.1) who is now acknowledged as the founder of modern histology. In addition to describing the Schwann cell, he made numerous contributions to the fields of biology, physiology and histology − not least as one of the instigators and main advocates of cell theory. The cell theory defined the cell as the base unit of all living organisms, and had great influence on the study of both plants and animals. The cell theory was radical for the time and irrevocably discredited Vitalism, the mainstream belief that life was attributed to a vital force. Among other things, Schwann is known for recognising that the crystals seen during fermentation, first reported by Leeuwenhoek in 1680, were in fact living organisms; although it was not until Pasteur in 1878 wrote to Schwann acknowledging this observation that Schwann's finding was accepted. In fact, Pasteur's germ theory stems from Schwann's work in which he showed that microorganisms were required for the putrefaction of meat.

Schwann spent his undergraduate years at the University of Bonn and then the equivalent of postgraduate study in Wuerzburg and Berlin. Schwann was appointed Professor of Anatomy at Louvain in 1839. In 1848 he moved to the Chair of Anatomy in Liege. In a biography of Schwann (Causey 1960), Causey reported that he avoided the strife of scientific controversy and appears to have risen above petty jealousies.

During his time in Louvain, Schwann described the nucleus in animal cells and defined the nucleus as being important in animal development. His studies were made in the dorsal horn cells. Schwann did not differentiate between the Schwann cell plasma membrane

Figure 1.1 Theodor Schwann.

and the Schwann cell itself. This has led to the interchangeable and confusing nomenclature of neurilemma and Schwann cell membrane. He worked at a time when microscopy was in its infancy, but with foresight recognised that his observations would in the future be verified or discounted. His contribution to neurohistology continued with his observations that nerve cells were ensheathed by cells he considered to be secondary nerve cells. Nevertheless, to this day they bear his name – the Schwann cell.

THE SCHWANN CELL – MORE THAN MEETS THE EYE

Throughout most of the history of neuroscience, neuroglial cells in both the peripheral (PNS) and the central (CNS) nervous systems have been regarded merely as the 'glue' that physically and metabolically holds the nervous system together. Neuroglia have historically been presented as passive bystanders; however, it is now clear that they are major players alongside the neurons of the PNS and the CNS. Ironically, it is the efficiency and blatancy with which the Schwann cell can produce the specialised unique and complex spirals of myelin membrane that has resulted in it being perceived as an axonal comfort blanket, with little regard for its more subtle but essential roles in the operation of the PNS. Such a biased view of the Schwann cell meant that for a long time our knowledge of its other equally important functions remained rudimentary. However, the interdependence between the Schwann cell and the neuron underpins the functioning of the entire PNS, and the fates of these two cell types are inextricably entwined (like star-crossed lovers) so that it is no longer valid to consider Schwann cells as passive support cells. This book challenges many of the preconceived ideas about the Schwann cell and highlights the importance of this cell to the functioning of the PNS in health, in disease and in repair following damage. In addition, as described in Chapter 7, current data suggest that Schwann cells can perform the entire spectrum of an immune response.

All neurons in the PNS are in intimate physical contact with Schwann and satellite cells, regardless of whether they are myelinated or unmyelinated, sensory or autonomic. All axons of the peripheral nerves are ensheathed by rows of Schwann cells, in the form of either one Schwann cell to each axonal length, or in Remak bundles, formed when an individual Schwann cell envelopes lengths of multiple unmyelinated axons (Figures 1.2, 1.3 and 1.4b).

There is now a large body of evidence that defines a multitude of Schwann cell functions that are not related to myelination (Lemke 2001). This uncoupling of myelin-associated functions from other Schwann cell roles emphasises the essentially symbiotic relationship between nerve cells and Schwann cells, where each is dependent on the other for normal development, function and maintenance. For example, it is the axon that controls the initiation of myelination, the number of myelin lamellae and the maintenance of the complex Schwann cell organisation (Michailov *et al.* 2004). However, it is the

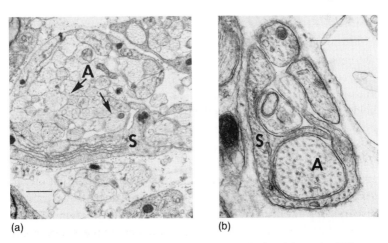

Figure 1.2 Ensheathed axonal bundles (Crawford and Armati, 1982).

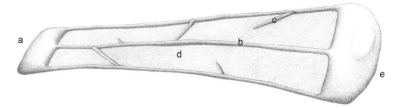

Figure 1.3 A diagrammatic representation of an unrolled Schwann cell showing (a) the inner mesaxon which ensheathes the axolemma, (b) the Schmidt Lanterman incisures, (c) The transverse processes, (d) the compacted regions of myelin, and (e) the outer mesaxon which rolls and ensheathes the whole Schwann cell/myelin complex and around which is the Schwann cell secreted basal lamina.

Schwann cell that regulates axonal diameter, neurofilament spacing and phosphorylation (Hsieh *et al.* 1994), and the clustering of ion channels at the node of Ranvier in myelinated axons (Poliak and Peles 2003). The Schwann cell and its extracellular basal lamina are also involved in axonal regeneration and the guidance of axons to their target destinations (Nguyen *et al.* 2002; Son and Thompson 1995b). Furthermore, Schwann cells have the capacity to interact with cells from outside the nervous systems, as evidenced by their well-established ability to communicate with cells of the immune system through the expression of MHC class II molecules (Armati *et al.* 1990; Armati and Pollard 1996).

Even though all Schwann cells are of neural crest lineage, Schwann cells in the mature PNS can be further categorised for convenience by their morphology, antigenic phenotype, biochemistry and anatomical location. These categories are (i) myelinating (MSCs), (ii) nonmyelinating (NMSCs), (iii) perisynaptic Schwann cells (PSCs) of the neuromuscular junction (Corfas *et al.* 2004) and (iv) satellite cells that ensheathe the cell bodies of sensory neurons (Hanani 2005). These different types of Schwann cells and their anatomical location are shown schematically in Figure 1.4.

Myelinating Schwann cells

Myelinating Schwann cells are the best-characterised cells of all Schwann cell categories, as there has been extensive research on myelination during development, and in demyelinating diseases where disruption of the specialised Schwann cell myelin membrane has been the defining characteristic. To some extent this concentration

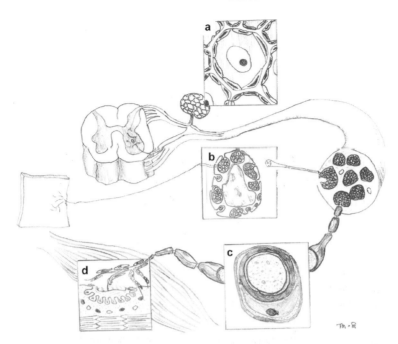

Figure 1.4 Locations of the Schwann cell subsets in the PNS. (a) Satellite Schwann cells are found in the dorsal root ganglia where they associate with the neuronal cell bodies of sensory neurons. (b) Non-myelin-forming Schwann cells are found in mixed peripheral nerve fibres and can ensheath up to ten axonal lengths. (c) Myelinating Schwann cells are also found in mixed peripheral nerve fibres but only myelinate one axonal length. (d) Perisynaptic Schwann cells are located at the neuromuscular junction where they not only enwrap the axonal length but also extend processes to encompass the synapse.

on myelin has diverted us from understanding this huge cell. Myelin-forming Schwann cells have a profound impact in both the healthy and diseased nervous system, not only on axonal conduction but also on many properties of the axons themselves. While Chapters 2 and 3 discuss in detail myelin-forming cells, it is salutary to consider the size and complexity of these Schwann cells.

Schwann cells in the human sciatic nerve can make as many as 100 spiral turns around an axonal length, so that their longitudinal length far exceeds the diameter of their associated axonal length (Webster 1971). To put this in perspective, if one unwrapped Schwann cell with 100 spirals of compact myelin membrane from a hypothetical axonal length (diameter 6 mm), the Schwann cell would be trapezoid and approximately 39 m in length. An axon of the leg, for example, could be 1 m in length and have as many as 100 000 Schwann cells associated with it. Clearly, the formation of so much myelin requires the synthesis of a vast amount of specialised membrane, the expression of a new array of genes, and complex cytoskeletal reorganisation, as discussed in Chapter 3.

With the advent of high throughput technologies, such as neuroproteomics (Kim *et al.* 2004), functional genomics (Bult *et al.* 2004), siRNA (Holen and Mobbs 2004) and improved methods for culturing neural cells, the workings of the molecular machinery required to achieve this intricate remodelling are only now beginning to be understood. Questions such as what determines whether an axon becomes myelinated by a Schwann cell (Taveggia *et al.* 2005) and how the number of myelin spiral lamellae is regulated (Michailov *et al.* 2004) are now being answered. Chan *et al.* (2006) have found that Par-3 (a member of the Par family of adaptor proteins) is localised within the myelin-forming Schwann-cell–axolemmal interface prior to the unidirectional spiralling of the Schwann cell around the axon. Inhibition of the asymmetrically expressed Par-3 inhibits myelination.

As well as the compacted myelin membranes of the Schwann cell, there are Schmidt–Lanterman incisures, cytoplasm-containing channels spiralling within the myelin membranes (Hall and Williams 1970). They contain the complete range of cytoplasmic components and are in continuity with the Schwann cell cytoplasm of the outer and inner mesaxonal regions (Fannon *et al.* 1995; Ghabriel and Allt 1981; Arroyo and Scherer 2000). These components of the myelin-forming Schwann cells are considered to be essential for development, growth, maintenance and turnover of myelin. They contain adherens junctions, including E-cadherin, catenins and F-actin. Tricaud *et al.* have further explored the role of adherens junctions, myelin architecture and their morphogenetic role in Schwann cells (Tricaud *et al.* 2005). They report that disturbance of the adherens junctions in mouse Schwann cells has a direct effect on the interface between compact and non-compact myelin, as discussed by Scherer and Arroyo in Chapter 3. These important but to date neglected components of the Schwann cell are also discussed in Chapter 8.

Non-myelin-forming Schwann cells

In a mixed peripheral nerve unmyelinated fibres outnumber myelinated fibres by a ratio of three or four to one (Jacobs and Love 1985). For example, a transverse section of a human sural nerve contains approximately 8000 myelinated fibres per mm^2, whereas the unmyelinated axons number 30 000 per mm^2. Most of these unmyelinated fibres are small-diameter axons of C fibres originating from sensory ganglia and axons from sympathetic neurons. Little attention has been given to the associated non-myelin-forming Schwann cells (Kennedy 2004). However, all of these unmyelinated fibres are in close contact with non-myelin-forming Schwann cells, which envelope a number of axonal lengths to form Remak bundles. The number of axons within a Remak bundle ranges from one to more than 10, and each axon is encircled by thin extensions of the Schwann cell (Berthold et al. 2005). Although the physiological roles of non-myelin-forming Schwann cells are poorly defined, it is becoming apparent that they are vital for the function and maintenance of unmyelinated axons and also necessary for pain sensation. This has been demonstrated in transgenic mice that have disrupted communication between unmyelinated axons and their non-myelin-forming Schwann cell partners. When communication between axons and adult non-myelinating Schwann cells was specifically disrupted, the mice began to lose their sensory perception and could no longer respond appropriately to hot and cold stimuli. Moreover, there was aberrant proliferation of non-myelinating Schwann cells as well as loss of unmyelinated C fibres (Chen et al. 2003). This study emphasises the fundamental need for communication between Schwann cells of all types and the axons they associate with. Although studies directed specifically towards non-myelin-forming Schwann cells are rare, they are becoming more frequent and important in uncoupling essential roles of the Schwann cell that are distinct from myelination.

Satellite cells

Satellite cells are of the same neural crest lineage as Schwann cells and the sensory neurons in the dorsal root ganglia (DRG), and have a distinct partnership with the neuronal cell they ensheathe when compared to other neuroglial interactions. Satellite Schwann cells associate directly with the cell body rather than the axon of a neuron, so that each neuronal cell body in the DRG is completely surrounded by a cap of several satellite Schwann cells to form a discrete anatomical unit. Although little is known about the physiology of satellite Schwann cells, they are likely to share many functions in common with the other classes of Schwann cells. Current information suggests that they have specific roles given their location and unique interaction with the sensory neuronal cell bodies they envelope. The proximal and distal arms of the peripheral nerves have a blood/nerve barrier (BNB) where the endothelial cells form tight junctions similar to that of the central nervous system blood/brain barrier (BBB). These barriers shield the nervous system from noxious blood-borne cells and molecules. However, the dorsal roots, the dorsal root ganglia and the neuromuscular junction (or tripartite synapse discussed in the next section) have an incomplete BNB. This lack of a BNB is in part compensated for by the satellite cells, which have the ability to regulate the neuronal environment by acting as a partial diffusion barrier. On the whole, satellite cells have barely been examined in terms of their specialisations and unique interactions with the neurons they support.

Schwann cells of the neuromuscular junction – the perisynaptic Schwann cells

The Schwann cell as an independent and essential player in the PNS is further exemplified at the neuromuscular junction (NMJ). The NMJ had long been considered as a synapse with two active components: the presynaptic nerve terminal and the postsynaptic muscle membrane. However, in recent years the perisynaptic Schwann cell (PSC) has gained recognition as the third component of what is now termed a tripartite synapse (Koirala *et al.* 2003). Although it is well established that Schwann cells regulate the extracellular neuronal environment by buffering ions and taking up neurotransmitters, it now appears that they can also contribute to the formerly archetypal neuronal function of neurotransmission (Robitaille 1998; Auld and Robitaille 2003a). PSCs encapsulate the motor end-plate (where the motor neurons synapse on the muscle and form an integral part of the neuromuscular synapse (Hughes *et al.* 2005a)), and not only are located in an ideal position to monitor synaptic activity but are also fully equipped to receive and respond to signals at this tripartite synapse (Araque *et al.* 1999; Auld and Robitaille 2003a). Although these specialised PSCs express many proteins in common with myelinating Schwann cells, they can be distinguished by their increased expression of neurotransmitter receptors and ion channels. Since PSCs are not directly connected to other synaptic elements their mode of action appears to be via a synapse–PSC–synapse regulatory loop (Auld and Robitaille 2003a), whereby PSCs are activated by synaptic activity to provide an inhibitory or excitatory feedback signal (Rochon *et al.* 2001). It is now accepted that PSCs make a substantial contribution to the NMJ, resulting in a more stable and efficient synapse. In addition to modulating NMJ function, PSCs are also essential for its formation and maintenance both during development and during repair following injury. Furthermore, denervated neurons can only reencounter the muscle end-plate when physically guided by Schwann cells.

SCHWANN CELLS RESPOND TO INJURY

Schwann cells have a pivotal role in response to PNS injury. The PNS regenerative powers are in part due to intrinsic properties of Schwann cells that encourage spontaneous regeneration and have been the focus of much investigation. PNS axonal regeneration occurs through the initiation of signalling cascades that activate Schwann cells to produce neurotrophic factors, cytokines, extracellular matrix and adhesion molecules, which aid in regrowth of the injured nerve. These abilities are in direct contrast to CNS neuroglia, particularly astrocytes, which produce a hostile environment for axonal regeneration in response to injury. In the CNS, damage results in extensive glial scarring, the production of inhibitory factors and the lack of axonal guidance, both physical and molecular. Due to these reparative qualities, Schwann cells are becoming candidates for use in cell transplantation to treat demyelinating diseases of the PNS and CNS and spinal cord injury. Schwann cells are considered a viable option for transplantation because they are not a target of immune attack in CNS autoimmune demyelinating disease, they can be cultured and expanded from adult biopsies for autologous engraftment (Rutkowski *et al.* 1995; Vroemen and Weidner 2003; Bachelin *et al.* 2005), and they can not only remyelinate but also promote nerve fibre regeneration. Moreover, because Schwann cells can be grown relatively easily in culture, they can be manipulated to enhance their regenerative properties or to act as vehicles for gene therapy. For example, non-human primate Schwann cells, genetically modified to overexpress the neurotrophic factors BDNF and NT-3 and transplanted into demyelinated spinal cord, promote not only axonal protection, Schwann cell differentiation and oligodendrocyte progenitor cell proliferation, but also clinical recovery (Girard *et al.* 2005).

However, if Schwann cells are to be used in spinal cord cell transplantation or genetic modification therapies, it becomes vital to recognise that PNS neuroglia are distinct from CNS neuroglia and cannot be used interchangeably. Although Schwann cells and oligodendrocytes are both myelinating cells, they often have different responses to the same stimulus, as in the case of nerve growth factor (NGF) (Chan *et al.* 2004). It has long been assumed that myelination in the CNS and PNS is regulated by a common axonal signal. However, a recent study (Chan *et al.* 2004) has shown that NGF is a potent regulator of myelination but has inverse effects in the PNS and CNS. This finding highlights the importance of studying the basic biology of Schwann cells in their own right, and offers a caveat against the extrapolation of CNS data to the PNS and vice versa. Moreover, differential responses between neurons and neuroglia should also be considered; for example, although neurotrophins have a number of effects on developing or regenerating neurons, including the promotion of cell survival, neurite outgrowth, differentiation and synaptic plasticity and function, they also affect the responses of myelinating neuroglia.

SCHWANN CELLS ORCHESTRATE INFLAMMATION IN THE PERIPHERAL NERVOUS SYSTEM

It is imperative to consider how the PNS interacts with other systems of the body during development and maintenance, and in response to damage. This is particularly pertinent to the immune and endocrine systems, which can have considerable impact on the function of the PNS. A balanced interplay between the PNS and the immune system is necessary for healthy function, and again the Schwann cell plays a critical role in maintaining this delicate equilibrium with involvement in the initiation, perpetuation and termination of immune responses in the PNS (Chapters 6 and 7). In view of the fact that Schwann cells respond to and produce an extensive collection of immunomodulatory factors (including neurotrophins, cytokines and chemokines), it is increasingly apparent that they are key orchestrators of the immune response in the PNS (Maurer *et al.* 2002). Indeed, the expression of MHC class 1 and class 11 molecules (Armati *et al.* 1990; Gold *et al.* 1995; Lilje and Armati 1997) and T cell activation accessory molecules (Van Rhijn *et al.* 2000) on the Schwann cell has also raised the possibility of its direct participation in the initiation or exacerbation of T-cell-mediated responses, with clear evidence that Schwann cell can act as antigen presenting cells (Argall *et al.* 1992b; Armati and Pollard 1996; Lilje and Armati 1997). However, the full ramifications of the Schwann-cell-mediated neuro-immune interactions in the PNS are as yet far from fully realised.

Current findings describe the extensive amount of cross-talk between the nervous system and the immune system, an overlap in communication exemplified by the changing role of the neurotrophins.

Neurotrophins are a group of potent growth factors that regulate development, survival, maintenance, plasticity and function in the nervous system (Huang and Reichardt 2001). Originally, neurotrophins such as NGF and BDNF were thought to act solely in the nervous system providing essential signals for neuronal survival and repair. However, it is now known that neurotrophins act outside the nervous system and are not only produced by cells of the immune system (including T cells, B cells and macrophages), but can also influence their development, proliferation and differentiation (Otten et al. 1989). In fact, there is so much cross-talk between neurotrophins and the immune system that they have also been alternately dubbed neurokines. It has now been shown in the CNS that immunity within the nervous system can be a double-edged sword – possessing both neuroprotective and neurodestructive effects (Hammarberg et al. 2000; Kerschensteiner et al. 2003). Given that the PNS was erroneously considered an immune-privileged site for so long, there is a lingering perception that any immune response within the nerve is harmful to neuronal/Schwann cell function. Although inappropriate inflammation of peripheral nerve does have catastrophic effects in some instances, the concept of neuroprotective immunity could have profound implications when considering the pathogenesis and treatment of inflammatory diseases of the PNS, as discussed in Chapters 9 and 10.

SCHWANN CELLS AND DISEASES OF THE PERIPHERAL NERVOUS SYSTEM

As this book will show, the communication between Schwann cells and neurons and its axons is so crucial that distinct clinical phenotypes result if it is perturbed or absent, as is the case in many hereditary peripheral neuropathies, such as the heterogeneous group of Charcot–Marie–Tooth (CMT) disorders (Chapter 8). The CMT disorders are caused by genetic mutations in Schwann cell and/or neuronal genes and can be divided into different classes depending on whether the clinical and electrophysiological picture is primarily demyelinating (CMT1, CMT3, CMT4) or axonal (CMT2) (Maier et al. 2002). These genetic disorders have revealed some key molecules that are vital to the Schwann cell/axon partnership and have dispelled the notion that peripheral neuropathies are distinct diseases that affect either the Schwann cell or the neuron. Analysis of the underlying gene mutations in CMT disorders has shown that primary loss of function in a Schwann cell gene can trigger a chain of events that can lead to secondary effects on axonal properties. It is common in these disorders to find that axons associated with a mutant Schwann cell can display secondary changes in neurofilament phosphorylation (Watson *et al.* 1994), axonal transport and even axonal degeneration. Further dissection of the biological mechanisms causing the CMT disorders will supply invaluable information regarding Schwann cell/axon relationships by giving detailed insight into the function of these mutated genes.

The prejudicial attitude regarding Schwann cells and the concentration on demyelination as a neuronal-based deficit has been the basis of study of the aetiology of the diseased PNS. There are many situations in diseases or injury to human peripheral nerve where myelinated axons are damaged or degenerate and the myelin breaks down, resulting in a histological picture of myelin devastation. This has led to grouping of such diseases as 'demyelinating neuropathies', which include Guillain–Barré syndrome (GBS) and chronic inflammatory demyelinating polyneuropathy (CIDP), discussed in Chapters 9 and 10. The picture of abnormal myelin is also seen in genetic diseases of peripheral nerves such as Charcot–Marie–Tooth disease (see Chapter 8). Although demyelination is the hallmark of these diseases, it is now recognised that this may represent secondary or endpoint pathology of a damaged Schwann cell or neuron.

THE 'SCHWANN SONG' FOR THE NEURONOCENTRIC NERVOUS SYSTEM

Research into the basic biology of the Schwann cell is revealing it to be complex, multifaceted and absolutely crucial to the PNS. Currently, almost every aspect of PNS function is being re-examined with respect to the role of the Schwann cell, and this work is exploding the myth of a passive, peripheral glial cell.

2

Early events in Schwann cell development

RHONA MIRSKY AND KRISTJÁN R. JESSEN

INTRODUCTION

The glial cells of adult peripheral nerves, myelinating and non-myelinating Schwann cells, are generated during development from neural crest cells. The protracted embryonic period of gliogenesis involves first the generation of Schwann cell precursors and subsequently the generation of immature Schwann cells. The signals controlling these early steps of gliogenesis from multipotent neural crest cells can now be analysed using transgenic and other molecular approaches, and the findings integrated with our knowledge of organogenesis of peripheral nerves. The subsequent postnatal generation of myelinating or non-myelinating Schwann cells from immature Schwann cells involves cessation of proliferation and resistance to cell death. The maturation of the nerve is likely to be regulated by a balance between signals that act as brakes to provide orderly timing for myelination and signals that actively promote it.

THE INITIAL DEVELOPMENT OF PERIPHERAL NERVES

Schwann cells originate from Schwann cell precursors, which in turn arise from multipotent neural crest cells that delaminate from the dorsal neural tube (reviewed in Jessen and Mirsky 2005a,b). The growing nerves consist initially of outgrowing axons and closely adherent Schwann cell precursors. At this stage the nerve lacks blood vessels, and there is no significant fibrous protective connective tissue around or within the nerve. Later in development, around the time that nerves establish stable contacts with their target tissues, blood vessels invade and the nerve acquires its characteristic connective tissue layers (Ziskind-Conhaim 1988; Jessen and Mirsky 2005a; Wanner et al. 2006).

Given the close association between axons and Schwann cell precursors in growing nerves, it is surprising that the Schwann cell precursors are not required for nerves to grow out to their targets (Grim et al. 1992; Riethmacher et al. 1997; Woldeyesus et al. 1999). Rather, Schwann cell precursors have five main functions. One important function is to give rise to Schwann cells. Schwann cell precursors represent an intermediate stage between the neural crest stem cells that they derive from and the Schwann cells that populate mature nerves. In addition, they provide essential trophic support for both sensory and limb level motor neurons (Garratt et al. 2000a). Schwann cell precursors are also necessary for normal nerve fasciculation (Morris et al. 1999; Wolpowitz et al. 2000; Lin et al. 2000). Furthermore, postjunctional folds are absent at the neuromuscular junction (NMJ) in those nerves that do form synapses, suggesting that perisynaptic Schwann cells (PSCs), discussed in detail in Chapter 5, are involved in this process (Wolpowitz et al. 2000). Finally, *in vivo*, there is evidence that Schwann cell precursors give rise to the small population of endoneurial fibroblasts within peripheral nerves (Joseph et al. 2004).

This result accords with previous work (see below), showing that *in vitro* Schwann cell precursors can generate non-glial lineages, including neurons (see below). In principle, it parallels the finding in the CNS that radial glia can give rise not only to astrocytes but also to unexpected cell types, including neurons and oligodendrocytes (Doetsch 2003; Gotz and Barde 2005). In both instances, cells that are unequivocally glia generate unexpected descendants that were previously thought to arise from other lineages.

OUTLINE OF THE SCHWANN CELL LINEAGE

As stated earlier, Schwann cells in spinal nerves arise from the neural crest. In mature nerves two morphological variants of Schwann cells, both associated closely with axons, exist throughout peripheral nerves (Figure 2.1). These are myelinating and non-myelinating Schwann cells, which surround large- and small-diameter axons, respectively (Bunge 1993b; Garbay et al. 2000; Jessen and Mirsky 1999; Jessen and Mirsky 2002; Lobsiger et al. 2002; Corfas et al. 2004; Jessen and Mirsky 2004; Sherman and Brophy 2005) (Figure 2.1). Other distinct glial cell types also exist within the mature PNS. These include olfactory cells, perisynaptic Schwann cells that cover the axon terminals at the skeletal NMJ, satellite cells that surround the cell bodies of sensory, sympathetic and parasympathetic neurons, enteric glia of the enteric

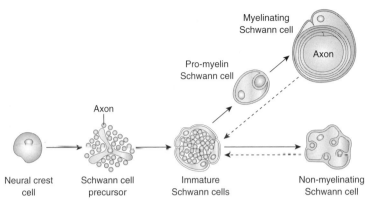

Figure 2.1 The Schwann cell lineage. Schematic illustration of the main cell types and developmental transitions in Schwann cell development. Dashed arrows indicate the reversibility of the final, and largely postnatal, transition that generates mature myelinating and non-myelinating cells. The embryonic phase of Schwann cell development involves three transient cell populations. First, migrating neural crest cells, second, Schwann cell precursors. These cells express a number of differentiation markers not found on migrating neural crest cells, including BFABP, P0 and Dhh (Figure 2.4). At any one time point, a rapidly developing population of cells − such as the glia of embryonic nerves − will contain cells that are somewhat more advanced than others. However, the cells isolated from E14 nerves (Morrison *et al.* 1999) as P0-negative using the P07 monoclonal P0 antibody (Archelos *et al.* 1993) referred to as 'neural crest stem cells', are unlikely to be significantly different from the bulk of the cells in the nerve referred to here as Schwann cell precursors (Morrison *et al.* 1999) (for a detailed discussion of this point see Jessen and Mirsky (2004) and Mirsky and Jessen (2005)). Third, immature Schwann cells. All of these cells are considered to have the same developmental potential and their fate is dictated by the axons with which they associate. Myelination occurs only in Schwann cells that by chance envelope large-diameter axons − Schwann cells that ensheath small diameter axons progress to become mature non-myelinating cells. Reproduced with permission from *Nature Reviews Neuroscience* (Jessen and Mirsky 2005a) © 2005 Macmillan Magazines Ltd.

nervous system (which are similar to astrocytes), and neuroglial cells that form part of the core of specialised nerve endings, including Pacinian corpuscles.

During progression along the Schwann cell lineage, two intermediate cell types are generated. Namely, Schwann cell precursors, which are the neuroglial cells of embryo day (E) 14/15 rat nerves

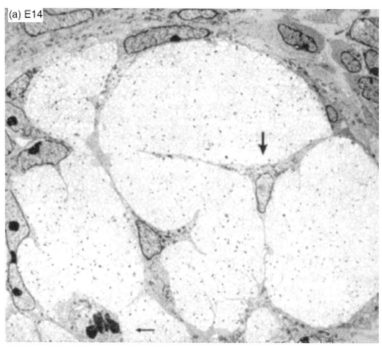

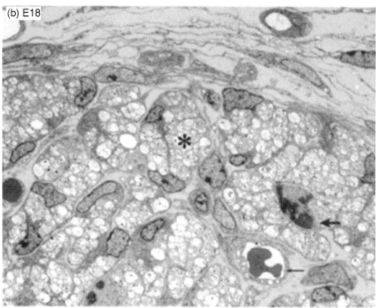

(mouse E12/13), and immature Schwann cells, which are generated from Schwann cell precursors, essentially from E15−E17, and which are the neuroglial cells found in rat nerves from E17/18 until about the time of birth (Figure 2.2) (Jessen et al. 1994; Dong et al. 1995). The postnatal fate of immature Schwann cells is subsequently determined by the axons with which they randomly associate, the myelination programme being selectively activated around birth in cells that associate with the larger axons. The final differentiation of non-myelinating Schwann cells starts about two weeks postnatally (Jessen and Mirsky 1999).

The generation of Schwann cells from neural crest cells is defined therefore by three major transitions: (1) from neural crest cells to Schwann cell precursors; (2) from Schwann cell precursors to immature Schwann cells; and (3) the divergence of this population into the two mature Schwann cell types (Figure 2.1). Most of these events hinge on axonal signals. These include survival signals, mitogenic signals and differentiation signals from the axons with which the Schwann cell precursors and Schwann cells continuously associate (Figure 2.3) (Jessen and Mirsky 1999; Jessen and Mirsky 2004;

Figure 2.2 The appearance of early cells in the Schwann cell lineage. (a) An electron microscopic image of a transverse section of a nerve in the hind limb of a rat embryo at E14. Schwann cell precursors branch among the axons inside the nerve (for example, the large arrow) and are also found in close apposition to axons at the nerve surface. One precursor cell is undergoing mitosis (small arrow). Extracellular connective tissue space (turquoise), which contains mesenchymal cells, surrounds the nerve but is essentially absent from within the nerve itself. These nerves are also free of blood vessels and the axons also have a smaller and more uniform diameter than those seen in mature nerves. Magnification ×2000.
(b) Schwann cells in a transverse section of the sciatic nerve of a rat embryo at E18, shown at the same magnification. In marked contrast to the E14 nerve, connective tissue spaces now branch throughout the nerve among compact bundles of immature Schwann cells and associated axons ('Schwann cell families') (Webster and Favilla 1984) (for example, see asterisk). Blood vessels (small arrow) and fibroblasts (for example, directly above the vessel) have also appeared inside the nerve. One Schwann cell is in mitosis (large arrow). Outside the nerve (in the uppermost part of the picture) connective tissue, containing flattened fibroblasts of the early developing perineurium and two blood vessels can be seen. Reproduced with permission from *Nature Reviews Neuroscience* (Jessen and Mirsky 2005a) © 2005 Macmillan Magazines Ltd. A colour version of this Figure is in the Plate section.

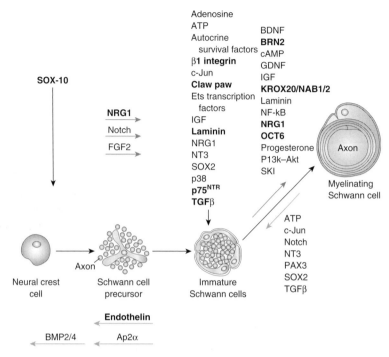

Figure 2.3 Some of the factors that have been implicated in the control of early Schwann cell development and myelination. Evidence for molecules shown in bold is based on *in vivo* observations of mutant animals. The other molecules have been implicated *in vitro*. In some cases the *in vitro* evidence is substantially more complete than in others. Sox-10 is required for the generation of all peripheral glia from the neural crest (Britsch et al. 2001), while BMPs inhibit glial differentiation (Shah et al. 1996). Axon-derived NRG1, in particular the type III isoform, is necessary for the survival of Schwann cell precursors in embryonic nerves both *in vivo* and *in vitro* (Dong et al. 1995; Wolpowitz et al. 2000; Garratt et al. 2000a). NRG1, FGF2 and Notch accelerate the Schwann cell precursor/Schwann cell transition (Dong et al. 1999; Brennan et al. 2000; Morrison et al. 2000; Leimeroth et al. 2002; A. Woodhoo, R. Mirsky and K. R. Jessen, unpublished observations), whereas the transcription factor AP2α and endothelins delay it (Brennan et al. 2000). In immature Schwann cells, survival is supported by autocrine survival factors (see text), NRG1, Ets transcription factors and laminin (Grinspan et al. 1996; Trachtenberg and Thompson 1996; Syroid et al. 1996; Meier et al. 1999; Parkinson et al. 2002; Yu et al. 2005), whereas TGFβ and p75 NTR induce Schwann cell death (Syroid et al. 2000; Parkinson et al. 2001; D'Antonio et al. 2006). *In vitro* experiments indicate that NRG1 is an axon-associated Schwann cell mitogen, but proliferation is also supported by TGFβ, laminin and a number of other

Arroyo and Scherer, Chapter 3). The cells of the lineage are also characterised by high levels of plasticity. Much of the developmental sequence is readily reversible. Notably, mature myelinating and non-myelinating Schwann cells de-differentiate and revert to a phenotype similar to that of immature Schwann cells when contact with axons is lost during nerve injury or after Schwann cell dissociation and cell culture. Furthermore, Schwann cell precursors can be diverted or reprogrammed, at least *in vitro*, to generate neurons, neuroglial–melanocytic precursors or melanocytes, and other neural crest derivatives after prolonged culture in complex media (Sherman et al. 1993; Hagedorn et al. 1999; Dupin et al. 2003; Morrison et al. 1999). Only the middle transition – from Schwann cell precursors to immature Schwann cells – seems to be irreversible.

(cont.) factors (Eccleston 1992; Morrissey et al. 1995; Yang et al. 2005; Yu et al. 2005; D'Antonio et al. 2006). The transcription factors Sox-2 and c-Jun support proliferation, although c-Jun is also required for cell death (Parkinson et al. 2004; Le et al. 2005a). ATP and adenosine, however, inhibit Schwann cell division (Fields and Stevens 2000; Stevens et al. 2004). NRG1, BDNF, NT3, IGFs and the p38 pathway function in Schwann cell migration and/or association with axons prior to myelination (Meintanis et al. 2001; Cheng et al. 2000; Yamauchi et al. 2004; Fragoso et al. 2003). Radial sorting is impaired in clawpaw, laminin and β1 integrin mutant mice (Xu et al. 1994; Feltri et al. 2002; Darbas et al. 2004; Pietri et al. 2004; Yu et al. 2005). Myelination is promoted by the transcription factor Krox-20 acting with NAB proteins, and by Oct-6, Brn-2 and NFkB (Topilko and Meijer 2001; Jaegle et al. 2003; Nickols et al. 2003; Le et al. 2005b), but inhibited by c-Jun, Pax-3 and Sox-2 (Kioussi et al. 1995; Parkinson et al. 2004; Le et al. 2005a; D. Parkinson, R. Mirsky and K.R. Jessen, unpublished observations). Cell-extrinsic signals that promote myelination include GDNF, NRG1, IGFs, BDNF, progesterone and laminin (Bunge 1993b; Koenig et al. 1995; Stewart et al. 1996; Chan et al. 2001; Hoke et al. 2003; Michailov et al. 2004). Intracellular PI3 kinase/Akt and cAMP-activated pathways also promote myelination, whereas it is retarded by Notch activation, NT3 and ATP (Maurel and Salzer 2000; Fields and Stevens 2000; Chan et al. 2001; Jessen and Mirsky 2004; Stevens et al. 2004; A. Woodhoo, R. Mirsky and K.R. Jessen, unpublished observations). TGFβ also inhibits myelination, while Ski, which suppresses TGFβ, stimulates it (Einheber et al. 1995; Guenard et al. 1995a; Atanasoski et al. 2004). Reproduced with permission from *Nature Reviews Neuroscience* (Jessen and Mirsky 2005a) © 2005 Macmillan Magazines Ltd.

Neural crest cells, Schwann cell precursors, and immature Schwann cells all divide rapidly *in vivo*, and the onset of myelination is the only step in the lineage that is clearly linked to cell cycle exit (Stewart et al. 1993). Nevertheless, even this exit is reversible because myelinating cells re-enter the cell cycle as they de-differentiate in response to nerve injury (Scherer and Salzer 2001). Developmental cell death also occurs in neural crest cells, Schwann cell precursors, and immature Schwann cells, whereas myelinating Schwann cells in early postnatal nerves and all Schwann cells in mature nerves are relatively resistant to cell death even after denervation. An important reason for this is likely to be the existence of autocrine survival circuits, discussed below (Homma et al. 1994; Grinspan et al. 1996; Meier et al. 1999; Winseck et al. 2002).

MOLECULAR PROFILE, SIGNALLING RESPONSES AND TISSUE RELATIONSHIPS DEFINE EACH STAGE OF THE LINEAGE

Each stage of the Schwann cell lineage can be characterised by a partially overlapping set of molecular differentiation markers (Figure 2.4). Formerly, analysis of lineage progression was hindered by a lack of markers other than S100 and glial fibrillary acidic protein (GFAP), both of which appear late in embryonic nerve development. Recently a number of markers that allow a more detailed analysis of earlier steps in the lineage have been identified (Jessen and Mirsky 2005a). For the first three lineage stages these markers fall into five main groups: (1) markers present on neural crest cells and PNS neuroglia, exemplified by Sox-10 (Britsch et al. 2001); (2) markers expressed by neural crest cells and Schwann cell precursors, but not, or at least at much lower levels, by immature Schwann cells, exemplified by the transcription factor AP2a and $\alpha 4$ integrin (Stewart et al. 2001; Bixby et al. 2002; V. Sahni, R. Mirsky and K. R. Jessen, unpublished observations); (3) markers present only on Schwann cell precursors, of which at present cadherin 19 (Cad19) is the only example (Takahashi and Osumi 2005); (4) markers present in Schwann cell precursors and Schwann cells but not in migrating neural crest cells, such as brain fatty acid binding protein (BFABP) and protein zero (P0) (Lee et al. 1997; Britsch et al. 2001); (5) markers that are expressed by immature Schwann cells, but are absent or present at very low levels in precursors and neural crest cells, such as S100 and GFAP (Jessen and Mirsky 2005a) (Figure 2.4).

A related set of markers has been used to delineate differentiation from immature Schwann cells to myelinating and non-myelinating Schwann cells and the reverse process, the de-differentiation that follows nerve injury or dissociation when Schwann cells are placed in culture. Myelination is characterised by up-regulation of the transcription factor Krox-20 and myelin-related proteins such as P0, peripheral myelin protein 22 (PMP22), myelin basic protein (MBP), proteolipid protein (PLP), myelin and lymphocyte (MAL) protein and periaxin, together with the lipids galactocerebroside and sulfatide. There is concomitant down-regulation of a set of markers that characterise immature Schwann cells and non-myelinating Schwann cells, including p75 neurotrophin receptor (NTR), L1, and neural cell adhesion molecule (N-CAM) (Jessen and Mirsky 2004; Jessen and Mirsky 2005a). Mature non-myelinating cells also express galactocerebroside and sulfatide, and possess a protein profile similar but not identical to that of immature Schwann cells (Figure 2.5).

Additional criteria can also be used to identify each stage in the lineage, as shown in Figure 2.4. Schwann cell precursors, unlike migrating neural crest cells, are intimately associated with axons, a characteristic of neuroglial cells (Figures 2.1 and 2.2). Furthermore, cultured Schwann cell precursors show numerous differences from migrating neural crest cells in response to survival factors, in particular in the ability to survive in β-neuregulin1 (NRG1) in the absence of laminin or fibronectin, conditions under which neural crest cells die (Figure 2.4) (Jessen and Mirsky 1999; Jessen and Mirsky 2004; Woodhoo et al. 2004). Also, when compared with neural crest cells, Schwann cell precursors are relatively insensitive to the neurogenic action of bone morphogenetic protein-2 (BMP2) and strongly biased towards Schwann cell generation (White et al. 2001; Kubu et al. 2002).

Additional differences between immature Schwann cells and Schwann cell precursors include the basal lamina, which starts to form soon after Schwann cells are generated (A. Kumar, R. Mirsky and K.R. Jessen, unpublished data). But perhaps the most striking difference between these cells is the ability of Schwann cells to help ensure their own survival using autocrine survival circuits (Meier et al. 1999). These are missing in Schwann cell precursors, leaving these cells wholly dependent on survival signals from axons such as neuregulin-1 (NRG1) (see below).

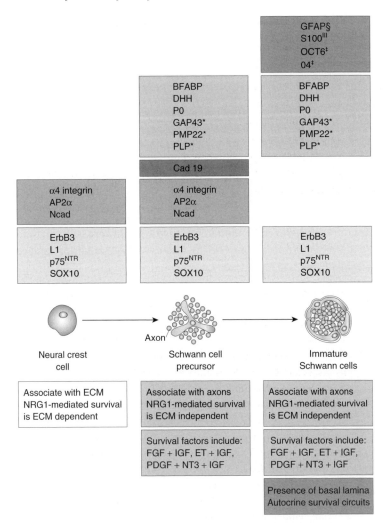

Fig 2.4 Changes in phenotypic profile as cells progress through the embryonic Schwann cell lineage. Shared profiles are indicated by distinct colours. The boxes above the lineage drawing indicate changes in gene expression that take place during embryonic Schwann cell development. The gene expression shown here is based on observations of endogenous genes rather than on observations of reporter genes in transgenic animals. Note that Cadherin 19 (Cad 19) is exclusively expressed in Schwann cell precursors (Takahashi and Osumi 2005). Each developmental stage also involves characteristic relationships with surrounding tissues, and distinctive signalling properties (boxes below lineage drawing). For instance, neural crest cells migrate through extracellular matrix. By contrast, Schwann cell precursors and Schwann

MOLECULES THAT CONTROL GLIOGENESIS FROM THE NEURAL CREST

Whether glial differentiation from the neural crest represents the default mode of differentiation, as has been suggested for CNS gliogenesis (Doetsch 2003; Gotz 2003), remains to be determined. The *in vitro* data on the action of NRG1, BMP2/4 and Notch activation, three major signals implicated in this process, are however consistent with this idea. NRG1 suppresses neuronal development, an action also shared by Notch activation, whereas BMP2/4 activates neuronal development and overrides the action of NRG1 (Shah et al. 1994; Shah et al. 1996;

> (cont.) cells are embedded among neurons (axons) with minimal extracellular spaces separating them from nerve cell membranes, a characteristic feature of glial cells in the CNS and PNS. Basal lamina is absent from migrating crest cells and Schwann cell precursors but appears on Schwann cells. In vitro NRG1 only supports neural crest survival in the presence of extracellular matrix, although this is not required for the NRG1-mediated survival of Schwann cell precursors and Schwann cells (Woodhoo et al. 2004). Migrating neural crest cells also fail to survive in the presence of several factors that support the survival of Schwann cell precursors and Schwann cells, including combinations such as FGF plus IGF, endothelin (ET) plus IGF, and PDGF plus NT3 and IGF (Meier et al. 1999; Woodhoo et al. 2004). Schwann cells also have autocrine survival circuits that are absent from Schwann cell precursors (Meier et al. 1999). * Proteins that also appear on neuroblasts/early neurons. ‡ Markers that are acutely dependent on axons for expression. § GFAP is a late marker of Schwann cell generation, as significant expression is not seen until about the time of birth. GFAP is reversibly suppressed in myelinating cells. The early expression of GFAP has not yet been examined carefully in mice. Eleven Schwann cell precursors have been shown to be S100-negative and Schwann cells S100-positive using routine histochemical methods – however, low levels of S100 are detectable in many mouse Schwann cell precursors when the sensitivity of the assay is increased. a4 integrin (Bixby et al. 2002; Joseph et al. 2004; V. Sahni, R. Mirsky and K.R. Jessen, unpublished observations); AP2a (Stewart et al. 2001); BFABP (Britsch et al. 2001); Dhh (Bitgood and Mcmahon 1995; Parmantier et al. 1999); ErbB3 (Garratt et al. 2000b); GAP43 (Jessen et al. 1994); L1 (Jessen and Mirsky 2004); N-Cad (Wanner et al. 2006); Oct-6 (Blanchard et al. 1996); O4 (Dong et al. 1999); PLP (Griffiths et al. 1998); PMP22 (Hagedorn et al. 1999); P0 (Lee et al. 1997); p75NTR (Jessen and Mirsky 2004); Sox-10 (Britsch et al. 2001). Reproduced with permission from *Nature Reviews Neuroscience* (Jessen and Mirsky 2005a) © 2005 Macmillan Magazines Ltd.

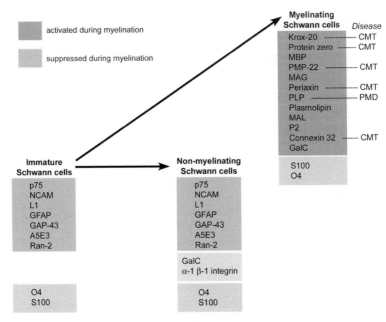

Figure 2.5 The main changes in protein and lipid expression that take place as the population of immature Schwann cells diverges to generate myelinating and non-myelinating cells during development. Myelination involves a combination of down-regulation and up-regulation of key molecules, most of which are associated with formation of the myelin lamellae. Boxes group molecules that are expressed in a similar pattern. Darkest grey box: molecules that are strongly up-regulated on myelination. Medium/dark grey box: molecules associated with immature Schwann cells that are down-regulated when cells myelinate. The generation of mature non-myelinating cells from immature cells involves many fewer molecular changes, and most of the molecules typical of immature Schwann cells are also expressed by non-myelinating cells, as indicated (Jessen et al. 1990; Salzer et al. 1998; Jessen and Mirsky 2005a). Lightest grey box: galactocerebroside is expressed by both myelinating and non-myelinating Schwann cells, but not by immature cells, while a1b1 integrin is up-regulated specifically on non-myelinating cells as they mature (Jessen and Mirsky 2005a). Medium/light grey box: 04 antigen and S100 are expressed by all Schwann cells in peripheral nerves, although 04 is down-regulated in the absence of axons (Jessen and Mirsky 2005a). Hereditary demyelinating neuropathies that result from abnormal expression of some of the myelin-associated proteins are indicated. CMT: Charcot–Marie–Tooth neuropathy; PMD: Pelizaeus–Merzbacher disease. A colour version of this Figure is in the Plate section.

Morrison *et al.* 2000). Thus the evidence that these signals have positive or negative effects on the appearance of neurons is convincing. It has proved more problematic to demonstrate that NRG1 or Notch actually initiate glial development from the crest. Similarly, Sox-10, an HMG box transcription factor expressed by all PNS glia that is required for glial development from the neural crest, would not appear to be part of a classical inductive signalling cascade, because it is expressed by all neural crest cells prior to neuroglial induction (Britsch *et al.* 2001).

A default model would be consistent with the known sequence of events during gangliogenesis, during which neurons develop first, followed subsequently by neuroglia. It would be sufficient to envisage neurogenic signals acting on neural crest cells, which would then be tempered by signals from early neurons (e.g. NRG1 or Delta/Notch signalling; see below) that would suppress excessive neurogenesis in neighbouring cells, thereby allowing neuroglial cells to emerge (Shah *et al.* 1994; Morrison *et al.* 2000). The possible roles of these signals in gliogenesis will be discussed individually below.

Sox-10

The role of the transcription factor Sox-10 will be considered first. This is the only molecule known to be essential for generation of the neuroglial lineage from the trunk neural crest (Southard-Smith *et al.* 1998; Britsch *et al.* 2001; Paratore *et al.* 2001). It is present in essentially all migrating crest cells and persists in all peripheral neuroglia and melanocytes. It is down-regulated in developing neurons. In mice in which Sox-10 has been inactivated all peripheral neuroglia are absent, while neurons are initially generated in normal numbers (Britsch *et al.* 2001). At early times the DRG contain neural crest-like cells in place of BFABP+ satellite cells and developing nerves contain a few crest-like cells that lack BFABP. Thus, in the absence of Sox-10, neuroglial specification is blocked, while crest cells survive and can generate neurons. *In vitro* experiments are also consistent with this view (Paratore *et al.* 2001). One function of Sox-10 might be to maintain levels of the NRG1 receptor erbB3, which is required for Schwann cell precursor survival and proliferation *in vivo* (Britsch *et al.* 2001).

Notch

Notch activation is involved in fate determination in the CNS, where enforced Notch activation *in vivo* promotes neuroglial cell generation

(Wang and Barres 2000). In the PNS, parallels exist between Notch and NRG1 action (see below). In crest cultures, Notch activation inhibits neuronal generation and increases the number of GFAP+ Schwann cells (Morrison et al. 2000; Wakamatsu et al. 2000; Kubu et al. 2002). Like NRG1, it stimulates the formation of immature Schwann cells from Schwann cell precursors and stimulates Schwann cell proliferation (A. Woodhoo, R. Mirsky and K.R. Jessen, unpublished data). A precedent for this exists in the CNS, where cooperative interactions between Notch and NRG1 have been noted in astrocyte development (Schmid et al. 2003). Whether Notch acts instructively to promote gliogenesis from uncommitted neural crest cells is still unclear, since alternative possibilities exist to explain the positive effects of Notch activation on Schwann cell development (Morrison et al. 2000). These include indirect effects due to inhibition of neurogenesis, stimulation of the Schwann cell precursor/Schwann cell transition and stimulation of Schwann cell proliferation.

BMP2 and 4

In vivo, BMPs are involved in sympathetic neuron generation and in cultures of neural crest cells they stimulate the formation of neurons (Shah et al. 1996; Schneider et al. 1999) and block glial differentiation (Shah et al. 1996). Involvement in controlling gliogenesis in vivo has not been shown.

NRG1

This molecule and its receptors erbB3/erbB2 play crucial roles at multiple stages of Schwann cell development. In neural crest cell cultures, NRG1 strongly inhibits neurogenesis (Shah et al. 1994). Whether it plays a comparable role in vivo is not clear, because overproduction of neurons is not seen in NRG mutant mice. And, although NRG1 promotes gliogenesis relative to neurogenesis due to suppression of neuronal development, NRG1 signalling does not appear to be obligatory for glial differentiation from the neural crest. In neural crest clonal cultures, GFAP+ Schwann cells appear readily even in the absence of added NRG1 (Shah et al. 1994), and the same is true if appearance of P0, an earlier marker of Schwann cell precursors, is used (N. Kazakova, R. Mirsky and K.R. Jessen, unpublished results), implying that NRG1 does not directly promote gliogenesis. Furthermore, in mice that lack NRG1 or its receptors erbB2 or erbB3, a major population of neural crest-derived glia, DRG satellite cells, develop normally; although Schwann cell precursors are lost, presumably due to the role of NRG1 in migration and survival of early neural crest derivatives (see below) (Garratt et al. 2000a). The NRG1 mutants suggest two possible functions for NRG1 in the development of neural crest derivatives. In these mutants sympathetic ganglia fail to form properly, particularly in caudal positions, probably due to impaired ventral migration of neural crest cells past the location of the DRG. This is inferred from the observation that sympathetic ganglia are severely hypoplastic in the absence of NRG1 signalling (i.e. in NRG1, erb2 and erbB3 mutants), whereas neural crest cells are generated in normal numbers and accumulate in dorsal locations (Britsch et al. 1998). The importance of NRG1 for survival is seen in developing spinal nerve trunks, where Schwann cell precursors, and later Schwann cells, are severely depleted in the absence of NRG1 signalling (Dong et al. 1995; Riethmacher et al. 1997; Woldeyesus et al. 1999; Wolpowitz et al. 2000). *In vitro* NRG1 is an essential survival signal from DRG neurons that rescues Schwann cell precursors from death and stimulates proliferation (Dong et al. 1995). *In vivo*, the major NRG1 isoform of importance for Schwann cell survival is the transmembrane isoform III, the major isoform expressed by sensory and motor neurons at the appropriate developmental time, where it accumulates along axonal tracts (Marchionni et al. 1993; Loeb et al. 1999; Wang et al. 2001; Leimeroth et al. 2002). In mice lacking this isoform, Schwann cell precursors initially populate the nerves at E11, presumably utilising other NRG1 isoforms for initial survival and migration. But by E14, a stage when Schwann cell precursors are converting rapidly to Schwann cells, cell numbers are severely depleted, indicating that isoform III is required *in vivo* for Schwann cell precursor survival. Nerves lacking isoform III are also defasciculated, particularly within the target field (Loeb et al. 1999; Wolpowitz et al. 2000). In contrast, Schwann cell precursors develop normally in mice lacking isoforms I and II (Meyer et al. 1997; Garratt et al. 2000a). In support of this, Schwann cell precursors depend on axons for survival *in vivo*, and the Schwann cell precursor death that follows nerve injury is prevented by application of NRG1 (Winseck et al. 2002).

At later stages in the lineage, NRG1 promotes the transition from Schwann cell precursors to Schwann cells (Dong et al. 1995; Dong et al. 1999; Brennan et al. 2000). It also promotes Schwann cell survival, acting in concert with autocrine survival signals (see below) to prevent the death of neonatal Schwann cells within nerve trunks and also at the

NMJ after injury (Morrissey et al. 1995; Trachtenberg and Thompson 1996; Grinspan et al. 1996). Furthermore, it is likely to be an important component of the axon-associated signal that drives immature Schwann cell proliferation prior to myelination, although this has not been demonstrated in vivo (Morrissey et al. 1995).

Therefore, NRG1 promotes the generation and expansion of the pool of immature Schwann cells by promoting survival, proliferation and progression within the Schwann cell lineage. Additionally, the inhibitory effect of NRG1 on neurogenesis (Shah et al. 1994) might act indirectly to increase glial cell production by prolonging the time available for uncommitted neural crest cells to adopt a glial fate. It is likely that a combination of these mechanisms results in the increased Schwann cell generation from neural crest cells seen in response to NRG1 in various in vitro experiments.

Most recently, a new role for NRG1 in regulating myelination has been revealed. Three papers underlie the importance of NRG1 type III in control of myelination (Garratt et al. 2000a; Michailov et al. 2004; Taveggia et al. 2005). Selective inactivation of the NRG1 receptor erbB2 or underexpression of NRG1 in heterozygous mice results in the formation of fewer myelin lamellae, whereas overexpression of NRG1 isoform III in neurons results in increased spirals of myelin lamellae (Garratt et al. 2000a,b; Michailov et al. 2004). Similarly DRG neurons from NRG1 type III null mice fail to induce proper Schwann cell ensheathment and myelination in neuron–Schwann cell cocultures, whereas overexpression leads to hypermyelination. The results suggest that, acting in concert with other signals, high levels of NRG1 type III provide an instructive signal for myelination, and that low levels are required for proper ensheathment of unmyelinated axons (Taveggia et al. 2005).

THE GENERATION OF IMMATURE SCHWANN CELLS

The transition from Schwann cell precursors to immature Schwann cells involves a major change in cellular architecture within the nerve (Figure 2.6). As described earlier, peripheral nerves at E14 in rat (E12 in mouse) consist of axons and Schwann cell precursors, which are found both at the surface and within the developing nerve. The Schwann cells are connected by adherens junctions and form a network of long sheet-like processes that communally envelop large groups of axons, dividing the nerve into territories longitudinally (Ziskind-Conhaim 1988; Wanner et al. 2006). The nerve is devoid of

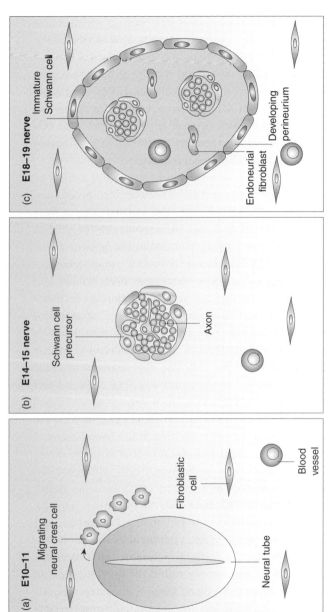

Figure 2.6 Cell and tissue relationships at key stages of Schwann cell development in rodents. There is a simple relationship between the main stages of embryonic Schwann cell development and organogenesis of spinal nerves. (a) Migrating neural crest cells move through immature connective tissue before the time of nerve formation. (b) Schwann cell precursors are tightly associated with axons and are found in early nerves that are still compact and do not contain vessels or connective tissue. (c) Immature Schwann cells are found in nerves that have acquired the basic tissue relationships of adult nerves. By this time, the developing perineurium defines the endoneurial space that now contains, in addition to axon/Schwann cell units, blood vessels, endoneurial fibroblasts and extracellular matrix. Reproduced with permission from *Nature Reviews Neuroscience* (Jessen and Mirsky 2005a) © 2005 Macmillan Magazines Ltd. A colour version of this Figure is in the Plate section.

blood vessels and connective tissue. By E18/19, when immature Schwann cells populate the nerve, the overall architecture is transformed and is essentially as described by Webster and Favilla (1984) for newborn nerve. The nerves consist of Schwann cell families, each enclosing smaller groups of axons than at E14, each of which is surrounded by basal lamina and connective tissue spaces containing collagen fibrils, endoneurial fibroblasts and blood vessels. By this stage the nerve is also enclosed by several sheaths of immature perineurial cells, and an epineurial sheath of collagen fibrils is forming. The basic relationship between nervous tissue, connective tissue and blood vessels is similar to that seen in adult nerves. The Schwann cell precursor/Schwann cell transition therefore correlates with an important step in the organogenesis of peripheral nerves.

A crucial change in cell survival mechanisms also occurs at this transition. Schwann cells, in contrast to Schwann cell precursors, can support their own survival using autocrine circuits. They secrete a cocktail of survival factors, that *in vitro* includes insulin-like growth factor 2 (IGF2), neurotrophin 3 (NT3), platelet-derived growth factor BB (PDGF-BB), leukaemia inhibitory factor (LIF) and a lysophosphatidic-acid-like molecule (Dowsing et al. 1999; Meier et al. 1999; Weiner and Chun 1999). These survival circuits are probably important in maintaining Schwann cell survival in injured nerves even after axons degenerate. The switch from paracrine dependence on axonal signals to autocrine survival support is likely to match biological need, the former providing a mechanism for matching the numbers of axons and Schwann cell precursors, and the latter ensuring Schwann cell survival in injured postnatal nerves.

Little is known about the transcription factors that control the Schwann cell precursor/Schwann cell transition. It has been suggested that AP2α provides negative regulation because it is strongly down-regulated both in rats and mice, as Schwann cells are generated *in vivo* and enforced expression in Schwann cell precursors delays Schwann cell generation *in vitro* (Stewart et al. 2001). Another negative regulator of Schwann cell generation is endothelin. Endothelins and their receptors are present in embryonic nerves, and in endothelin-B-receptor-deficient rats Schwann cells form prematurely (Brennan et al. 2000). Evidence for a positive regulation of the Schwann cell precursor/Schwann cell transition comes from *in vitro* experiments, indicating that NRG accelerates the conversion of Schwann cell precursors to Schwann cells, in addition to promoting Schwann cell precursor survival and proliferation (Brennan et al. 2000).

Notch activation also promotes the generation of Schwann cells from Schwann cell precursors *in vitro* (Morrison et al. 2000; A. Woodhoo, R. Mirsky and K. R. Jessen, unpublished observations).

FUNCTIONS OF SCHWANN CELL PRECURSORS AND THEIR DERIVATIVES

Because of the importance of *Sox-10*, isoform III of NRG or the NRG1 co-receptors *erbB2* or *erbB3*, Schwann cell precursors and Schwann cells are absent from the major limb nerves of mouse embryos lacking these genes (see above). In these mutants, most of the DRG neurons and the motor neurons that project into limb nerves die by E14 and E18 respectively, although they are initially generated in normal numbers. This implies that an important function of Schwann cell precursors and immature Schwann cells is to provide essential survival signals for developing neurons, perhaps even using bidirectional NRG1 signalling from Schwann cells to axons as anti-apoptotic signals (Riethmacher et al. 1997; Morris et al. 1999; Garratt et al. 2000a; Britsch et al. 2001; Bao et al. 2003; Falls 2003). In addition, impaired axon–target contacts probably contribute to sensory and motor neuron death (Wolpowitz et al. 2000). Taken together with the finding that axons control the survival of Schwann cell precursors, these studies identify a defined period in early nerve development when neurons and glia are mutually dependent on each other for survival.

Recently, a number of classical ideas about developmental potential and lineage restrictions have been challenged. A notable example from the developing and adult CNS is the finding that neurons can arise from cells that show obvious similarities to astrocytes (Doetsch 2003; Alvarez-Buylla and Lim 2004; Gotz and Barde 2005). Another surprising lineage relationship involving neuroglial cells has been noted in embryonic nerves (Joseph et al. 2004). Here, genetic lineage tracing suggests that the small population of endoneurial fibroblasts that provide collagen for mechanical strength originate from cells in the nerve that are derived from the neural crest, that express desert hedgehog (Dhh) and are therefore presumably Schwann cell precursors. The observation that both cell types appear in the nerve at the Schwann cell precursor/Schwann cell transition (A. Kumar, R. Mirsky and K. R. Jessen, unpublished observations), fits well with the idea that Schwann cell precursors generate some fibroblasts in addition to Schwann cells. It is also consistent with the previous findings that early PNS neuroglia from rodents and birds can generate cells other than Schwann cells. *In vitro*, in the quail, P0+ cells from embryonic nerves can generate melanocytes (Ciment 1990; Sherman *et al.* 1993). More recently, chick Schwann cell precursors expressing the glial-specific Small Myelin Protein, were induced to generate melanocytes in the presence of endothelin, a signal that also acts more broadly to promote plasticity of crest derivatives (Dupin *et al.* 2000; Dupin *et al.* 2003). Melanocytes have also been induced in injured nerves of adult mice, particularly in lines that are heterozygous for the *neurofibromin-1* gene (Rizvi *et al.* 2002), and in the rat, P0+ cells from early nerves or DRGs have the potential to generate neurons and fibroblasts (Hagedorn *et al.* 1999; Morrison *et al.* 1999).

If confirmed, the generation of fibroblasts from Schwann cell precursors *in vivo* would bring the trunk neural crest lineage in line with the cardiac and cephalic neural crest lineages, both of which are known to generate connective tissue (Le Douarin and Kalcheim 1999). They also accord with the emerging concept that early glia can act as multipotent progenitors in the developing nervous system (Doetsch 2003; Gotz 2003).

Direct evidence for the obvious idea that Schwann cell precursors generate Schwann cells includes the following. (1) It is the simplest explanation of the finding that in limb spinal nerves, as Schwann cells appear at E15–E17, Schwann cell precursors disappear, and substantial numbers of both cell types are present only at around E16. (2) In Schwann cell precursor cultures prepared from E14 nerves, the ongoing appearance of cells with the phenotype of Schwann cells can be observed directly as Schwann cell precursors disappear. *In vitro*, the conversion of most of the Schwann cell precursors to Schwann cells is completed in 3–4 days, which is close to the time course of the Schwann cell precursor/Schwann cell transition *in vivo* (Jessen *et al.* 1994; Dong *et al.* 1995). (3) In mouse mutants that lack Schwann cell precursors, Schwann cells are not generated (Garratt *et al.* 2000a; Britsch *et al.* 2001).

At the Schwann cell precursor/Schwann cell transition, significant narrowing of developmental options occurs. De-differentiation of immature Schwann cells to Schwann cell precursors has not yet been observed, and *in vitro*, immature Schwann cells are resistant to signals, including BMP2 and FGF2, that can induce the generation of other neural crest derivatives from Schwann cell precursors (Sherman *et al.* 1993; Morrison *et al.* 1999; Morrison *et al.* 2000). Their single option seems to be the reversible generation of myelinating or non-myelinating cells, a fate choice determined by axon-associated signals (as discussed above).

EVENTS JUST PRIOR TO MYELINATION

Myelination occurs over an extended period during the first three weeks of postnatal life in rodents, and in humans during a period that begins in embryonic life and ends with puberty (Berthold et al. 2005). Considerable evidence now exists for the idea that positive and negative signals are involved (Jessen and Mirsky 2005a). Just before the onset of myelination at E18, significant changes occur in the relationship between Schwann cells and axons. They involve radial sorting, a radical change in cellular relationships that allows Schwann cells to start myelinating single large diameter axons (Webster and Favilla 1984). At the same time Schwann cell numbers are adjusted by controlling survival and proliferation and premature myelination appears to be prevented by the activity of signalling systems that function as 'myelination brakes'.

Radial sorting

This change starts in the late embryonic period and continues postnatally in rodents. Individual Schwann cells start to segregate from Schwann cell families at around E18 and form a 1:1 relationship with large diameter axons – the pro-myelin stage – and subsequently myelinate them, starting 2–3 days later. The molecular control of this critical step, which involves important rearrangements of axonal/axonal, Schwann cell/axonal and Schwann cell/Schwann cell interactions is not well understood. In Schwann cells in which β1integrin, laminins or erbB2 have been conditionally ablated, and in the clawpaw mutant mouse, which has a mutation in the Lgi4 gene, it is impaired (Xu et al. 1994; Feltri et al. 2002; Darbas et al. 2004; Pietri et al. 2004; Yang et al. 2005; Yu et al. 2005; Bermingham, et al. 2006) (see also Chapter 4).

Additionally, factors that affect Schwann cell migration in cell culture might govern cell movements during radial sorting *in vivo*. These include NRG1, IGF, neurotrophin-3 (NT3) and brain-derived neurotrophic factor (BDNF) (Cheng et al. 2000; Meintanis et al. 2001; Yamauchi et al. 2004). There is also evidence that p38MAP kinases, important in the migration of other cell types, are required prior to myelination, perhaps to attain the correct alignment between axons and Schwann cells (Fragoso et al. 2003).

CONTROL OF SCHWANN CELL NUMBERS: PROLIFERATION
AND DEATH

Myelination takes place at a stage when the young animal is growing rapidly. In developing nerves, the number of Schwann cells must be matched with the number of axons. As the period of neuronal cell death is largely over, the matching process is controlled by Schwann cell proliferation and death.

From studies using co-cultures of neurons and Schwann cells, the major axonal mitogen appears to be NRG1, although *in vivo* evidence for this is lacking (Salzer *et al.* 1980; Morrissey *et al.* 1995). Other potential mitogens include platelet-derived growth factor (PDGF), fibroblast growth factor (FGF) and transforming growth factor beta (TGFβ), all of which stimulate Schwann cell proliferation *in vitro* (Eccleston 1992; Einheber *et al.* 1995; Guenard *et al.* 1995b; Mirsky and Jessen 2005). Recently, conditional ablation of the TGFβ type II receptor in late embryonic mouse nerves has been shown to reduce Schwann cell proliferation *in vivo* at E19. This indicates that TGFβ is directly or indirectly involved in controlling Schwann cell division *in vivo* (D'Antonio *et al.* 2006).

The survival of immature Schwann cells in late embryonic and perinatal nerves is likely to be governed by a balance between survival and death signals. Survival is probably promoted by autocrine survival circuits, neuronally derived NRG1 and signals from the basal lamina (Grinspan *et al.* 1996; Syroid *et al.* 1996; Trachtenberg and Thompson 1996; Dowsing *et al.* 1999; Meier *et al.* 1999; Weiner and Chun 1999). Death signals so far identified include a signal, perhaps NGF, acting via the p75 NTR, which promotes the increased cell death seen in newborn nerves after transection (Soilu-Hanninen *et al.* 1999; Syroid *et al.* 2000), and TFGβ (Parkinson *et al.* 2002). In late embryonic and early postnatal nerves, death is also induced by TGFβ, since deletion of type II TGFβ receptors suppresses normal developmental cell death in E18-newborn nerves, and also inhibits the increased cell death seen in the distal stump following nerve transection (D'Antonio *et al.* 2006).

Myelination brakes

Signalling pathways that inhibit myelin differentiation are active in immature Schwann cells, and these pathways are suppressed at the onset of myelination. For example, the Jun-NH2-terminal kinase

(JNK)/cJun pathway is active in Schwann cells in mouse E18-newborn nerves, where it is required for NRG1-induced proliferation and TGFβ-induced death (Parkinson et al. 2001; Parkinson et al. 2004). As individual Schwann cells start to myelinate, this pathway is inactivated by a mechanism that depends on Krox-20 (Egr-2) and its associated proteins NAB1 and 2 (Parkinson et al. 2004; Le et al. 2005a). If the JNK/c-Jun pathway remains active, myelination in neuron–Schwann cell co-cultures is blocked, and myelin gene expression in response to pro-myelin signals such as Krox-20 or cAMP elevation is inhibited (Parkinson et al. 2004; Parkinson et al. 2005). Similarly, Notch signalling promotes proliferation of immature Schwann cells, but is suppressed as cells start to myelinate, and if suppression is prevented, myelination is blocked (A. Woodhoo, R. Mirsky and K.R. Jessen, unpublished data). The transcription factors Pax-3 and Sox-2 act in an analogous way. Both are expressed in proliferating Schwann cells prior to myelination, and induce proliferation when adenovirally expressed. They are down-regulated in myelinating Schwann cells and exert a negative effect on myelin differentiation (Kioussi et al. 1995; Le et al. 2005a; D.B. Parkinson, R. Mirsky and K.R. Jessen, unpublished observations). In neuron–Schwann cell co-cultures, axon-derived ATP delays myelination, acting via P2Y receptors, while both adenosine and ATP inhibit Schwann cell proliferation (Fields and Stevens 2000; Stevens et al. 2004). NT3 also acts to delay myelination in these cultures (Chan et al. 2001).

These studies are starting to define the signals that promote the immature Schwann cell state by acting as inhibitors of further differentiation, including myelination, and by stimulating, or being required for, the proliferation of these cells.

SIGNALS THAT PROMOTE MYELINATION

Myelination involves inactivation of myelin-inhibitory pathways together with activation of pro-myelin pathways (Figure 2.3). Space precludes a full description of this process, discussed further in Jessen et al. (2004) and Jessen et al. (2005). Pro-myelin signals include the transcription factors Krox-20, Oct-6 and Brn-2, Sox-10 and NFkB (Wegner 2000a; Wegner 2000b; Topilko and Meijer 2001; Jaegle et al. 2003; Nickols et al. 2003). Of these, Krox-20 is the most crucial for myelination because *in vivo*, Schwann cells fail to myelinate in the absence of Krox-20. Furthermore, mutations in Krox-20 are associated with Charcot–Marie–Tooth neuropathy (Topilko et al. 1994; Le et al. 2005a), further discussed in Chapter 9. Many of the actions of Krox-20 in myelination are mediated through the binding of the repressor proteins NAB1 and 2 (Le et al. 2005b). Absence of Oct-6 leads to delayed myelination and failure to up-regulate Krox-20 at the appropriate time (Bermingham et al. 1996; Jaegle et al. 1996). The related POU-domain transcription factor Brn-2 can partially compensate for the absence of Oct-6, resulting in delayed up-regulation of Krox-20 and subsequent myelination (Jaegle et al. 2003). Schwann-cell-specific enhancer regions in both Oct-6 and Krox-20 genes have been described (Ghazvini et al. 2002; Ghislain et al. 2002).

The positive effects of NRG1 signalling on myelination have already been discussed, but other factors are known to promote both the timing and extent of myelination *in vitro*. These include BDNF, GDNF, IGFs, PI3 kinase signalling, the proto-oncogene Ski and progesterone (Koenig et al. 1995; Stewart et al. 1996; Cheng et al. 1999; Maurel and Salzer 2000; Chan et al. 2001; Hoke et al. 2003; Sereda et al. 2003; Atanasoski et al. 2004; Melcangi et al. 2005; Taveggia et al. 2005). For further discussion of this topic see Jessen and Mirsky (2004); and Mirsky and Jessen (2005).

3

The molecular organisation of myelinating Schwann cells

EDGARDO J. ARROYO AND STEVEN S. SCHERER

INTRODUCTION

Myelinating Schwann cells have unique structural and molecular adaptations that promote saltatory conduction. The myelin lamellae itself can be divided into several domains — compact myelin, regions of specialised junctions between the layers of the myelin lamellae ('non-compact myelin'), the abaxonal/outer membrane, and the adaxonal/inner membranes, and the regions at the node, paranode, and juxtaparanode. P0, myelin basic protein (MBP), and peripheral myelin protein 22 kD (PMP22) are the main proteins of compact myelin. Incisures and paranodes contain the molecular components of adherens junctions, tight junctions and gap junctions. Nodal microvilli contain cytoskeletal components, dystroglycan, syndecan-3 and -4, and possibly cell adhesion molecules (CAMs) that may interact with CAMs on the nodal axolemma (Nr-CAM and neurofascin 186 kD; NF186) to cluster voltage-gated Na^+ (Na_V) channels. The nodal axolemma also contains voltage-gated K^+ channels (KCNQ2 and Kv3.1b). Na_V channel α subunits are directly linked to the spectrin cytoskeleton by ankyrin$_G$, and indirectly by their β subunits. The paranodal glial loops contain NF155, which interacts directly with contactin, and indirectly with contactin/Caspr heterodimers, forming septate-like junctions. In the juxtaparanodal region, connexin29 (Cx29) may form hemi-channels on the adaxonal Schwann cell membrane; these directly appose a complex of Kv1.1/Kv1.2 K^+ channels and Caspr2 on the axonal membrane; *trans*-interacting, transiently expressed axonal surface glycoprotein-1 (TAG-1) dimers may join the two apposed membranes. Kv1.1, Kv1.2 and Caspr2 each have a PDZ binding site and may interact with the same PDZ protein. Caspr and Caspr2 bind to band 4.1B, which links them to the spectrin cytoskeleton. Mutations that abolish septate-like junctions allow Kv1.1 and Kv1.2 into the paranodal region, leading to altered conduction. The adaxonal Schwann cell membrane contains myelin-associated glycoprotein (MAG), which interacts with gangliosides on the axonal membrane. Understanding how axon–Schwann interactions create the molecular architecture of myelinated axons is fundamental and is involved in the pathogenesis of peripheral neuropathies.

RECIPROCAL RELATIONSHIPS FORM MYELINATED AXONS

Myelinating Schwann cells differentiate from immature Schwann cells in response to axonal signals, of which neuregulin-1 may be key (see Chapter 2 by Mirsky and Jessen). These axonal signals cause Schwann cells to express the transcription factors Egr2/Krox-20, Brn-5 and Oct6, thereby altering the expression of numerous genes, including the components of the myelin lamellae. The maintenance of the myelinating phenotype, moreover, appears to depend on a continuous relationship with an axon, as axotomy results in the down-regulation of myelin-related genes and the dedifferentiation of previously myelinating Schwann cells.

Myelinating Schwann cells, in turn, organise the axonal membrane. The molecular components of nodes, including Na_V channels, ankyrin$_G$, NF186 and Nr-CAM accumulate at the ends of developing myelin lamellae, and as two adjacent internodes elongate, two clusters, or hemi-nodes, fuse to form a node of Ranvier (Lambert et al. 1997; Vabnick and Shrager 1998). The clustering of these nodal components appears to be organised by Schwann cell processes (Melendez-Vasquez et al. 2001). *Shaker*-type K^+ channels, Kv1.1 and Kv1.2, are subsequently excluded from the nodal axolemma and sequestered beneath the myelin lamellae by the developing paranode (Vabnick et al. 1999).

In this review, we consider some of the recent findings relating to the structure and function of myelinated axons in the peripheral nervous system (PNS). Various aspects of this topic are considered in other reviews (Marcus and Popko 2002; Girault et al. 2003; Poliak and Peles 2003; Salzer 2003; Rasband 2004; Scherer et al. 2004), which give additional references to the original literature, as space limitations preclude us from doing so here. These data refute the commonly repeated notion that myelin is a passive insulator of axons. Rather, the emerging view is that the molecular adaptations of myelinating Schwann cells facilitate saltatory conduction, and likely other aspects of axon–Schwann cell interactions, even in unanticipated ways. We emphasise the roles of junctional specialisations in the myelin lamellae, the structure of the nodal region, the molecular interactions of key molecules and disease-related alterations. A closely related topic – the laminin receptors of myelinating Schwann cells is discussed in Chapter 4 by Feltri and Wrabetz.

MYELIN LAMELLAE

The organisation of a myelinated axon is shown schematically in Figure 3.1a. It depicts two internodes; one has been unrolled to reveal its trapezoidal shape. The myelin lamellae can be divided into two domains – compact and non-compact myelin – each containing a non-overlapping set of proteins with specialised functions (Figure 3.1b). The lateral edges of the trapezoid spiral around axon, defining the paranodal regions of both the axon and the Schwann cell. Nodes of Ranvier are located between two adjacent paranodes. The juxtaparanodal region is on the opposite side of the paranode, and has distinct molecular features from the rest of the internode. The adaxonal/outer Schwann cell membrane apposes the basal lamina. Some of these features are also shown in Figure 3.2 – a transverse section through the internodal region of a myelinated axon.

Compact myelin forms the bulk of the myelin lamellae and its molecular organisation is depicted in Figure 3.3. Compact myelin is highly enriched in lipids, including the glycolipids galactocerebroside and sulphatide (Trapp and Kidd 2004). P0 is the most abundant protein in PNS myelin (Kirschner et al. 2004), forming tetramers that interact in *cis* and *trans* – the molecular glue that holds together the extracellular space (between the intraperiod lines) of compact myelin. In *Mpz*-null mice, myelin lamellae form but the intraperiod lines are widened. Even mice that are heterozygous for a null *Mpz* mutation have focal areas of widened intraperiod lines; this is also a feature of some human *MPZ* mutations. Hence, both *MPZ/Mpz* alleles may be required for proper myelination; the loss of one allele causes haplotype insufficiency.

Even though peripheral myelin protein 22 kDa (PMP22) is much less abundant than P0, duplication and deletion of *PMP22* cause the most common inherited demyelinating neuropathies (Suter 2004; Wrabetz et al. 2004). Three copies of *PMP22* cause Charcot–Marie–Tooth disease type 1A (CMT1A), whereas one copy causes a different phenotype – hereditary neuropathy with liability to pressure palsies (HNPP).

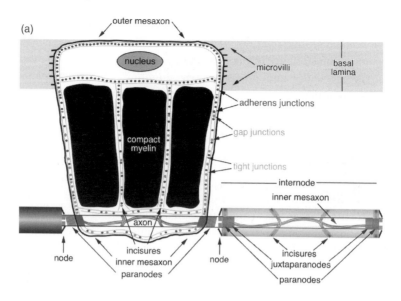

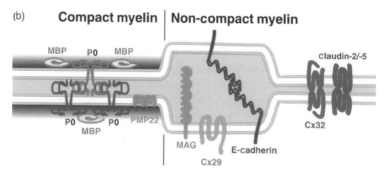

Figure 3.1 The organisation of a myelinated axon. (a) Depicts an 'unrolled' myelinating Schwann cell, revealing the regions that form compact and non-compact myelin. Tight junctions are depicted as two continuous (green) lines; these form a circumferential belt and are also found in incisures. Gap junctions are depicted as orange ovals; these are found between the rows of tight junctions, and are more numerous in the inner aspects of incisures and paranodes. Adherens junctions are depicted as purple ovals; these are more numerous in the outer aspects of incisures and paranodes. The nodal, paranodal, and juxtaparanodal regions of the axonal membrane are coloured blue, red and green, respectively. (b) The proteins of compact and non-compact myelin. Compact myelin contains P0, PMP22, and MBP; non-compact myelin contains E-cadherin, MAG, Cx32, Cx29, and claudins 1&5. Modified from Scherer and Arroyo (2002), with permission of Springer-Verlag. A colour version of this Figure is in the Plate section.

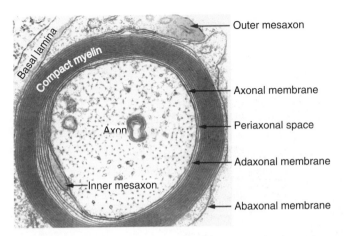

Figure 3.2 The ultrastructure of a myelinated axon. This is an electron micrograph of a transversely sectioned myelinated axon. Multiple layers of closely apposed, specialised cell membrane comprise compact myelin. The abaxonal/outer cell membrane apposes the basal lamina. The periaxonal space separates the adaxonal/inner cell membrane from the axonal membrane.

These results provide strong evidence that the amount of PMP22 in compact myelin is critical, slightly too much causes CMT1A and slightly too little causes HNPP (Vallat et al. 1996). In spite of its clinical importance, the function of PMP22 in the myelin lamellae remains unknown. PMP22 probably forms dimers, which may interact with P0 tetramers, but how overexpression and underexpression of PMP22 leads to demyelination remains unclear. If compact myelin is a liquid crystal composed of highly ordered proteins and lipids, then perhaps perturbations in the stoichiometry of either P0 or PMP22 alter its stability. How other dominant *PMP22* mutations cause even more severe forms of inherited demyelinating neuropathies appears to be related to the intracellular retention of the mutant proteins, which likely give rise to aberrant protein–protein interactions between mutant and wild type PMP22, or other toxic effects (Suter and Scherer 2003).

JUNCTIONAL SPECIALISATIONS

As depicted in Figures 3.1 and 3.4, non-compact myelin contains junctional specialisations between the layers of the myelin lamellae – so-called 'reflexive' or 'autotypic' junctions (Scherer et al. 2004).

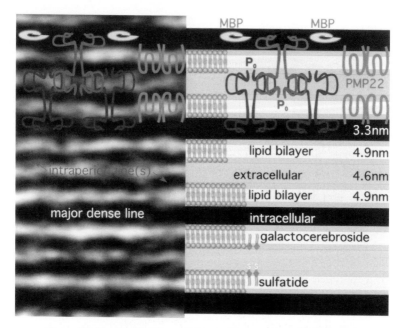

Figure 3.3 The localisation of PNS myelin proteins in compact myelin. The left panel is an electron micrograph of compact myelin, which consists of alternating layers known as the intraperiod line (which is actually a double line) and the major dense line. The right panel is a schematic depiction of how apposed cell membranes create the intraperiod and major dense lines. The disposition of P0 tetramers, PMP22 dimers, and MBP monomers, as well as the glycolipids (galactocerebroside and sulphatide) are shown. The approximate thickness of the lipid bilayers, as well as the intracellular and extracellular spaces is indicated. Reproduced with permission of Springer-Verlag (Scherer and Arroyo 2002). A colour version of this Figure is in the Plate section.

Some of these reflexive junctions are also found in inner and outer mesaxons. Adherens junctions are most numerous in the outer mesaxon and in the outermost layers of the paranodes and incisures; these contain E-cadherin, α-catenin, and β-catenin, and are linked to the actin cytoskeleton. Eliminating adherens junctions by conditionally deleting E-cadherin in Schwann cells has surprisingly little effect on the structure of the myelin lamellae (Young et al. 2002). Claudin-1 and -5 probably form the tight junctions at the paranodes and incisures, respectively, but their functional importance remains to be determined. Presumably, these tight junctions limit the diffusion of molecules between the layers of the myelin lamellae – at the inner

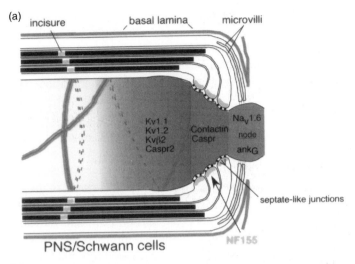
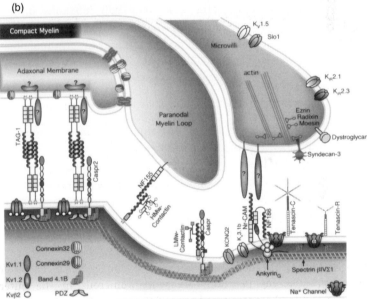

Figure 3.4 Nodal specialisations in PNS myelin lamellae. (a) Schematic depiction of the node, paranode and juxtaparanode. (b) is a schematic drawing of possible *cis* and *trans* interactions between the molecular components of nodes. Modified from Scherer and Arroyo (2002), with permission of Springer-Verlag. A colour version of this Figure is in the Plate section.

and outer mesaxons, as well as at Schmidt–Lanterman incisures and paranodes.

Tight junction strands enclose gap junction plaques, some of which are Cx32-positive (Meier et al. 2004). A role for gap junctions in the myelin lamellae was not established until mutations in the gene encoding Cx32, *GJB1*, were found to cause X-linked Charcot–Marie–Tooth disease (CMT1X) (Wrabetz et al. 2004). Dye transfer studies in living myelinating Schwann cells provide functional evidence that gap junctions mediate a radial pathway of diffusion across incisures. A radial pathway would be advantageous as it provides a much shorter pathway (up to 1000-fold), owing to the geometry of the myelin lamellae. A disrupted radial pathway may be the reason that *GJB1* mutations cause CMT1X. However, the pathway and even the rate of 5,6-carboxyfluorescein diffusion in *Gjb1/cx32*-null mice did not appear to be different than in wild-type mice, implying that another connexin(s) forms functional gap junctions in PNS myelin lamellae. This could be Cx29, which is also found in incisures, with the caveat that Cx29 by itself does not appear to form functional gap junctions (Altevogt et al. 2002).

SPECIALISATIONS AT NODES OF RANVIER

Nodes of Ranvier are the sites of action potential propagation, and share many characteristics with axon initial segments, the site of action potential initiation. By restricting currents to nodes, myelin lamellae facilitate saltatory conduction. The idea that myelin contains functional gap junctions is inconsistent with the notion that myelin 'insulates' axons. Rather, myelin lamellae facilitate saltatory conduction by reducing the capacitance of the internode, and by organising axonal ion channels.

In spite of the differences between myelinating Schwann cells and oligodendrocytes and their myelin lamellae, the organisation of the axon itself is similar in the PNS and CNS (Scherer et al. 2004; Trapp and Kidd 2004). Na_V channels are highly concentrated in the nodal axolemma (Figures 3.4 and 3.5). The α subunits are glycoproteins that form the pore; these belong to a multi-gene family, of which $Na_V1.1$, $Na_V1.2$, $Na_V1.3$, $Na_V1.6$, $Na_V1.7$, $Na_V1.8$ and $Na_V1.9$ are expressed by neurons, of which some but not all have been localised to nodes. The intracellular loop between the second and third domains harbours the ankyrin$_G$ binding site that is presumed to localise the channels to nodes and axon initial segments. Each Na_V channel

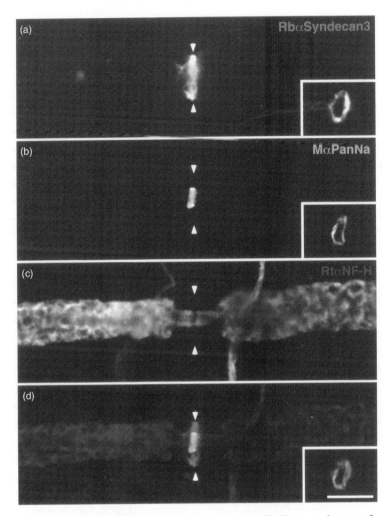

Figure 3.5 N-syndecan in Schwann cell microvilli. These are images of paraformaldehyde fixed fibres from adult rat sciatic nerve, immunostained for N-syndecan (red), Na$^+$ channels (green), and neurofilament-heavy NF-H (blue). N-syndecan is localised to Schwann cell microvilli situated around the node while Na$^+$ channels localise to the node on the axolemma. Note that in cross-section Na$^+$ channels lie inside N-syndecan staining (inset). Scale bar: 5 μm. A colour version of this Figure is in the Plate section.

is associated with two β subunits, which are encoded by four distinct genes. Na$_V$ channels are properly targeted in β1- and β2-null mice (Chen et al. 2002; Chen et al. 2004), so that the functions of β subunits remain to be elucidated, but include at least one important electrophysiological characteristic – the resurgent current of $Na_V1.6$ (Grieco et al. 2005).

$Na_V1.6$ is the main channel in mature mammalian nodes, but in the CNS and perhaps in the PNS many nodes initially express $Na_V1.2$. The gene encoding $Na_V1.6$ is mutated in mice with motor endplate disease (*med*), a recessively inherited disease that causes death by respiratory paralysis around post-natal day 20, demonstrating that $Na_V1.6$ is essential for motor axon conduction and that another Na_V channel is expressed prior to $Na_V1.6$. Further, because the conduction velocity of a mixed nerve is only minimally slowed in *med* mice, some myelinated axons (presumably sensory) must express other Na_V channels. Demyelination and remyelination alter the repertoire of nodal Na_V channels. $Na_V1.8$ is prominently expressed at nodes *Mpz*-null and *TremblerJ* mice (Devaux and Scherer 2005; Ulzheimer et al. 2004) both of which have a severe demyelinating neuropathy. The functional significance of this aberrant $Na_V1.8$ expression is unclear, as tetrodotoxin affects conduction more in *TremblerJ* nerves than in wild type mice, even though $Na_V1.8$ is relatively resistant to tetrodotoxin (Devaux and Scherer 2005).

Two voltage-gated potassium channels, Kv3.1b and KCNQ2, have also been identified at PNS nodes (Devaux et al. 2003; Devaux et al. 2004). Kv3.1b is localised to about 20% of nodes in normal adult rodent nerve, but is much more prominent in *TremblerJ* nodes. The functional significance of nodal Kv3.1b in mature nerves is unknown, as 4-aminopyridine, which blocks Kv3.1b channels (and other Kv channels, but not KCNQ channels), does not effect conduction. Based on the observation that a dominant KCNQ2 mutant causes neuromyokymia (muscle cell activity resulting from spontaneous activity of motor axons), KCNQ2 was localised to nodes (Figure 3.6). In keeping with the idea that KCNQ2 probably forms the 'slow' nodal current, which helps to repolarise axons, treating mature nerves with retigabine, a KCNQ channel 'opener', slightly slows conduction (Devaux et al. 2004). Conversely, the dominant KCNQ2 mutant associated with neuromyotonia has dominant-negative effects on co-expressed wild type KCNQ2 (Dedek et al. 2001), so that decreased slow nodal current causes excessive excitability.

Ankyrins are a family of three genes encoding adaptor proteins that link intrinsic membrane proteins to the spectrin cytoskeleton (Bennett and Baines 2001). Two splice variants of ankyrin$_G$, 270 and 480 kD, anchor Na_V channels at nodes. These isoforms have a membrane-binding domain composed of ANK repeats,

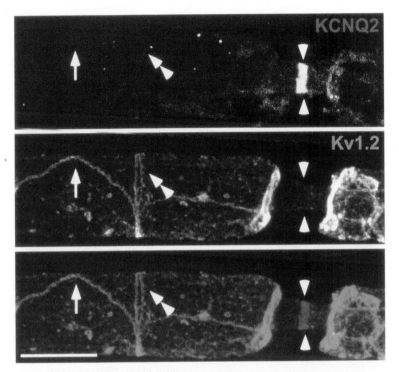

Figure 3.6 Distinct localisation of KCNQ2 and Kv1.2 channels in myelinated axons. These are images of unfixed teased fibres from adult rat sciatic nerve, immunostained for KCNQ2 (red) and Kv1.2 (green). KCNQ2 (facing arrowheads) is localised to the nodal axolemma, and Kv1.2 is localised to the juxtaparanodal region, as well as the 'juxta-mesaxons' (arrow; aligned with the glial inner mesaxon) and the 'juxta-incisures' (double arrowhead; aligned with the inner aspect of incisures). With permission of the Society for Neuroscience (Devaux et al. 2004). A colour version of this Figure is in the Plate section.

a spectrin-binding domain, and a serine/threonine-rich domain. As depicted in Figure 3.4b, the membrane-binding domain interacts with the cytoplasmic domains of NF186, Nr-CAM, and the α and β subunits of Na_V channels, thereby linking them to the spectrin cytoskeleton. Deletion of the *ank3/ankyrin$_G$* gene in mice reduces the amount of Na_V channels and NF186 in axon initial segments (Zhou et al. 1998). Because initial segments and nodes share many molecular characteristics, nodal membranes are probably similarly affected.

Spectrins are extended, flexible cytoskeletal molecules, 200–260 nm in length, that form a planar hexagonal lattice under cell membranes (Bennett and Baines 2001). The basic subunit is a heterotetramer comprised of two dimers linked head-to-tail; each dimer contains one α and one β spectrin in antiparallel orientations. Two α and four β genes have been described in mammals, and there are various alternatively spliced isoforms. Initial segments and nodes contain splice variants of spectrin IV/β4, and nodal structure and the clustering of Na_V channels are affected by mutations of the β4 spectrin gene (Lacas-Gervais et al. 2004; Yang et al. 2004).

Lambert and colleagues (Lambert et al. 1997) proposed that axonal NF186 and Nr-CAM have heterophilic interactions with other CAMs on Schwann cell microvilli. This suggestion is in accord with the ultrastructural data showing tethering of microvilli to the nodal axolemma, whereby Schwann cell processes appear to contact Na_V clusters (Ichimura and Ellisman 1991; Melendez-Vasquez et al. 2001). Furthermore, the clustering of Nr-CAM and NF186 appears prior to the clustering of Na_V channels (Lambert et al. 1997; Custer et al. 2003), and disrupting Nr-CAM in myelinating co-cultures affects Na_V clustering (Lustig et al. 2001). The putative partner on the microvilli remains to be determined. Dystroglycan may play a role, as it is localised to microvilli, and mice in which myelinating Schwann cells lack dystroglycan have dispersed nodal clusters of Na_V channels and slowed conduction (Saito et al. 2003). Microvilli also contain syndecan-3 and -4 (Figure 3.5) (Goutebroze et al. 2003), heparan sulphate proteoglycans that bind to a novel collagen in the extracellular matrix (Erdman et al. 2002), but their roles at nodes are not known. Syndecans bind to ezrin, radixin and moesin (the defining members of the ERM family of proteins), which are actin-binding proteins found in microvilli. The regulation of ERM proteins/actin appears to be involved in the formation of nodal microvilli and indeed of nodes themselves (Gatto et al. 2003; Melendez-Vasquez et al. 2004).

SPECIALISATIONS AT PARANODES

In freeze-fracture, the paranodal loops of the myelin lamellae contain rows of large particles that are in register with a double row of smaller particles on the axolemma (Ichimura and Ellisman 1991). Together, these particles connect the terminal loops to the axons, and correspond to the so-called 'terminal bands' seen by transmission electron microscopy, more recently termed septate-like junctions because they resemble invertebrate septate junctions (Figure 3.4a). Septate junctions may function similarly to vertebrate tight junctions, forming intercellular junctions that prevent the diffusion of small molecules and ions. Septate-like junctions, however, do not prevent the diffusion of lanthanum or even microperoxidase (molecular mass 5 kDa) into the periaxonal space (Feder 1971); hence, they are not 'tight' in the conventional sense. In addition, septate-like junctions are thought to act like a 'fence', separating nodal from juxtaparanodal channels (see below).

The molecular organisation of the paranode is depicted in Figure 3.4b. The paranodal axolemma contains contactin, a GPI-linked CAM that can either dimerise with Caspr (in which case it is glycosylated in the endoplasmic reticulum) or remain 'free' (in which case it is glycosylated differently, probably in the Golgi). Caspr (also known as paranodin) is a transmembrane glycoprotein belonging to the neurexin superfamily, and must heterodimerise with contactin in the endoplasmic reticulum to reach the cell membrane, and it apparently does so without passing through the Golgi (Bonnon et al. 2003). NF155 is an alternatively spliced isoform of neurofascin, and is expressed by myelinating neuroglial cells. NF155 binds in *trans* to 'free' contactin but to not contactin-Caspr heterodimers (Gollan et al. 2003). Although NF155 is partitioned into lipid rafts, it is stabilised from Triton-X100 extraction by its molecular interaction in paranodes (Schafer et al. 2004).

As might be expected, the absence of either contactin or Caspr results in the loss of septate-like junctions (Poliak and Peles 2003). An NF155-null mouse has not yet been reported, mice lacking CD9 or the genes required to synthesise galactocerebroside and sulphatide (glycolipids found in the myelin lamellae; Figure 3.3) provides a glimpse of the likely phenotype. Like *contactin* or *Caspr*-null mice, these mice lack septate-like junctions, and in all of these mutant mice, Kv1.1, Kv1.2 and Caspr2 are mislocalised to the paranodal axonal membrane, thereby directly apposing nodal Na_V channels. This mislocalisation likely shunts current, thereby slowing conduction, so that Kv1.1/Kv1.2-blockers (dendrotoxin and 4-aminopyridine) improve conduction velocity in these mutant mice (whereas they have no effects in normal mice). Finally, Na_V channels are less well confined to the nodal region in mutant mice lacking septate-like junctions (Rios et al. 2003; Rosenbluth et al. 2003).

SPECIALISATIONS AT JUXTAPARANODES

By freeze-fracture electron microscopy, the axolemma in the region extending 10–15 μm from the paranode – the juxtaparanode – contains clusters of 5–6 particles (Stolinski et al. 1985). The distribution of these juxtaparanodal particles corresponds to the distribution of Kv1.1 and Kv1.2 and their associated β2 subunit. Kv1.1 and Kv1.2 subunits can freely mix in varying proportions to form tetramers, the functional channels, and the size of the particles (10 nm in diameter) compares well to the expected size of a tetramer (Kreusch et al. 1998). Although Kv1.1 and Kv1.2 channels appear to be concealed under the myelin lamellae, juxtaparanodal K^+ channels are thought to dampen the excitability of myelinated fibres. The finding that *Kv1.1/Kcna1*-null mice have abnormal impulse generators near the NMJs supports this idea (Smart et al. 1998). Similarly, mutations in the human Kv1.1 gene, *KCNA1*, cause a form of familial episodic ataxia that is associated with ectopic impulse generators in the intramuscular branches of myelinated axons.

TAG-1 is a GPI-linked CAM that forms dimers in *cis*, and *trans* interactions of these dimers appear to link the glial and axonal membranes at juxtaparanodes (Poliak et al. 2003; Rasband 2004). Both neurons and Schwann cells express TAG-1. On the axonal side, TAG-1 interacts directly with Caspr2, a homologue of Caspr that is localised to the juxtaparanodes of myelinated fibres. Caspr2 and Caspr have similar domain structures, including a binding site for band 4.1B, an adaptor protein that is specifically localised to paranodes and juxtaparanodes. Caspr2 also has an intracellular PDZ domain binding site, as do Kv1.1 and Kv1.2, and all three proteins form a complex, probably mediated by a protein with multiple PDZ domains. So far, only one such PDZ protein, PSD95, has been found at the juxtaparanodal region, but it is not required for the formation of Caspr2/Kv1.1/Kv1.2 complexes, or for their clustering in myelinated axons (Rasband et al. 2002). The localisations of Caspr2 and TAG-1 at the juxtaparanodes are interdependent, and both are essential for clustering of Kv1.1/Kv1.2 channels.

Freeze-fracture electron microscopy reveals hexamers on the adaxonal Schwann cell membrane, facing the periaxonal space (Stolinski et al. 1985). These hexamers are comprised of Cx29 (Altevogt et al. 2002; Li et al. 2002b). How Cx29 is localised to the juxtaparanodal region is unknown. It is unlikely to interact directly with TAG-1, so additional molecular interactions are required as depicted in Figure 3.4. Because myelinating Schwann cells are not gap-junction-coupled to axons (Balice-Gordon et al. 1998), Cx29 appears to form hemi-channels. These findings raise the possibility that Cx29 and Cx32 provide pathways for the spatial buffering of K^+: the repolarisation of axons causes K^+ to diffuse from the axoplasm into the periaxonal space through

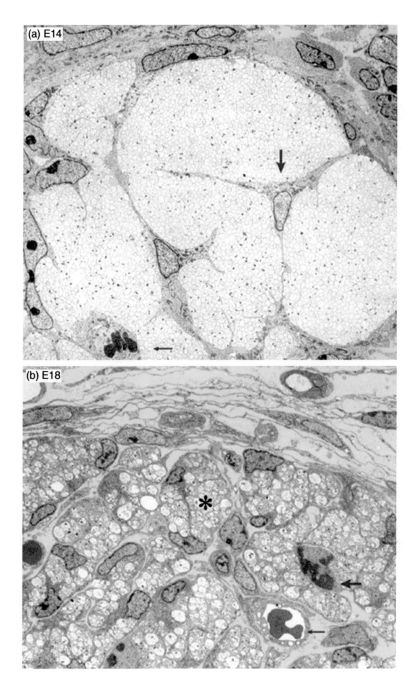

Figure 2.2 See page 17 for caption.

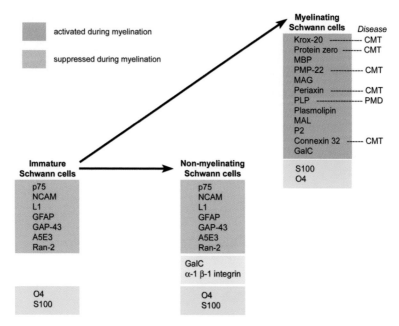

Figure 2.5 See page 24 for caption.

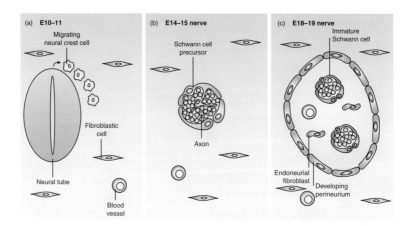

Figure 2.6 See page 29 for caption.

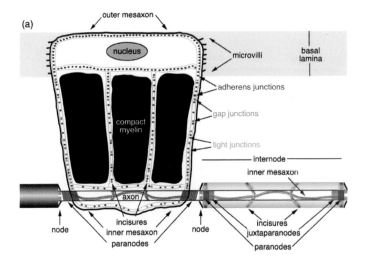

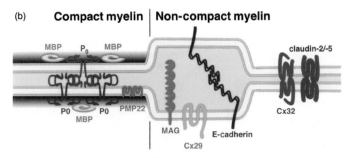

Figure 3.1 See page 40 for caption.

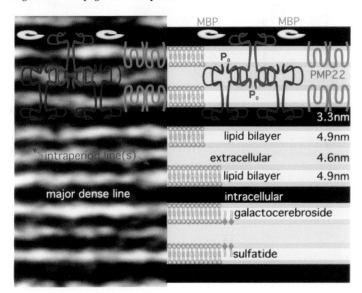

Figure 3.3 See page 42 for caption.

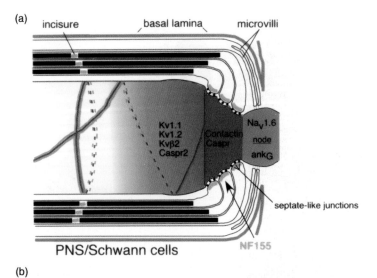
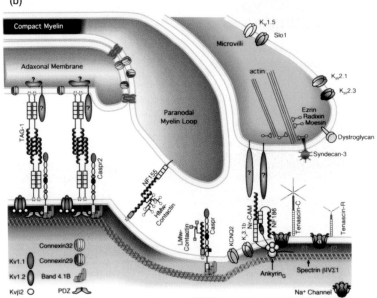

Figure 3.4 See page 43 for caption.

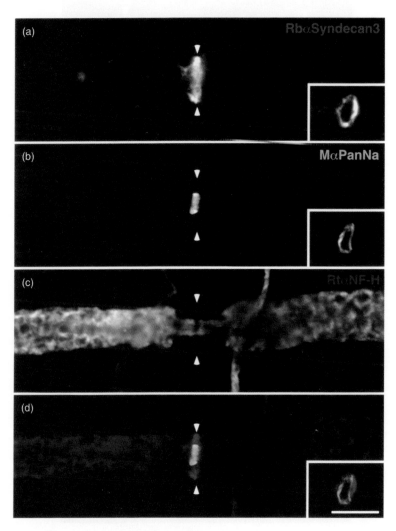

Figure 3.5 See page 45 for caption.

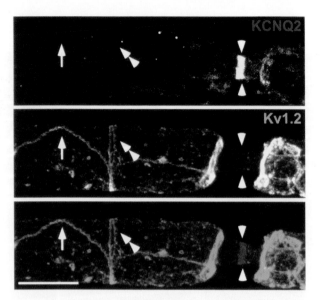

Figure 3.6 See page 47 for caption.

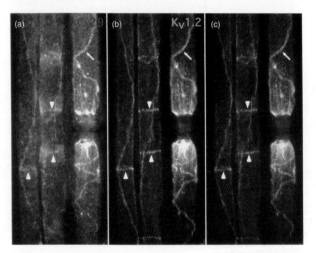

Figure 3.7 See page 51 for caption.

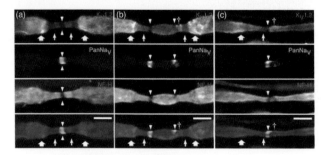

Figure 3.9 See page 54 for caption.

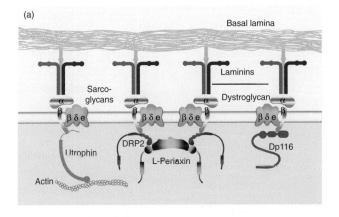

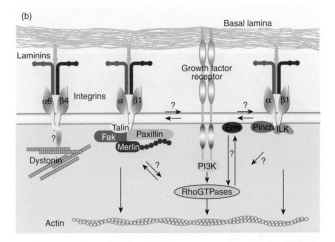

Figure 4.1 See page 65 for caption.

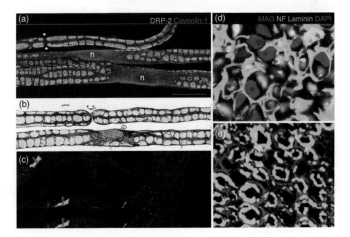

Figure 4.2 See page 66 for caption.

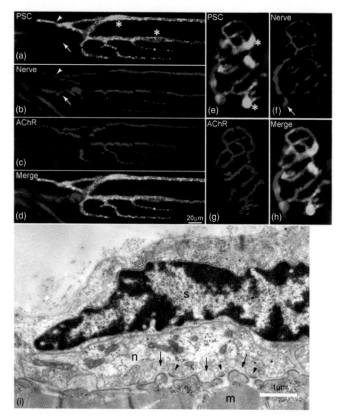

Figure 5.1 See page 74 for caption.

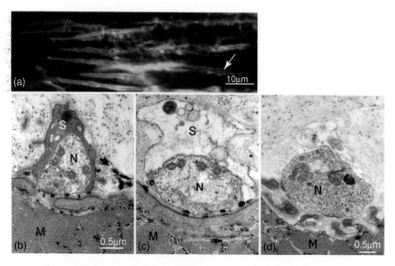

Figure 5.2 See page 77 for caption.

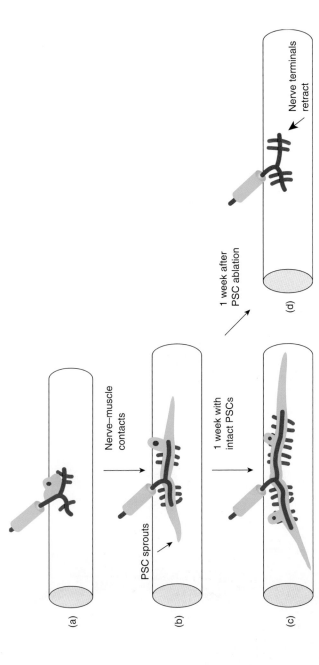

Figure 5.3 See page 83 for caption.

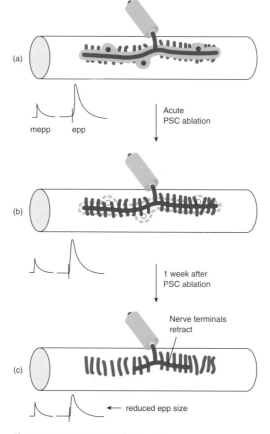

Figure 5.4 See page 90 for caption.

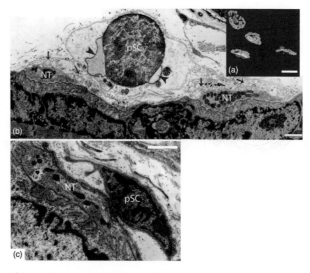

Figure 5.5 See page 92 for caption.

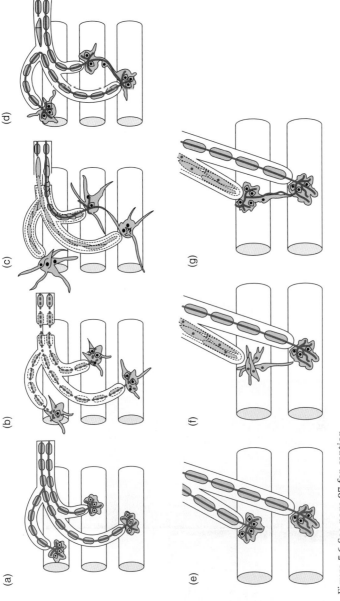

Figure 5.6 See page 97 for caption.

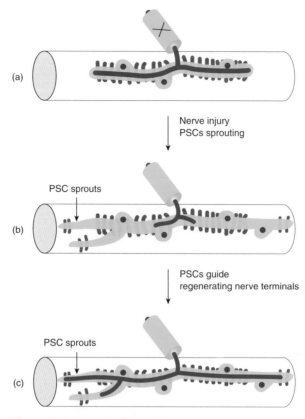

Figure 5.7 See page 98 for caption.

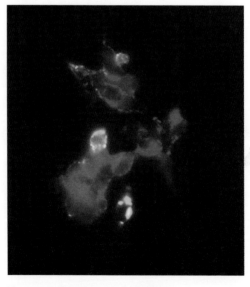

Figure 6.1 See page 115 for caption.

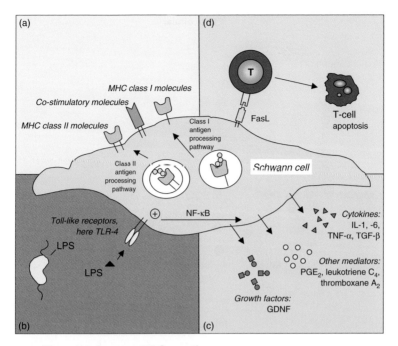

Figure 7.1 See page 125 for caption.

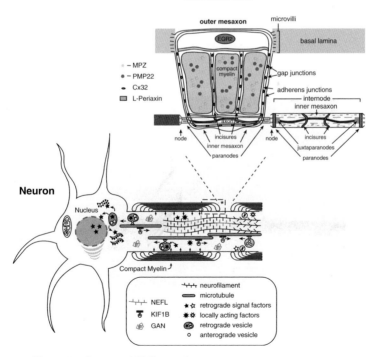

Figure 8.1 See page 130 for caption.

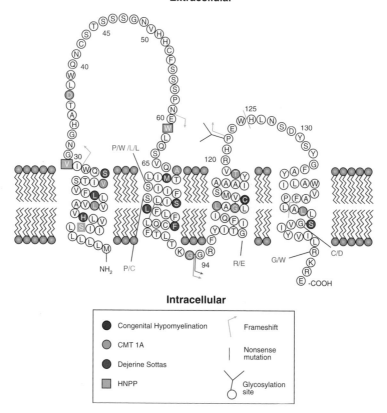

Figure 8.2 See page 133 for caption.

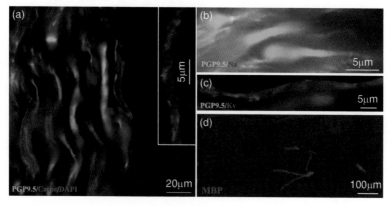

Figure 8.3 See page 137 for caption.

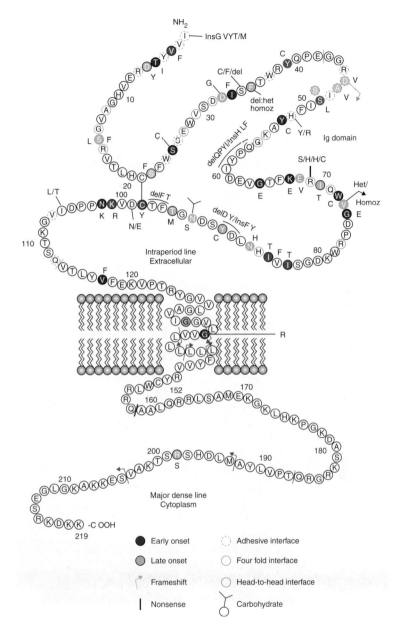

Figure 8.5 See page 141 for caption.

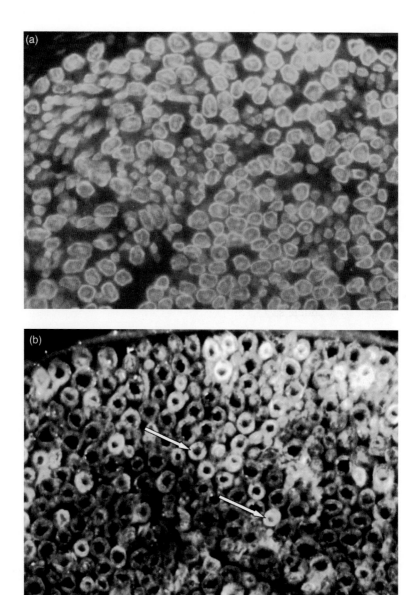

Figure 10.9 See page 183 for caption.

Kv1.1/Kv1.2 channels, from the periaxonal space into the adaxonal Schwann cell cytoplasm via Cx29 hemi-channels, and from the adaxonal cytoplasm to the abaxonal cytoplasm through paranodal Cx32 channels. The abaxonal membrane contains a variety of K^+ channels (Kv1.5, Slo1, Kir2.1 and Kir 2.3) that could provide an egress for K^+ into the extracellular space.

THE INTERNODAL REGION

In transmission electron microscopy, the internodal neuroglial and axonal membrane lacks the conspicuous specialisations of the nodal region. Nevertheless, freeze-fracture reveals similar intramembranous particles on both the neuroglial and axonal membrane as seen in the paranodal and juxtaparanodal regions (Stolinski et al. 1985). In accord with these findings, the adaxonal membrane of myelinating Schwann cells has a strand of NF155/MAG at the inner mesaxon, flanked by a double strand of Cx29 (Figure 3.7) (Altevogt et al. 2002).

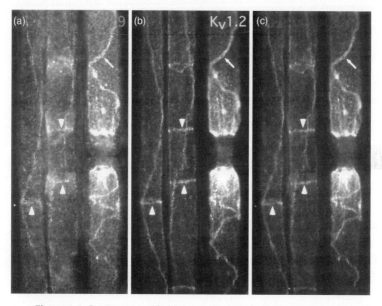

Figure 3.7 Cx29 apposes *Shaker*-type channels. These are images of fixed teased fibres from rat sciatic nerve, double-labelled for Cx29 (red) and Kv1.2 (green). Cx29 and Kv1.2 are strikingly co-localised in the juxtaparanodal region, and aligned both at the inner mesaxon (arrow) and at the innermost aspect of incisures (arrowheads). With permission of the Society for Neuroscience (Altevogt et al. 2002). A colour version of this Figure is in the Plate section.

A tripartite ring of Cx29-MAG/NF155-Cx29 staining is also found at the innermost aspect of incisures. The internodal axonal membrane has a complementary organisation: a tripartite strand – consisting of a central strand of Caspr/contactin, flanked by strands of Kv1.1/Kv1.2/Kvβ2/Caspr2 – apposes the inner mesaxon ('juxta-mesaxonal') and the innermost aspect of incisures ('juxta-incisural'). Presumably *trans*-interactions between NF155 and contactin, and TAG-1 dimers provide the molecular basis of these internodal localisations (Figure 3.4 and 3.8).

MAG is a member of the siglec family of immunoglobulin super-family of sialic acid binding lectins (Schnaar 2004). It is localised on the adaxonal glial membrane all along internodes, including para-nodes. The gangliosides GD1a and GT1b appear to be high-affinity ligands for MAG, and are components of the axolemma. Mice that lack all complex gangliosides, including GD1a and GT1b, have

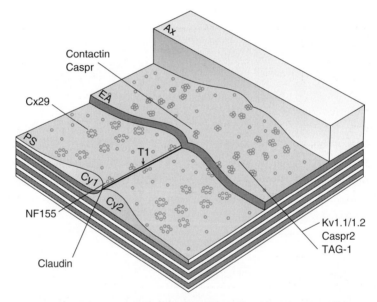

Figure 3.8 Internodal specialisation of myelinating Schwann cells. This is a schematic summary of the putative identities and distributions of intramembranous particles on the axonal and adaxonal Schwann cell membranes in the internodal region, as visualised by freeze-fracture electron microscopy. The location of the tight junction (TJ) at the inner mesaxon is indicated. Modified from Stolinski *et al.* (1985), with permission of Elsevier Science.

diminished levels of MAG, supporting the idea that gangliosides interacts with MAG *in vivo*, but the much more pronounced phenotype of these mice compared to *Mag*-null mice (Trapp and Kidd 2004) remains to be clarified.

DEMYELINATION ALTERS THE MOLECULAR ARCHITECTURE OF AXONS

A myriad of genetic and acquired perturbations cause demyelination. In nerves acutely demyelinated by lysolecithin, the paranodal (clustered Caspr and contactin), juxtaparanodal (clustered Kv1.1/Kv1.2), and internodal (juxta-incisural and juxta-paranodal localisation of Caspr and Kv1.2) axonal specialisations disappear in conjunction with the breakdown of the previously myelinated internode (Arroyo et al. 2004). In contrast, myelinated axons immediately adjacent to demyelinated internodes maintain these specialisations, including the hemi-nodes that abut the demyelinated internodes. Thus, the local differentiation of the axon depends on the myelin lamellae. In this model, Kv1.1 and Kv1.2 K^+ channels are found throughout the demyelinated internode and appear to accumulate adjacent to hemi-nodes. Further, even subtler disruptions of paranodes are often accompanied by the infiltration of Kv1.1 and Kv1.2 into the paranodal region (Figure 3.9). The distribution of Kv1.1 and Kv1.2 in *TremblerJ* mice are similarly redistributed – they are diffusely localised along demyelinated internodes and infiltrate paranodes (Devaux and Scherer 2005). Because Kv1.1/Kv1.2 have fast kinetics, these aberrantly localised channels would be expected to contribute to conduction slowing and even block, and represent a potential therapeutic target.

CONCLUSION

The intricate molecular organisation of the axonal membrane is highly related to the structure of the overlying myelin lamellae, and the molecular interactions that generate these relationships are being illuminated. Demyelination completely disrupts the molecular organisation of the axonal membrane at the level of individual myelin lamellae, and the redistribution of Kv1.1/Kv1.2 channels may contribute to conduction slowing and block. Even more discrete disruptions of myelinated axons, such as altered septate-like

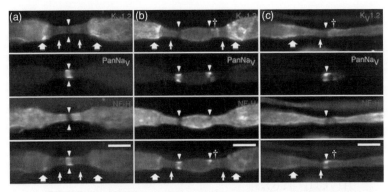

Figure 3.9 Redistribution of Kv1.1 in paranodal and segmental demyelination. These are images of teased fibres 5–7 days after intraneural injection of lysolecithin, immunolabelled for Kv1.2 (red), Na_V channels (green), and neurofilament-heavy (NF-H). Panels a, b, and c show examples of normal appearing node, paranodal demyelination and segmental demyelination (to the right of the hemi-node), respectively. Apposed pairs of arrowheads indicate the locations of nodes (a), single arrowheads indicate the locations of hemi-nodes (b,c), arrows indicate the locations of paranodes, and large arrows indicate the positions of juxtaparanodes. In panel b, note that Kv1.2 has infiltrated into the paranodal region on the right; in panel c, note that Kv1.2 apposes the hemi-node. Scale bars: 10 μm. From Arroyo *et al.* (2004) with permission of Wiley-Liss. A colour version of this Figure is in the Plate section.

junctions, may have deleterious consequences. Understanding how the molecular architecture of myelinated fibres is disrupted in inherited and acquired neuropathies will provide insight into their pathogenesis, as well as novel therapeutic opportunities.

4

The role of the extracellular matrix in Schwann cell development and myelination

MARIA LAURA FELTRI AND LAWRENCE WRABETZ

INTRODUCTION

The cellular components of the peripheral nervous system (PNS) — neurons and their axons, Schwann cells, perineurial cells, endoneurial fibroblasts and vessel endothelia — are surrounded by extracellular components such as collagens, fibronectin, laminins and proteoglycans. These components are either part of extracellular matrix or organised in the basal lamina. These molecules are recognised by cellular receptors on Schwann cells, such as integrins and dystroglycan. Engagement of matrix components by cognate receptors modulates nearly all aspects of Schwann cell development including: proliferation, survival, migration, interaction with axons, differentiation, myelination, and formation of the nodes of Ranvier. After a general review of the extracellular matrix molecules and their known functions in peripheral nerve, we will focus on the role of laminins in Schwann cell differentiation, Schwann cell/axon interactions, organisation of the axonal membrane and myelination.

EXTRACELLULAR MATRIX IN ENDONEURIUM

Every nerve is characterised by: the epineurium, surrounding the whole nerve; the perineurium, surrounding fascicles of nerve fibres; and the endoeurium, defined as the area contained within the perineurium, and in which the single fibres are contained. The adult endoneurium contains abundant amounts of extracellular matrix, which can be present in extracellular spaces among nerve fibres, associated with fibroblasts and organised in basal laminae around Schwann cells or endothelia; or anchored or spanning cell membranes in Schwann cells, endothelia or fibroblasts. The non-cell-associated matrix is mainly composed of fibrillary collagen, while most of the other extracellular matrix components are associated with fibroblasts (Joseph *et al.* 2004). Most extracellular matrix components have potentially important roles in Schwann cell development, as seen by analogy to other cells and as suggested by *in vitro* experiments. However, so far, an important role in peripheral nerve development has been confirmed *in vivo* only for laminins, derived from observations in both human hereditary diseases and from animal models. Before reviewing these recent discoveries, we will briefly describe other extracellular components for nerves in which a potential role in Schwann cells has been described.

COLLAGENS

Collagens are triple helical molecules formed by the assembly of three different α chains of procollagen. Each type of collagen fibril (e.g. type IV) forms from multiple combination of α chains. For example collagen IV, abundant in basal laminae, is formed by trimers of six possible α chains, encoded by six different genes. The fibrils can be connected by small proteoglycans and by intra-fibrillary collagen (reviewed in Ottani *et al.* (2002)). Collagens can be present in several varied conformations and cellular locations and have many functions such as the formation of fibrils, networks or beaded-like structures. Notably, certain collagens have not only extracellular matrix properties, but can also be trans-membrane receptors, or shed by proteases to release peptides with biological activities. For example, the C-terminal domain of the proteoglycan collagen XVIII is called endostatin and has anti-angiogenic activity (O'Reilly *et al.* 1997). Peripheral nerves contain fibrillary collagens type I, III, V, and XI; intrafibrillary collagen X (Seyer *et al.* 1977; Shellswell *et al.* 1979; Fujii *et al.* 1986; Jaakkola *et al.* 1989; Bunge 1993; Chernousov *et al.* 2000); collagens associated with endoneurial and perineurial basal laminae such as type IV, V, VI (Jaakkola *et al.* 1989; Peltonen *et al.* 1990; Chernousov *et al.* 2000) and the proteoglycan collagen XVIII (Halfter *et al.* 1998); non-fibrillary collagens types IX and XIII (Sund *et al.* 2001). The latter class of collagens are particularly interesting, as they are type II transmembrane protein, that can act as laminin receptors or be shed by a disintegrin and metalloprotease (ADAM) proteases from the cell surface, and behave as extracellular matrix components (reviewed in Franzke et al. 2003).

Among nerve collagens, collagen V has been extensively characterised. It contains the ubiquitous α1 and α2 chains, and the nerve-specific α4 chain, suggesting that it may play an important role in nerve development (Chernousov et al. 2000). Indeed, collagen V *in vitro* promotes Schwann cell migration and axonal fasciculation, and inhibits axonal outgrowth through distinct domains (Chernousov et al. 2001). Schwann cells adhere to collagen V through syndecan 3, and this adhesion activates Erk1/2 protein kinase and migration (Erdman et al. 2002). Finally, the N-terminal domain of α4 collagen V is released by Schwann cells and binds perlecan and glypican 1 (Rothblum et al. 2004). The function of α4 collagen V *in vivo* has yet to be investigated.

PROTEOGLYCANS

Proteoglycans – core proteins linked to glycosaminoglycan chains such as heparan, chondroitin, keratan or dermatan sulphate – can be extracellular, transmembrane or membrane-associated through GPI-linkage. More than 25 core proteins are found in the nervous system, such as syndecan, glypican, agrin, perlecan, brevican, neurocam, aggrecam, versican, phosphocan, collagen IX and XVIII, amyloid precursor protein (APP), tenascin and decorin. They can bind growth factors and extracellular matrix components such as fibroblast growth factor (FGF), laminins, hyaluronan, and heparin. Proteoglycans can act as co-receptors with adhesion molecules (e.g. integrins, transiently expressed axonal surface glycoprotein-1 (TAG-1), and neural cell adhesion molecule (N-CAM)) and can modulate the activation of integrins and bind the actin cytoskeleton (reviewed in Carey (1997)). A condroitin sulphate proteoglycan, neuroglycan C, is a member of the neuregulin family (Kinugasa et al. 2004). Schwann cells normally synthesise syndecans 3 (Carey et al. 1992), syndecan 4 (Goutebroze et al. 2003) and glypican 1 (Carey et al. 1993). Glypican on Schwann cells has been shown to cooperate with integrins to promote spreading on laminin surfaces (Carey et al. 1990; Carey et al. 1993). When syndecan 1 is ectopically expressed in Schwann cells, it mediates morphological changes (Carey et al. 1994). Syndecan 3 and glypican 1 in Schwann cells bind collagen V (Erdman et al. 2002; Rothblum et al. 2004), while syndecan 3 can also be shed by metalloproteinases (Asundi et al. 2003), and has been postulated to mediate actin rearrangements during axonal sorting

(Erdman et al. 2002). All these data suggest that syndecans and glypicans may cooperate with integrins in Schwann cells to mediate actin rearrangement during peripheral nerve development. It will be very interesting in the future to study the interactions between signals initiated by collagen V/syndecan 3 or glypican1 and laminins/ α6β1 integrins in axonal fasciculation.

Some authors have reported that AN2/NG2 is synthesised by immature non-myelinating Schwann cells, where it may have a role in Schwann cell migration (Schneider et al. 2001), whereas other authors report that fibroblasts, but not Schwann cells, produce AN2/NG2 (Martin et al. 2001). Many proteoglycans, syndecan 3, 4 (Goutebroze et al. 2003), NG2 (Martin et al. 2001) glypican (Carey et al. 1993) and versican/hyaluronectin (Delpech et al. 1982; Apostolski et al. 1994), are found at the nodes of Ranvier, either in microvilli such as syndecans, or as part of the nodal substance. Here they may perform several roles. By binding both to other extracellular components and to cell surface molecules, they may contribute to the nodal region organisation, or to voltage-gated sodium channel (Nav) clustering in cooperation with dystroglycan (see below); by virtue of their negative charges they may buffer Na ions, and being anti-adhesive (NG2, versican), they may prevent lateral extension of Schwann cells or inappropriate axonal sprouting. Finally, some proteoglycans, like agrin (which binds dystroglycan and laminin) (Yamada et al. 1996), perlecan (Yang et al. 2005), collagen XVIII (Halfter et al. 1998; Rothblum et al. 2004), decorin and versican (Braunewell et al. 1995) are present in the Schwann cell basal lamina.

TENASCIN

Tenascins are a family of glycoproteins which in general have antiadhesive properties. They have a characteristic structure formed by oligomers of six polypeptides, arranged in symmetrical branched structures, where the six arms emanate from a central core and they are bound at their amino-termini (reviewed in Jones and Jones (2000)). Tenascins can bind other extracellular matrix (ECM) molecules and a variety of cell surface receptors, therefore generating responses that can be opposite to each other. These responses may be due directly to tenascins, or to the modulation of the response of other adhesion molecules (reviewed in Jones and Jones (2000)). Tenascin C, R and X are present in the PNS. In most tissues, tenascins are expressed during development and down-regulated in the adult differentiated tissues.

However, they are then re-expressed in pathological states such as vascular disease and tumorigenesis, and during tissue remodeling (wound healing, involution). Similarly, tenascin C is expressed in the PNS during development, and is strongly up-regulated after Wallerian degeneration (Sanes et al. 1986; Daniloff et al. 1989; Martini et al. 1990). At the cellular and sub-cellular level, tenascin C is expressed in perineurial fibroblasts and pre-myelinating Schwann cells before birth; after birth its expression decreases in perineurium and becomes restricted in Schwann cells: it is present in nodes of Ranvier in myelinated fibres and is absent in non-myelinated fibres (Rieger et al. 1986). Mice in which the tenascin C gene was inactivated showed hypomyelination and myelin degeneration in peripheral nerves (Cifuentes-Diaz et al. 1998), a result that has not been confirmed by other investigators (reviewed in Mackie and Tucker (1999)).

Tenascin R has been mostly studied in the central nervous system (CNS), where it is produced by oligodendrocytes, binds F3/contactin and the $\beta1$ subunit of sodium channels and myelin-associated glycoprotein (MAG) (Xiao et al. 1998; Xiao et al. 1999; Yang et al. 1999). In the mouse, Tenascin R is only transiently expressed in Schwann cells between E14 and E18. Here it may counteract the permissive role of fibronectin in Schwann cell migration and neurite outgrowth (Probstmeier et al. 2001).

Tenascin X is expressed by Schwann cells and perineurial cells (Tucker et al. 2001; Matsumoto et al. 2002). In chickens it is mostly localised in the proximal roots, where it may contribute to the barrier between the PNS and the CNS, as it inhibits both Schwann cell migration and neurite outgrowth *in vitro* (Tucker et al. 2001). Tenascin X *in vivo* may be redundant, as deletion of Tenascin X does not produce peripheral nerve abnormalities (Matsumoto et al. 2002).

VITRONECTIN

Vitronectin is normally present only on the basal lamina of perineurial cells, but is widely expressed in Schwann cells in sural nerves of Charcot–Marie–Tooth-1 patients (Palumbo et al. 2002).

FIBRONECTIN

Fibronectins are a family of glycoproteins with affinities to both other ECM components, such as collagen and fibrin, and to several integrin receptors such as $\alpha3\beta1$, $\alpha4\beta1$, $\alpha5\beta1$, and $\alpha v\beta1$ (reviewed in Hynes (1992)).

Fibronectin is encoded by one gene, but then undergoes extensive alternative splicing to generate multiple isoforms. The active protein comprises a dimer of two covalently linked subunits. Fibronectin exists in plasma, or can be cell-associated.

Fibronectin and its main receptor, $\alpha5\beta1$ integrin, are found in the basal lamina of Schwann cells and perineurial cells (Palm and Furcht 1983; Tohyama and Ide 1984; Lefcort et al. 1992). The strongest fibronectin expression is found during embryonic development in axons and Schwann cells, whereas it is down-regulated in the adult and strongly re-expressed after injury (Lefcort et al. 1992). Fibronectin is believed to have a role during development in Schwann cell precursor migration (Baron Van Evercooren et al. 1982; Milner et al. 1997), and in nerve regeneration through facilitation of axonal outgrowth and Schwann cell proliferation (Baron Van Evercooren et al. 1982). This hypothesis has so far not been tested genetically, as fibronectin and $\alpha5$ integrin null mice die after gastrulation (George et al. 1993; Yang et al. 1993; Georges-Labouesse et al. 1996).

FIBRINOGEN AND FIBRIN

Fibrinogen in plasma can extravasate in tissues after damage and be transformed in fibrin. In peripheral nerve damage therefore, fibrin becomes an additional ECM component (Akassoglou et al. 2000), which has been shown to have important functional effects on Schwann cells. Schwann cells bind fibrin through an integrin receptor, $\alpha v\beta8$ (Chernousov and Carey 2003), and activate intracellular signalling leading to Schwann cell de-differentiation (Akassoglou et al. 2002). Thus, fibrin is useful after acute nerve damage, when Schwann cells de-differentiate and proliferate to mediate removal of myelin debris and facilitate axonal regrowth. However, fibrin must be degraded at the time of remyelination, as it inhibits Schwann cell migration and differentiation.

BASAL LAMINA AND SCHWANN CELL MYELINATION

Each Schwann cell in the endoneurium is surrounded by a basal lamina. The basal lamina is first deposited by immature Schwann cells that form "families" around bundles of axons, at E15.5 in the mouse. Mary and Richard Bunge were the first to show that a basal lamina was important for myelination, and to draw an analogy between epithelia and Schwann cells, as both cell types polarise to have a surface facing the basal lamina and a surface facing the lumen or the axon, respectively (Bunge and Bunge 1983). *In vitro* co-cultures of rat dorsal root ganglion (DRG) neurons, seeded with Schwann cells, demonstrated that with the addition of serum and ascorbic acid, triple helical collagen could form, a basal lamina could be assembled, and myelination could proceed (Carey et al. 1986; Eldridge et al. 1987; Eldridge et al. 1989). Later, similar studies showed that partial myelination could be achieved in the absence of serum and ascorbic acid, as long as a defined medium (B27) containing antioxidant was provided (Podratz et al. 1998; Podratz et al. 2004). In this system, although myelin was formed in the absence of basal lamina, laminin was still present around Schwann cells (Podratz et al. 2001). This suggests that laminins themselves, but not the assembly of a complete basal lamina, are required for myelination. This conclusion is supported by several examples *in vivo*, where myelin can be found in the absence of a continuous basal lamina *in vivo* (Madrid et al. 1975; Nakagawa et al. 2001a; Yang et al. 2005). Thus laminins and the signals that they generate seem to be the most important requirement for myelination. The peripheral nerve phenotype of patients and animals with mutated laminins confirms this view.

LAMININS AND LAMININ RECEPTORS

The first observation that signals from laminins were important for peripheral nerve development came from the discovery that mice (dystrophyc dy/dy and dy^{2J}/dy^{2J}) and patients with Congenital Muscular Dystrophy 1A (CMD1A – often called merosin-deficient CMD) that lack the α2 chain of laminin, manifest as a peripheral neuropathy (Shorer et al. 1995; Di Muzio et al. 2003; Quijano-Roy et al. 2004). Neurophysiological evidence of a neuropathy are found in most, if not all, of the CMD1A patients examined (Quijano-Roy et al. 2004). Lack of laminin signals in turn leads to abnormalities in Schwann cell interactions with axons. The defects observed in the animal model arise at different times in development and at different locations; an early arrest in the process of radial sorting of axons is detected mainly in the proximal PNS, while later abnormalities in myelinated internodes and nodes of Ranvier are found postnatally and distally (Shorer et al. 1995; Brett et al. 1998; Deodato et al. 2002; Di Muzio et al. 2003). Laminin signals are interpreted by Schwann cells through laminin receptors. Schwann cells express an array of laminin isoforms and receptors (reviewed in Feltri and Wrabetz (2005)), and the analysis of mice lacking one or more of these components shows how different phenotypes can arise due both to the presence of multiple and partially redundant laminins during development, and to activation of specific laminin/receptor pairs occurring only at specific locations and times in development.

MULTIPLE LAMININS IN SCHWANN CELLS

Laminins constitute a large family of trimers, formed by an α, β and γ subunit. Different paring of the five α, four β and three γ subunits give rise to at least 15 isoforms (Table 4.1). In mature peripheral nerves, laminin 2 and low levels of laminin 8 are found in the basal lamina surrounding single Schwann cells (Sanes et al. 1990; Patton et al. 1997; Wallquist et al. 2002). Laminin α5 (probably forming laminin 10)

Table 4.1. *Laminin trimer composition and adult PNS distribution**

Laminin	Subunit composition	DRG (satellite cells)	Perineurium	Endoneurium (Schwann cell basal lamina)	Node of Ranvier	NMJ (synaptic cleft)
1	α1β1γ1					
2	α2β1γ1	+		+	+	
3	α1β2γ1					
4	α2β2γ1		.			+
5	α3β3γ2					
6	α3β1γ1					
7	α3β2γ1		+/−[†]			
8	α4β1γ1	+		+		
9	α4β2γ1		+			+
10	α5β1γ1				+	
11	α5β2γ2		+			+
12	α2β1γ3			+[‡]		
13	α3β2γ3					
14	α4β2γ3					
15	α5β2γ3					

*Trimer distribution is assumed from coincident expression of individual isoforms
[†]Found by (Wallquist et al. 2002) but not (Patton et al. 1997)
[‡]Found by *in situ* hybridisation after crush injury (Wallquist et al. 2002)

was recently described in Schwann cell basal laminae only over nodes of Ranvier and paranodes (Vagnerova et al. 2003; Occhi et al. 2005), suggesting a specific role there. Interestingly this shows that the Schwann cell basal lamina is longitudinally polarised, in parallel with the underlying axon. During development, laminins 2 and 8 are expressed, with laminin 8 expressed at higher levels than in the adult (Lentz et al. 1997; Wallquist et al. 2002).

Laminins contribute to basal lamina assembly by interacting with other matrix components such as nidogen, fibulins, and can self-assemble in homo and hetero complexes causing spontaneous polymerisation of laminin molecules (Yurchenco et al. 1992; Odenthal et al. 2004). This polymerisation is enhanced after engagement of cellular receptors (Colognato et al. 1999). Finally, laminins interact with cellular receptors to activate outside-in signaling. Depending on subunit length, structure and interaction sites, laminins can be divided into functional groups (reviewed in Miner and Yurchenco (2004)). According to these criteria, peripheral nerve laminins can be divided functionally into three groups: laminins 2 and 4 (polymerising with other trimers, bind to $\alpha6\beta1$, $\alpha7\beta1$ integrin and dystroglycan), 10 and 11 (polymerising with other trimers, bind to $\alpha6\beta1$, $\alpha6\beta4$ integrin and dystroglycan), 8 and 9 (not polymerising with other trimers, bind to $\alpha6\beta1$ integrin and dystroglycan) (Miner and Yurchenco 2004).

MULTIPLE LAMININ RECEPTORS IN SCHWANN CELLS

Laminin receptors include molecules that belong to the integrin or collagen superfamily and dystroglycan. Integrins are transmembrane dimeric proteins composed of an α and a β subunit not covalently bound (Tamkun et al. 1986). Twenty-four α and nine β subunits have been identified in humans, from which 25 dimers have been characterised (reviewed in (Clegg et al. 2003)). The $\alpha\beta$ combination specifies ligand affinity and intracellular signaling. Thus far, known laminin receptors are $\alpha1\beta1$, $\alpha2\beta1$, $\alpha3\beta1$, $\alpha6\beta1$, $\alpha7\beta1$ and $\alpha6\beta4$ integrins, and they are all present in Schwann cells (reviewed in Previtali, 2001; Clegg, 2003). In some cell types also, integrins $\alpha v\beta8$ (Venstrom and Reichardt 1995), $\alpha9\beta1$ (Forsberg et al. 1994) and $\alpha v\beta3$ (Kramer et al. 1990), may act as laminin receptors. Whereas the expression of $\alpha9\beta1$ has not been investigated, integrins $\alpha v\beta3$ and $\alpha v\beta8$ are synthesised by Schwann cells (Milner et al. 1997); the role of αv integrin in Schwann cells is unclear, as conditional inactivation of αv in neural crest derivatives, including Schwann cells, did not show peripheral nerve alterations (McCarty et al. 2004).

Collagen XIII is a non-fibrillary collagen that can be shed by ADAM proteases and behave as extracellular matrix components, or can act as laminin receptors (reviewed in Franzke et al. (2003)) and is synthesised in peripheral nerves (Sund et al. 2001). As for other collagens, genetic experiments *in vivo* are so far lacking to determine a role for this interesting family of molecules.

Dystroglycan is formed by an extracellular α and a transmembrane β subunit (Ervasti and Campbell 1993). The extracellular portion can bind to laminin and agrin in Schwann cells, whereas the intracellular portion connects to the actin cytoskeleton through complexes that contain DP116, dystrophin-related protein 2 (DRP-2), periaxin, β, δ and ε sarcoglycans, utrophin, α-dystrobrevin (Byers et al. 1993; Saito et al. 1999; Imamura et al. 2000; Sherman et al. 2001). There are probably two or three separate dystroglycan–dystrophin complexes in Schwann cells, one containing DRP2 and periaxin, one containing DP116 and possibly a different one containing utrophin (Figure 4.1), as suggested by studies in mutant mice lacking single components of the dystroglycan–dystrophin complex, taken together with co-localisation and biochemical studies of normal nerves (Imamura et al. 2000; Sherman et al. 2001; Saito et al. 2003). These complexes are located in different sub-cellular compartments, with utrophin localised in outer cytoplasmic channels (also known as Cajal bands), DRP2 in non-channel-regions, and DP116 enriched in microvilli (Figure 4.2) (Sherman et al. 2001; Albrecht and Froehner 2004; Court et al. 2004; Occhi et al. 2005). Dystroglycan in peripheral nerve is glycosylated (mannose and sialic acid), and glycosylation is important for laminin binding (Yamada et al. 1996; Chiba et al. 1997). It has recently been shown that glycosylation of α-dystroglycan is important for many, if not all, dystroglycan functions in humans and mice. Five human congenital muscular dystrophies are caused by mutations in glycosyltrasferases that include α-dystroglycan as a substrate, (reviewed in Muntoni et al. (2004)). In addition the *Large* gene, coding for the LARGE α-dystroglycan glycosyltransferase, is mutated in myodystrophy (*Large^{myd}*) and *enr* mice (Grewal et al. 2001; Levedakou et al. 2005), both of which show defects in peripheral nerves (Levedakou et al. 2005). Surprisingly, peripheral nerve defects in *enr* mice only partially resemble defects in mice lacking Schwann cell dystroglycan (Levedakou et al. 2005), suggesting that either additional substrates for LARGE exist, or that different compensating mechanisms are available in the two mutants.

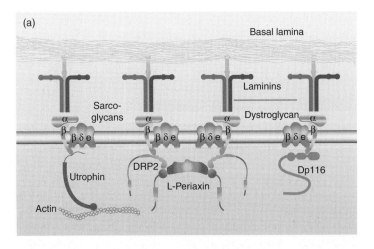

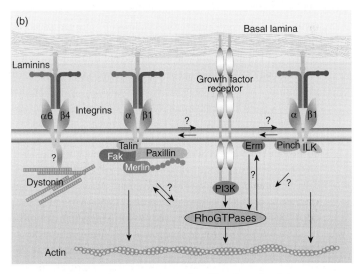

Figure 4.1 Linkages, interactors and signalling from laminin receptors. (a) The dystroglycan–dystrophin complex links laminins in the basal lamina to actin in the Schwann cell. α-dystroglycan is extracellular, binds laminin when glycosylated, and forms a complex with the transmembrane β-dystroglycan and β, δ and ε sarcoglycan. Intracellularly, β-dystroglycan probably forms three different complexes with utrophin, Dp116, or DRP2/periaxin. (b) Receptors containing β1 integrin, possibly α6β1, have been shown to form a complex with focal adhesion kinase (FAK), paxillin and merlin in Schwann cells, whereas other β1 integrin interactors, such as ILK and PINCH, are present in Schwann cells. These complexes possibly regulate actin rearrangements through RhoGTPases and ERMs, and in cooperation with growth factor receptors such as ERB2/3. A colour version of this Figure is in the Plate section.

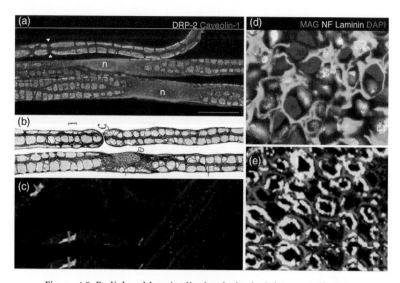

Figure 4.2 Radial and longitudinal polarity in Schwann cells. (a)–(c) Dystroglycan is linked to different dystrophins or dystrophin-related proteins in Schwann cells. The distribution of dystroglycan linkages differs along the longitudinal domains of myelinated axons: DRP2 (a, green) is in the internode, in areas of cytoplasmic apposition that are interspersed among cytoplasmic channels (stained with anti-caveolin 1 antibodies, red) previously described by Ramon y Cajal (b); in contrast Dp116 (c, red) is enriched in microvilli at nodes of Ranvier. Arrowheads in (a) indicate the position of the node of Ranvier, n= nuclei. (d)–(e) Laminin and its receptors are localised at the abaxonal, external side of Schwann cells: laminin (d, green) surrounds the outer Schwann cell surface, in contrast to myelin-associated-glycoprotein (MAG, red) which is found at the adaxonal surface that surrounds axons (stained for neurofilaments, NF, light blue). Similarly, β1 integrin (e, red) like other components of laminin receptors is found at the abaxonal surface, external to the myelin lamellae (stained with antibodies against MBP, green). (a) Modified with permission from Scherer and Arroyo (2002). (b) Modified with permission from Cajal *et al.* (1991). A colour version of this Figure is in the Plate section.

LAMININ RECEPTOR EXPRESSION IN SCHWANN CELL DEVELOPMENT

During development, Schwann cell precursors and embryonic Schwann cells express the dual laminin/collagen α1β1 integrin (Stewart *et al.* 1997), α6β1 integrin, and low levels of α2β1 and α3β1 (Hsiao *et al.* 1991; Previtali *et al.* 2003b). At birth α6β4 integrin and dystroglycan appear (Masaki *et al.* 2002; Previtali *et al.* 2003b). All of these receptors

Table 4.2. PNS phenotype of laminin or laminin receptor mutants

	Radial sorting defects	Hypermyelination, and/or myelin in/out-folding	Polyaxonal myelination	Node of Ranvier defects	Short internodes	Reference
Laminin γ1/P0cre	yes	no	no	yes	n.d.	(Chen and Strickland 2003)
Laminin α2-null or dystrophic	yes	yes	no	yes	yes	(Bradley and Jenkison 1973; Stirling 1975; Occhi et al. 2005)
Laminin α4-null	yes	yes	yes	n.d.	n.d.	(Wallquist et al. 2002; Yang et al. 2005)
Laminin α2/α4-null	yes	n.s.	n.s.	n.d.	n.d.	(Yang et al. 2005)
Integrin β1/P0cre	yes	no	no	n.d.	n.d.	(Feltri et al. 2002)
Dystroglycan/P0cre	rare	yes	yes	yes	n.d.	(Saito et al. 2003)
Integrin α7-null	no	no	no	n.d.	n.d.	(Previtali et al. 2003a)
Integrin αV/nestinCre	no	no	no	n.d.	n.d.	(McCarty et al. 2004)
MDC1A	n.s.	yes	n.s.	yes	yes	(Deodato et al. 2002; Brett et al. 1998; Di Muzio et al. 2003; Shorer et al. 1995)

n.d. = not determined
n.s. = not shown

are synthesised by both myelin-forming and non-myelin-forming Schwann cells (Feltri et al. 1994; Niessen et al. 1994; Masaki et al. 2000; Previtali et al. 2003a, 2003b), except for α1β1 integrin, whose expression is restricted to mature non-myelin-forming Schwann cells (Stewart et al. 1997). Schwann cells that are destined to form myelin start to synthesise dystroglycan at the promyelinating stage, when a Schwann cell has obtained a 1:1 relationship with an axon (Masaki et al. 2002; Previtali et al. 2003b). α7β1 integrin is expressed later by mature Schwann cells (Previtali et al. 2003a). All of these receptors are synthesised by both myelin-forming and non-myelin-forming Schwann cells (Feltri et al. 1994; Niessen et al. 1994; Masaki et al. 2000; Previtali et al. 2003a,b), except for α1β1 integrin, whose expression is restricted to mature non-myelin-forming Schwann cells (Stewart et al. 1997). In mature Schwann cells, laminin receptors are polarised to the abaxonal surface, facing the basal lamina. Dystroglycan is also present in microvilli at nodes of Ranvier.

MULTIPLE ROLES FOR LAMININ AND LAMININ RECEPTOR PAIRS IN SCHWANN CELLS (SEE ALSO TABLE 4.2)

Radial sorting

Normally each axon in adult nerves is surrounded by Schwann cell processes, but in many dystrophic humans and mice, laminin α2 deficient axons are naked and grouped in bundles (Bradley and Jenkison 1973; Stirling 1975). These naked axonal bundles represent Schwann cell development that has been arrested during radial sorting, a process normally taking place in embryonic nerves, when Schwann cells proliferate and extend processes to sort axons destined for myelination. Laminin-deficient Schwann cells are not able to sort axons, due to both a decrease in number (Bray et al. 1977; Jaros and Bradley 1978) and to an inability to extend their cellular processes (Feltri et al. 2002). These abnormalities are more severe in proximal roots than in distal nerves, possibly due to variable compensation and redundancy by other laminins. The hypothesis that this might reflect redundancy by other laminins is suggested by recent experiments with laminin-α4-null mice. These experiments show that the absence of laminin 8 causes mild defects in axonal sorting (Wallquist et al. 2005; Yang et al. 2005), whereas in the absence of both laminin 2 and 8 (mice null for laminin α2 and α4, or with Schwann cell inactivation of laminin γ1 using a PoCre line), all proximal and distal axons remain unsorted.

These data show that laminins 2 and 8 play complementary, rather than redundant, roles (Yang et al. 2005; Yu et al. 2005). In addition, compensating phenomena also occur as laminin α1 is up-regulated in distal nerves, but not roots, of *dystrophic* mice (Previtali et al. 2003b) and transgenic laminin α5 can substitute for loss of laminin α2 and α4 in distal nerves (Yang et al. 2005).

The β1 integrin subunit is found in all the known receptors present prenatally during radial sorting of axons (Previtali et al. 2003b), and has been shown in both *in vitro* and *in vivo* studies to be required for axonal sorting. Early studies using neuronal/Schwann cell co-cultures showed that anti-β1 integrin antibodies prevented myelination (Fernandez-Valle et al. 1994; Podratz et al. 2001). Later, Schwann-cell-specific inactivation of β1 integrin in mice, using the Cre/LoxP system, confirmed these results, and showed that the block was at the stage of axonal sorting (Feltri et al. 2002). Since the phenotype resembles that of laminin mutants, the receptor involved could be α1β1 or α3β1 integrins or, more probably, the abundant α6β1 integrin. The major effect of lack of β1 integrin seems to be on impaired signaling to the actin cytoskeleton, with consequent deficiency in process extension. In fact, β1 integrin-null nerves show evidence of reduced extension or even retraction of Schwann cell processes. α6β1 integrin in Schwann cells is known to interact with paxillin, merlin and FAK (Chen et al. 2000; Fernandez-Valle et al. 2002). Other α6β1 integrin interactors, such as Integrin Linked Kinase (ILK) and PINCH are also present in Schwann cells (Chun et al. 2003; Campana et al. 2003).

Differently from laminin mutant mice, Schwann cells lacking β1 integrin proliferate and survive normally (Feltri et al. 2002). In contrast, α2 and α2/α4 integrin mutants show a decreased rate of Schwann cell proliferation (Jaros and Bradley 1978; Yang et al. 2005), while laminin γ1/PoCre mice have a PI3K−mediated loss of Schwann cells due to increased death and decreased proliferation (Yu et al. 2005). It is not clear if this derives from the lack of a direct laminin signal, or whether it is an indirect consequence of impaired access to axonal neuregulins (Yu et al. 2005). Since, at birth, several laminin receptors are expressed by Schwann cells, normal death/proliferation in β1-null mice can be explained by redundancy with another receptor such as dystroglycan (Li et al. 2005c). In contrast, before birth, no receptors without β1 integrin are known to be expressed (Previtali et al. 2003b) or upregulated in β1 mice (Feltri et al. 2002), which may suggest that an unidentified receptor(s) mediates the survival/proliferation signals from laminins to Schwann cells. Thus, signals from laminins allow Schwann cells to radially sort axons during development in at least two ways: by promoting the extension of their processes via β1 integrin-mediated cytoskeletal rearrangements, and by matching the numbers of axons and Schwann cells via another receptor.

Postnatal abnormalities in the morphogenesis of the myelinated axon and nodes of Ranvier

In the absence of laminin α2 or β1 integrin, some Schwann cells are able to bypass the defective sorting and myelinate, probably because the loss of laminin signalling is compensated by laminin 8 (Yang et al. 2005) or by the onset at birth of α6β4 integrin and dystroglycan expression (Previtali et al. 2003b). However, the resulting myelinated internodes in *dystrophic* mice are not normal. Abnormalities include a reduced or increased number of myelin lamellae, the formation of short internodes, and abnormalities in nodes of Ranvier such as defective microvilli and wider nodes (Bradley et al. 1977; Bradley and Jenkison 1973; Jaros and Bradley 1979; Occhi et al. 2005).

Conditional ablation of dystroglycan in Schwann cells, using the Cre/LoxP technology, shows that these abnormalities reflect the loss of dystroglycan function during myelin lamellae elaboration. Mice lacking dystroglycan in Schwann cells have abnormally folded myelin lamellae, reduced nerve conduction velocity, and altered clusters of sodium channels at nodes of Ranvier (Saito et al. 2003). Schwann cell microvilli contact the axons at nodes, and are important for clustering of sodium channels. Microvilli contain dystroglycan (Saito et al. 2003) and Dp116 (Occhi et al. 2005) and are atrophic in dystroglycan mutant mice, suggesting that dystroglycan has a specific function in microvilli organisation or sodium channel clustering. In addition, dystroglycan may have a function also in internodes. Here dystroglycan is part of a dystrophin–glycoprotein complex that includes periaxin and DRP-2 (Sherman et al. 2001). The periaxin/DRP 2 complex is localised in cytoplasmic compartments, among outer channels known as Cajal-Bands (Court et al. 2004). This localisation and compartment formation is lost in periaxin-null mice (Sherman et al. 2001), in association with improper elongation of Schwann cells, possibly determining internodal length (Court et al. 2004). Thus, signals from laminin via Schwann cell dystroglycan are probably required for normal internode elongation, formation of microvilli and clustering of sodium channels (see also Chapter 3 by Arroyo and Scherer).

CONCLUSIONS

As the complexity of the expression of laminin isoforms and laminin receptors in nerve increases, the roles assigned to laminins expand correspondingly. Continued genetic exploration of their roles in vertebrates where Schwann cells and axons relate properly, and identification of downstream signalling, are two key objectives for future studies of laminin and laminin receptors pairs. This kind of analysis will need to extend also to other extracellular matrix molecules such as collagens and proteoglycans, to give us a deeper understanding of the many roles of extracellular matrix in nerves.

5

The biology of perisynaptic (terminal) Schwann cells

CHIEN-PING KO, YOSHIE SUGIURA AND ZHIHUA FENG

There are four classes of Schwann cells in the mature vertebrate nervous system: (1) myelinating Schwann cells, which wrap around large-diameter axons including motor axons; (2) non-myelinating Schwann cells, which associate with small-diameter axons of many sensory and all postganglionic sympathetic neurons; (3) satellite cells of peripheral ganglia; and (4) non-myelinating perisynaptic Schwann cells (PSCs), also known as terminal Schwann cells, which cap the nerve terminal at the neuromuscular junction (NMJ) (Corfas et al. 2004). While the role of motor and sensory axon-associated Schwann cells in saltatory conduction has been well-acknowledged and characterised, relatively little is known about the role of the synapse-associated Schwann cells. However, in the past decade, there has been widespread interest in unraveling the role of Schwann cells in peripheral synapses as well as the role of astrocytes in central synapses. Extensive studies on synapse–neuroglial interactions in both the peripheral nervous system (PNS) and central nervous system (CNS) have led to the concept of the tripartite synapse (Araque et al. 1999; Volterra et al. 2002; Kettenmann and Ransom 2005). The emerging concept suggests that neuroglia cells are active and essential participants in modulating synaptic function, promoting synapse repair and development and stabilising synapses. Thus, it is no longer tenable to view the neurochemical synapse as a synaptic contact made of only the presynaptic nerve terminal and the postsynaptic target, without taking into consideration the multiple active roles of the third element, neuroglia, specifically here the PSCs.

In this chapter, we will focus on the role of PSCs in NMJ synapse formation, function, maintenance, remodeling and regeneration at the NMJ. Similar topics on synapse–neuroglial interactions at the NMJ have been reviewed recently (Auld et al. 2003; Auld and Robitaille

2003a,b; Kang et al. 2003; Koirala et al. 2003; Colomar and Robitaille 2004; Corfas et al. 2004; Feng et al. 2005). In addition to reviewing the role of PSCs, we will highlight findings that are also relevant to the role of astrocytes in CNS synapses. For thorough insights into synapse–neuroglial interactions in the CNS, there are several reviews and books (Haydon 2001; Fields and Stevens-Graham 2002; Volterra et al. 2002; Auld and Robitaille 2003a; Nedergaard et al. 2003; Slezak and Pfrieger 2003; Hatton and Parpura 2004; Newman and Volterra 2004; Ullian et al. 2004a).

THE TRIPARTITE ORGANISATION OF THE NEUROMUSCULAR JUNCTION (NMJ)

The NMJ is arguably the best-studied synapse due to its large size, relatively simple organisation and easy accessibility for *in vivo* observations and experimental manipulations (Katz 1966; Salpeter 1987; Sanes and Lichtman 1999; Ko and Thompson 2003). The NMJ is a tripartite synapse composed of the motor nerve terminal, the postsynaptic specialisations with acetylcholine receptors (AChRs), and PSCs. The frog, *Rana pipiens* (Figure 5.1a–d) and mouse, *Mus musculus* (Figure 5.1e–h) are the best studied NMJs among all vertebrates.

At the frog NMJ, PSCs can be labeled with two vital probes: peanut agglutinin (PNA), which recognises the extracellular matrix associated with PSCs (Ko 1987), and the monoclonal antibody (mAb) 2A12, which binds to the external surface membrane of PSCs (Astrow et al. 1998). Figure 5.1a shows an example of mAb 2A12 immunostaining of PSC somata (asterisks) and processes at a frog NMJ, also fluorescently labelled with anti-neurofilament antibody for axons and anti-synapsin I antibody for nerve terminals (Figure 5.1b), and α-bungarotoxin (α-BTX) for AChRs on muscle fibres (Figure 5.1c). There are typically around 3–4 PSC somata per NMJ in frog muscles (Astrow et al. 1998; Herrera et al. 2000). The tripartite arrangement can be further appreciated in the merged image (Figure 5.1d). There is no Schwann cell labelling by mAb 2A12 along axons, including the pre-terminal axons (compare the arrows in Figure 5.1a and b), in whole-mount preparations, but there is labelling in axonal Schwann cells in cryosections. This is probably attributed to the impermeability of mAb 2A12 through the sheath of perineurim around nerve bundles in whole-mount preparations (Astrow et al. 1998). Since binding of PNA and mAb 2A12 is extracellular and does not alter synaptic function, these two probes reveal the dynamics of PSCs in relation to nerve terminals during synaptic

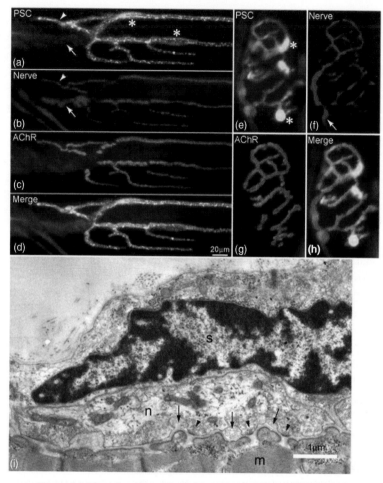

Figure 5.1 The tripartite organisation of vertebrate neuromuscular junctions (NMJs). (a)–(d) Frog, Rana pipiens, NMJ triple labelled with a monoclonal antibody, mAb 2A12, for perisynaptic Schwann cells (PSCs) (a, green), anti-neurofilament and anti-synapsin I antibodies for axons and presynaptic nerve terminals (b, blue), and α-bungarotoxin (α-BTX) for postsynaptic acetylcholine receptors (AChRs) (c, red). Antibody 2A12 labels PSC somata (asterisk in a) and processes (arrowhead in a) but does not label the Schwann cells along axons including pre-terminal axons (compare arrows in a and b). The tripartite organisation is further shown in the merged image (d). Scale bar in d applies to a–c. (e)–(h) An NMJ in a transgenic mouse that expresses green fluorescence protein (GFP) in Schwann cells (e, green) and transgenic cyan fluorescence protein (CFP) for nerve terminal (f, blue). The NMJ is also labelled with α-BTX for AChRs (g, red). Asterisks in e mark PSC somata, and the arrow in f points to a pre-terminal axon. The merged image is shown in h. (Reproduced from

development, sprouting and regeneration in living animals (see below). Unfortunately, PNA and mAb 2A12 do not recognise mammalian PSCs.

At the mammalian NMJ, PSCs can be labelled with antibodies, such as antibody to the Ca^{2+} binding protein S100 (Reynolds and Woolf 1992) or LNX-1 (an E3 ubiquitin ligase) (Young et al. 2005) in intact muscles, and antibodies to p75 neurotrophin receptor (Hassan et al. 1994) or GAP-43 (Woolf et al. 1992) in denervated muscles. Although anti-S100 also labels other cell types, it has been very useful in revealing the dynamic behaviour of mammalian PSCs, but requires membrane permeabilisation and thus cannot be used as a vital stain. A recent approach using transgenic mice that express green fluorescent protein (GFP) family, not only in axons (Feng et al. 2000; Lichtman and Sanes 2003), but also in Schwann cells (Kang et al. 2003; Zuo et al. 2004) has revolutionised our ability to view the dynamic behaviour of axons and PSCs in living animals (Figure 5.1e–h). GFP is expressed not only in PSC somata and processes but also in Schwann cells along the pre-terminal axon, which do not have AChR clusters underneath. The merged picture (Figure 5.1h) further illustrates the tripartite arrangement of the mammalian NMJ. Despite the compact size of mammalian NMJs (~20–40 μm in length), which are approximately 10-fold smaller in length than frog NMJs, there are also about 3–4 PSC somata at each NMJ, and the PSC number is correlated with the endplate size (Herrera et al. 1990; Love and Thompson 1998; Lubischer and Bebinger 1999; Jordan and Williams 2001).

The tripartite nature of the frog NMJ is seen in Figure 5.1i. The PSC(s) cap the motor nerve terminal, which makes synaptic contacts with the muscle fibre. The PSC also sends finger-like processes projecting into the synaptic cleft. These PSC 'fingers' interdigitate with active zones – sites of transmitter release. These PSC 'fingers' contain L-type calcium channels (Robitaille et al. 1996) and are colocalised with F-actin concentrated at the non-release domains of the frog nerve terminal (Dunaevsky and Connor 2000). The functional significance of

(cont.) Kang et al. (2003) with permission). (i) Electron micrograph of a longitudinally sectioned frog NMJ shows the PSC (s) with its electro-dense nucleus capping the nerve terminal (n) on muscle fibre (m). Fine PSC processes or 'fingers' (arrows) are seen in the synaptic cleft and interdigitate with active zones (arrowheads) at the presynaptic nerve terminal membrane. Scale bar: 1 μm. A colour version of this Figure is in the Plate section.

PSC fingers remains to be determined. In mammals, in contrast to frog, PSC 'fingers' are usually not found in the synaptic cleft of the NMJ. The exclusion of PSC processes from the cleft may be attributed to laminin 11 ($\alpha 5\beta 2\gamma 1$) in the synaptic cleft (Patton et al. 1998). PSCs are also covered with basal lamina, which appear to merge with the extra-synaptic basal lamina of the muscle fibre (Saito and Zacks 1969; Engel 1994). PSCs also contain some extracellular matrix molecules that are distinct from those in the extrasynaptic muscle surface and in the synaptic cleft (Ko 1987; Astrow et al. 1997; Patton et al. 1997; for a review see Patton (2003). Furthermore, the extracellular matrix associated with PSCs may be involved in guiding nerve terminal sprouts at the frog NMJ (Chen and Ko 1994; Ko and Chen 1996) (see below).

Why is the NMJ organised in such a manner with PSCs capping the nerve terminal? Does the tripartite organisation imply an essential role of PSCs at the NMJ? To address these questions, Reddy et al. (2003) used mAb 2A12 and complement-mediated lysis (Walport 1998; Reddy et al. 2003) to selectively ablate PSCs in frog muscles (Figure 5.2a). In contrast to the normal appearance of PSCs in control muscles without any treatment (Figure 5.2b), PSCs become electron-lucent and swollen with debris of damaged membranes seen in muscle (Figure 5.2c). The PSC ablation does not cause any acute damage to the nerve terminal and muscle fibre. One week later, PSCs remain absent and there are 'naked' nerve terminals apposed to the muscle fibre (Figure 5.2d). We used this approach to 'knock out' PSCs en masse and to examine the essential roles of PSCs in synaptic function, growth and maintenance in living frog muscles. A similar approach using antibodies against certain epitopes of gangliosides seen in Miller–Fisher syndrome has also been applied to ablate PSCs in rats (Halstead et al. 2005) (also see below).

Despite the juxtaposition of PSCs with the nerve–muscle junction, most studies of the NMJ have focused on the nerve terminal and the muscle fibre. Although there are studies of the synaptic structure, function, development and plasticity at the NMJ (Katz 1966; Salpeter 1987; Sanes and Lichtman 1999; Ko and Thompson 2003), our knowledge of PSCs in this model synapse remains rudimentary.

BRIEF HISTORY OF PERISYNAPTIC SCHWANN CELLS

While Theodor Schwann was credited for observing rows of his namesake cells along peripheral nerve fibres, the first hint of the existence of PSCs can be attributed to Louis-Antoine Ranvier (1878),

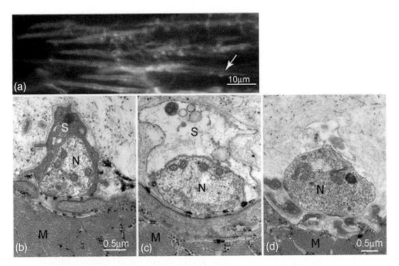

Figure 5.2 Selective ablation of frog PSCs *in vivo*. (a) Fluorescent image of frog NMJs (labeled with PNA, green) treated with mAb 2A12 and complement. The complement-mediated lysis of PSCs is confirmed with positive staining of ethidium homodimer-1 (red), which labels nucleic acids in the cell bodies (arrow) only when the cell membranes are damaged. (b)–(d) Electron micrographs of NMJs in a normal muscle (b), and in PSC-ablated muscles, which were examined acutely (~2 h) (c), and one week after PSC ablation (d). Similar to the longitudinal section in Figure 5.1(i), the cross-section of the normal NMJ (b) shows the typical tripartite arrangement with the PSC (s) overlying the nerve terminal (n) on the muscle (m). In (c), PSCs appear damaged with swollen cytoplasm and vacuoles of fragmented PSC membranes about 2 h after mAb 2A12 plus complement treatment. Despite the severe damage in the PSC, the ultrastructure of the nerve terminal (n) and the muscle fibre (m) appears normal. One week after PSC ablation (d), PSCs remain absent from NMJs, while 'naked' nerve terminals (n) maintain the normal synaptic contact with muscle fibres (m). Electro-dense crystals seen in (b) and (c) are reaction products from acetylcholinesterase staining used to facilitate the identification of NMJs. Scale bar = 0.5 μm. A colour version of this Figure is in the Plate section.

who identified clusters of 'arborisation nuclei', which were accompanied by the motor nerve terminal arbors and were distinct from the muscle fibre nuclei (Couteaux 1973). Through developmental studies and improved histological staining, these arborisation nuclei were later recognised as nuclei of 'teloglia' or terminal Schwann cells (Couteaux 1938; Tello 1944; Boeke 1949; Couteaux 1973). The non-neuronal nature of these cells was further implicated by the observation that these cells survived after nerve degeneration (Tiegs 1953; Birks et al. 1960b). Electron microscope studies performed in the 1950s and 1960s provided high-resolution images of vertebrate NMJs (Beams and Evans 1953; Palade and Palay 1954; Reger 1955; Robertson 1956), and confirmed the intimate contacts of Schwann cells with the nerve terminals (Birks et al. 1960a; Couteaux 1973). The juxtaposition of terminal Schwann cells with the nerve terminal at the NMJ was further confirmed by scanning electron microscopy and freeze-fracture studies (Heuser et al. 1976; Desaki and Uehara 1981; Ko 1981).

The intimate association of PSCs with the neuromuscular synapse was also shown at denervated NMJs. Shortly after nerve injury, PSCs become phagocytic as they engulf degenerating nerve terminals and occupy the synaptic cleft (Birks et al. 1960b; Miledi and Slater 1968). Surprisingly, miniature endplate potentials can be recorded even after nerve injury at denervated NMJs (Birks et al. 1960b; Miledi and Slater 1970; Dennis and Miledi 1974). The recording of these small potentials, which were thought to be caused by spontaneous release of acetylcholine (ACh) from PSCs, further implicates the proximity of PSCs to the synapse. However, decades after Schwann cell miniature endplate potentials were described, we still do not know their physiological significance.

The proximity of PSCs to the neuromuscular synapse has begged a battery of questions regarding the potential role of PSCs in synaptic function, formation and maintenance. With advances in calcium imaging and probes for labelling PSCs, there was a flurry of studies in the early 1990s that galvanised interest in addressing these questions. One of the seminal findings was that PSCs can sense synaptic activity by raising their intracellular calcium (Jahromi et al. 1992; Reist and Smith 1992), and that PSCs can modulate synaptic transmission at the NMJ when G-protein pathways or intracellular calcium levels in PSCs are modified pharmacologically (Robitaille 1998; Castonguay and Robitaille 2001). Thus, PSCs not only can 'listen' to nerve activity but also can 'talk' back to nerve terminals under certain conditions (Colomar and Robitaille 2004). Another key discovery is that PSCs sprout profusely after nerve injury at vertebrate NMJs and that PSC sprouts guide regenerating and sprouting nerve terminals (Reynolds and Woolf 1992; Son and Thompson 1995a; Koirala et al. 2000). Using vital probes for PSCs, we have shown that PSC processes are dynamic and may also guide nerve terminal sprouts during synaptic remodeling at the frog NMJ (Chen et al. 1991; Chen and Ko 1994; Ko and Chen 1996). As further described in this chapter, a barrage of recent studies on synapse–neuroglial interactions at the vertebrate NMJs have led to our appreciation and current understanding of the diverse roles of PSCs in synapse formation, maturation, maintenance and function.

ARE PERISYNAPTIC CELLS AT THE NMJ OF SCHWANN CELL ORIGIN?

Although it has been widely accepted that the motor nerve terminal is capped with non-neuronal cells, whether these synapse-associated cells indeed originate from Schwann cells was questioned in the late 1960s. Based on electron microscope observations, Shanthaveerappa and Bourne (1967) used the term 'PSC' as an abbreviation for 'perisynaptic cells', which they believed were perineural epithelial cells, rather than Schwann cells (Shanthaveerappa and Bourne 1966; Shanthaveerappa and Bourne 1967). However, this idea was later refuted by Saito and Zacks (1969), who found no evidence of perineural epithelial cells covering the nerve terminal, and reconfirmed with electron microscopy that the nerve terminal was covered with non-myelinating Schwann cells (Saito and Zacks 1969). Recent studies using molecular approaches have put this controversy largely to rest, demonstrating that these perisynaptic cells are indeed of neural crest/Schwann cell origin as they express many neuroglial markers (see also Chapter 2). For example, PSCs can be labelled with antibodies to the Ca^{2+} binding protein S100 (Reynolds and Woolf 1992) and the glial fibrillary acidic protein (GFAP) (Georgiou et al. 1994), both of which are abundant in neuroglia, albeit in other cells as well. In addition, transgenic mice that express GFP under control of the human S100B gene show GFP expression in myelinating Schwann cells as well as PSCs at the NMJ (Zuo et al. 2004). Furthermore, PSCs express Schwann cell molecular markers such as protein zero (P0) as well as myelinating glial markers myelin-associated glycoprotein (MAG), galactocerebroside and 2′,3′-cyclic nucleotide 3′-phosphodiesterase (Georgiou and Charlton 1999). Thus, there is little doubt that the perisynaptic cells are of Schwann cell origin, and PSCs can be appropriately referred to as 'perisynaptic Schwann cells', as dubbed by Jahromi et al. (1992). However, whether and how PSCs differentiate directly from Schwann cell precursors or from immature Schwann cells during development is not known. Furthermore, it remains a mystery why the non-myelinating PSCs express proteins thought to be critical for myelination and what role these myelinating proteins might play in the biology of PSCs.

ROLES OF PSCs IN SYNAPTOGENESIS

Are Schwann cells necessary for axon outgrowth to target muscles?

Although the differentiation of PSCs during embryonic development is not well understood, the involvement of PSCs in the development of NMJs has been investigated. To address the roles of PSCs in the formation of the NMJ, the first question one would like to raise is whether developing Schwann cells, which are in close association with growing axons, are necessary for the axon to grow out to target muscles. Harrison (1908, 1924) addressed this question by surgical removal of the neural crest in tadpoles, and found that the absence of Schwann cells did not prevent axonal outgrowth to target muscles. Although this finding was later confirmed in the chicken (Lewis et al. 1983), other studies have shown that migrating Schwann cells apparently lead axonal growth cones (Noakes and Bennett 1987) and the nerves would not grow into the limb in the absence of Schwann cells (Noakes et al. 1988; Yntema 1943). However, the controversy as to whether Schwann cells are necessary, and whether Schwann cells are leaders or followers during the initial navigation of axons to their target muscles (Keynes 1987) has largely been resolved by recent studies of ErbB2 (Morris et al. 1999; Woldeyesus et al. 1999; Lin et al. 2000), ErbB3 (Riethmacher et al. 1997) and Splotch (Grim et al. 1992) mutant mice, all of which lack Schwann cells in both the peripheral nerves and nerve–muscle contacts. Despite the absence of Schwann cells through this 'genetic ablation', motor axons still project to their target muscles and even form nerve–muscle contacts, albeit only transiently. These studies suggest that Schwann cells are dispensable for axonal navigation to target muscles during development.

Are PSCs necessary for the development of NMJs?

The observation of transient nerve–muscle contacts in mutant mice lacking Schwann cells suggests that PSCs are also not essential for the initial formation of nerve–muscle contacts. In addition, functional nerve–muscle contacts can be formed without Schwann cells in cultures of frog tissue (Kullberg et al. 1977; Chow and Poo 1985). Thus, similar to axonal pathfinding, the initiation of nerve–muscle contact formation does not require the involvement of Schwann cells. However, as described below, increasing evidence suggests that

Schwann cells play essential roles in the subsequent growth, maturation and maintenance of developing NMJs.

The importance of neuroglial cells in promoting NMJ development has been indicated by the observation that Schwann cells are present at developing NMJs soon after initial nerve–muscle contacts are formed (Kelly and Zacks 1969; Linden et al. 1988; Herrera et al. 2000). In addition, the number of PSCs per NMJ increases and correlates with the length of NMJ during the period of intensive synaptic growth (Hirata et al. 1997; Love and Thompson 1998; Herrera et al. 2000). Moreover, a possible maintenance role of PSCs is indicated by the findings mentioned above that NMJs are transiently formed, but not maintained, in mutant mice lacking Schwann cells (Morris et al. 1999; Woldeyesus et al. 1999; Lin et al. 2000; Wolpowitz et al. 2000). However, myelinating and non-myelinating Schwann cells are also absent in these mutants, which die at birth. Thus, these mice cannot be used to address the specific role of PSCs in synapse growth, maturation and maintenance following the initial formation of nerve–muscle contacts.

To directly test the essential role of Schwann cells in promoting neuromuscular development, we have used two strategies. First, we examined the dynamic behaviour of PSCs in relation to nerve terminal growth throughout the course of NMJ development in frog muscles. Fluorescent triple-labelling revealed PSCs, nerve terminals, and AChRs (Herrera et al. 2000; Reddy et al. 2003). At the earliest discernible nerve–muscle contacts, PSCs are already present but only partially cover the nerve terminals. PSCs then quickly occupy the full junctional area, and extend profuse sprouts, ranging from a few microns to hundreds of microns, beyond the borders of nerve–muscle contacts as development progresses (Herrera et al. 2000). The period of extensive sprouting of PSC processes coincides with the period of the most vigorous synaptic growth. The number of PSCs per NMJ correlates with the extension of junctional size and increases from approximately 0.5 before metamorphosis to almost 3 in the adult *Xenopus*. As development advances, nerve terminals extend along PSC processess. Shortly after metamorphosis, these PSC processes retract and NMJs acquire the tripartite organisation as seen in adult NMJs (Herrera et al. 2000). This series of static images from *Xenopus* at various developmental stages suggest that PSC sprouts lead nerve terminal growth during synaptogenesis. To further reveal this dynamic relationship, we have repeatedly imaged *in vivo* both PSCs and nerve terminals at identified NMJs (Reddy et al. 2003). At first, profuse PSC processes extending beyond nerve terminals were routinely seen at developing NMJs in tadpole muscles. At the same junctions observed 8–12 days later, PSCs extend even longer processes and subsequent growth and addition of nerve terminals follows the contours of the preceding PSC processes in tadpole muscles. Taken together, these results suggest that, while PSCs are not required for the initiation of nerve–muscle contacts in frogs, PSCs and their processes lead and guide nerve terminal growth during NMJ development.

The second strategy to unravel the necessary role of PSCs in synaptogenesis was to test whether and how the removal of PSCs alters synaptic growth and maintenance *in vivo* (Reddy et al. 2003). By combining PSC ablation in frog tadpoles with repeated *in vivo* imaging of identified NMJs, Reddy et al. (2003) were able to compare the morphology of the same NMJs before and at 8 and 12 days after PSC ablation. In control muscles, treated with mAb 2A12 or complement alone, PSCs were intact, and over 50% of tadpole NMJs showed nerve terminal extension and fewer than 3% retraction after 8–10 days. In contrast, very few NMJs (6.7%) showed nerve terminal extension, but almost half (44%) underwent retraction in PSC-ablated tadpole muscles. Among those nerve terminals that changed in length, terminal length increased by about 70–80% in control PSC-intact muscles, but decreased by about 42% in PSC-ablated muscles. Furthermore, while ~10% of new NMJs were added between the first and the second observation periods in PSC-intact controls, no new NMJs were seen in PSC-ablated muscles. These results suggest that PSCs promote synaptic growth and are necessary for the maintenance of developing amphibian NMJs *in vivo* (Figure 5.3).

In rats, the essential role of PSCs in developing NMJs has also been shown. At birth, approximately half of all NMJs have only one PSC soma, while the remainder show only PSC processes arising from somata and arranged along the pre-terminal axon (Love and Thompson 1998). As development progresses, the number of PSC somata increases and NMJs show an additional two to three PSCs by two months of age. The increase, through cell migration and proliferation, occurs largely during the first two postnatal weeks, and precedes the increase in endplate size. There is a similar increase in PSC numbers at the postnatal rat NMJs (Hirata et al. 1997). In contrast to the survival of PSCs after axotomy in the adult, PSCs in neonatal denervated muscles undergo apoptosis, which can be prevented by injection of a neuroglial growth factor, neuregulin 1 (NRG1), which is present in developing motor neurons (Trachtenberg and Thompson 1996). In addition, NRG1

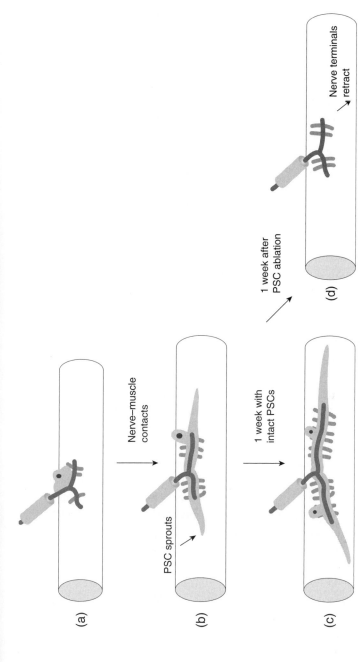

Figure 5.3 A schematic diagram summarising the role of PSCs in the development of frog tadpole NMJs. While PSCs are not required for the formation of the initial nerve–muscle contact (a), soon after the contact is formed, PSCs extend profuse processes ahead of growing nerve terminals (b). As development progresses, nerve terminals grow along the preceding PSC sprouts (c). (d) One week after PSC ablation, nerve terminals stop growing and some also retract, suggesting that PSCs promote synaptic growth and are essential for the maintenance of developing NMJs. In Figures 5.3, 5.4 and 5.7, myelinating Schwann cells and PSCs are shown in green; motor axons and nerve terminals in blue; and clusters of AChRs in red. A colour version of this Figure is in the Plate section.

or Schwann cell transplants can alter PSC morphology and synaptic structure in developing rat muscles (Trachtenberg and Thompson 1997). Thus, NRG1 and ErbB receptor signalling can regulate the development of not only axonal Schwann cells but also nonmyelinating Schwann cells including PSCs (Chen et al. 2003; Corfas et al. 2004). The necessary role of PSCs in synaptic growth has further been illustrated by the observation that partial denervation in neonatal, but not adult, rat muscles results in apoptosis of PSCs and absence of nerve terminal sprouting in neonatal muscles (Lubischer and Thompson 1999). Moreover, the withdrawal of NMJs observed in ErbB2 and ErbB3 mutant mice may be in part attributed to the lack of PSCs (Riethmacher et al. 1997; Morris et al. 1999; Woldeyesus et al. 1999). Taken together, these studies suggest that mammalian PSCs also play an essential role in the growth and maintenance of developing nerve terminals at the NMJ. Similarly, astrocytes in the CNS promote synaptogenesis and maintain developing synapses (Pfrieger and Barres 1997; Ullian et al. 2001; Christopherson et al. 2005). Similar results were obtained when examining the effects of postnatal rat Schwann cells on these CNS synapses (Ullian et al. 2004b).

Whether PSCs also play an active role in pruning excess nerve terminals at multiply innervated NMJs in postnatal muscles is not known. Culican et al. (1998) have shown that retraction of nerve terminals is accompanied, over a similar time course, by withdrawal of PSC processes from the sites of synapse elimination (Culican et al. 1998). Based on the spatial relationship between retracting axons and Schwann cells, Hirata et al. (1997) suggest that Schwann cells do not play an active role in axon pruning at developing NMJs (Hirata et al. 1997). However, a recent study has shown that retracting axons shed cellular fragments, axosomes, which are phagocytosed by adjacent Schwann cells and assimilated into their cytoplasm during synapse elimination (Bishop et al. 2004). While neuroglial cells are believed to play an active and necessary role in axon pruning in the brain of the invertebrate Drosophila (Awasaki and Ito 2004; Watts et al. 2004), whether Schwann cells play a similar role at vertebrate NMJs remains to be investigated (see also Koirala and Ko (2004)). The recent availability of transgenic mice that express fluorescent reporter genes in both axons and Schwann cells may provide a new tool to study the involvement of PSCs in axon pruning at developing NMJs *in vivo* (Kang et al. 2003; Zuo et al. 2004).

ROLES OF SCHWANN-CELL-DERIVED FACTORS IN
SYNAPTOGENESIS

It has been known that NRG1 supplied by developing axons is required for the survival of the Schwann cell precursors (Dong et al. 1995), as well as the survival of axonal Schwann cells and PSCs in neonatal animals (Grinspan et al. 1996; Trachtenberg and Thompson 1996). Conversely, Schwann cells may be required for the survival of developing motor neurons as shown by the widespread motor neuron death in ErbB receptors-null or NRG1 null mutant mice that lack Schwann cells (Morris et al. 1999; Riethmacher et al. 1997; Woldeyesus et al. 1999; Wolpowitz et al. 2000). Thus, NRG1-ErbB signalling is essential for the reciprocal interactions between Schwann cells and motor neurons and their survival in neonates (Corfas et al. 2004). However, the molecular mechanisms by which Schwann cells promote neuromuscular synaptic growth and maintenance are not well understood.

To identify potential molecules through which Schwann cells promote NMJ development, we examined the effects of Schwann-cell-conditioned medium on synapse formation and differentiation in frog nerve–muscle cultures (Peng et al. 2003). Embryonic frog spinal neurons usually die within 3–4 days in culture, but can be kept alive for up to 10 days by the exogenous application of a mixture of the growth factors: brain-derived neutrophic factor (BDNF), glial-derived neurotrophic factor (GDNF), neurotrophin-3 (NT3), neurotrophin-4 (NT4), plus forskolin to elevate cAMP. As expected, this trophic treatment also induces extensive neurite outgrowth. However, the number of nerve–muscle contacts marked by AChR clusters is greatly reduced. The paucity of neuromuscular synaptic contacts may be attributed to the downregulation of agrin, a protein essential for clustering of AChRs (McMahan 1990), in cultured spinal neurons following treatment with the mixture of trophic factors. Importantly, the addition of conditioned medium restored agrin expression, and greatly increased the number of nerve–muscle contacts showing AChR clusters. The result suggests that, while trophic factors promote a neurite 'growth mode', Schwann-cell-derived factors switch this mode to a 'synaptogenic mode', in which neurite growth stops and synaptogenesis ensues.

We have recently found that transforming growth factor (TGF)-β1 can mimic the effect of conditioned medium alone, in increasing the number of nerve–muscle contacts associated with AChR as well as upregulating agrin expression in spinal neurons (Feng and Ko 2004).

This synaptogenesis-promoting role of conditioned medium is absent if the culture is treated with conditioned medium immunoprecipitated with an antibody to TGF-β1. The results suggest that TGF-β1 is a key molecule mediating the conditioned-medium-induced synaptogenesis in frog nerve muscle cultures. Whether TGF-β1 is essential for synaptogenesis *in vivo* remains to be examined. It is interesting to note that conditioned medium also increases the number of synapses formed between mammalian spinal motor neurons in culture, although those synapses are glutamatergic (Ullian *et al.* 2004b). Whether TGF-β1 also promotes synaptogenesis in mammalian nerve–muscle culture is not known. Christopherson *et al.* (2005) have recently identified thrombospondins as astrocyte-secreted proteins that promote synaptogenesis in the developing CNS. It would be interesting to test if thrombospondins also play a role in the formation of NMJs.

In addition to promoting synaptogenesis, we have also found that conditioned medium greatly enhances synaptic transmission in frog nerve–muscle culture (Cao and Ko 2001). The frequency of spontaneous synaptic currents was dramatically increased by approximately 150-fold. The presynaptic enhancement of transmitter release is an acute effect, starting within 5–10 min of conditioned-medium addition and was not attributable to any known factors, including neurotrophins, glutamate and ATP, that modulate synaptic efficacy. We have found that the molecular weight of the active factor(s) in Schwann-cell-conditioned medium responsible for enhancing transmitter release is smaller than 5 kDa, although the exact identity of the molecule(s) remains unknown. Recent studies have demonstrated that mammalian astrocytes and astrocyte-secreted factors can also enhance transmitter release (Allen and Barres 2005). Thus, Schwann cells and astrocytes of the CNS share many similar roles in promoting synaptogenesis.

ROLES OF PSCs IN SYNAPTIC TRANSMISSION

Can PSCs 'listen' to nerve terminals?

The first indication that PSCs might be involved in modulating NMJ transmission came from the seminal work by Jahromi *et al.* (1992) and Reist and Smith (1992), who demonstrated that high-frequency stimulation of peripheral nerve increases intracellular calcium levels in PSCs at frog NMJs (Jahromi *et al.* 1992; Reist and Smith 1992). A similar increase in PSC calcium level has also been seen at mouse

NMJs (Rochon et al. 2001). The calcium response is limited to PSCs and is not detected in myelinating Schwann cells lined up and ensheathing axons (Jahromi et al. 1992; Reist and Smith 1992). Despite the presence of L-type calcium channels (Robitaille et al. 1996) and the alpha1B subunit (encoding the N-type) of the voltage-dependent calcium channel (Day et al. 1997) on PSCs, the increase of intracellular calcium in PSCs is believed to be caused by calcium released from internal calcium stores. This is because the removal of extracellular calcium does not block the calcium response (Jahromi et al. 1992; Rochon et al. 2001).

What is the mechanism by which neuronal activity elevates intracellular calcium levels in PSCs? It has been shown that the increase in PSC calcium level is prevented by ω-conotoxin, which blocks transmitter release, but not by α-BTX, which blocks AChRs (Jahromi et al. 1992; Rochon et al. 2001). Thus, substances released from the nerve terminal during high-frequency stimulation probably mediate the calcium increase in PSCs. Indeed, local application of ACh can mimic nerve stimulation in increasing PSC calcium levels, probably through muscarinic AChRs (Jahromi et al. 1992; Robitaille et al. 1997; Rochon et al. 2001). Exogenous applications of ATP, adenosine and substance P can also induce the calcium response in PSCs (Bourque and Robitaille 1998; Jahromi et al. 1992; Robitaille 1995; Rochon et al. 2001). Taken together, these studies suggest that PSCs can detect or 'listen' to synaptic activities at the NMJ.

Can PSCs 'talk' to nerve terminals?

As indicated by the studies described above, PSCs contain receptors, such as muscarinic ACh receptors, ATP receptors, adenosine receptors and substance P receptors. Since most of these receptors are G-protein-coupled receptors, activation of the G-protein signalling pathways in PSCs may in turn influence NMJ activities. Robitaille (1998) has tested this idea by microinjection of GTP-γS, a non-hydrolysable analogue of GTP, into frog PSCs and found that transmitter release in response to nerve stimulation is reduced. Conversely, microinjection of GDP-βS, a non-hydrolysable analogue of GDP, into frog PSCs inhibits NMJ depression during high-frequency stimulation, but has no effect on tonic release of transmitter. Thus, activation of G-protein pathways in PSCs may depress synaptic transmission at the NMJ.

As discussed above, high-frequency nerve stimulation increases the intracellular calcium level in PSCs. This raises another possibility, that neuronal activity may also be modulated by this PSC calcium increase. Castonguay and Robitaille (2001) have shown that indeed increasing intracellular calcium level in frog PSCs by treatment with thapsigargin, a Ca^{2+}-ATPase pump inhibitor, or microinjection of inositol 1,4,5-triphosphate (IP3) can potentiate the evoked release of neurotransmitter ACh at the NMJ. Furthermore, microinjection of BAPTA, a calcium chelator, into frog PSCs can enhance synaptic depression induced by high-frequency nerve stimulation. This study suggests that the increase in PSC calcium level during high-frequency nerve stimulation may in turn potentiate transmitter release at the NMJ.

Thus, based on the above two studies in the frog (Castonguay and Robitaille 2001; Robitaille 1998), PSCs can both enhance and reduce evoked-transmitter release depending on which pathway is manipulated pharmacologically. Whether PSCs potentiate or depress NMJ transmission may depend on the net outcome of these pathways (Colomar and Robitaille 2004). It has been shown that PSCs express enzymes required for synthesis of nitric oxide (NO) (Descarries *et al.* 1998; Rothe *et al.* 2005). Whether NO is released by PSCs and modulates synaptic transmission at the NMJ remains to be studied. Furthermore, while PSCs have the ability to modulate transmitter release, it is unclear whether and how PSCs influence NMJ activities under physiological conditions. Our recent work using complement-mediated lysis to ablate PSCs *in vivo* suggests that PSCs, in frogs at least, do not play an acute role in modulating synaptic transmission at the NMJ (Reddy *et al.* 2003) (see below).

ROLES OF PSCs IN THE MAINTENANCE OF THE ADULT NMJ

The development studies discussed above suggest that PSCs play an essential role in the maintenance of developing NMJs. Are PSCs critical for the maintenance of structure and function at the adult NMJ as well? To address this question, we took advantage of complement-mediated cell lysis to selectively ablate PSCs from frog NMJs *in vivo*, and examine the acute and chronic effects of PSC removal on NMJ structure and function. Despite osmotic damage to PSCs following mAb 2A12 and complement treatment, the ultrastructure of nerve–muscle contacts appear normal following acute (within 5 h) PSC ablation (Figure 5.2). This acute PSC ablation also does not cause any significant changes in synaptic transmission, including the amplitude and frequency of miniature endplate potentials, the amplitude of endplate potentials, and quantal contents. In addition, paired-pulse facilitation and synaptic depression in response to high-frequency stimulation are not altered. Thus, PSCs do not seem to play an *acute* role in modulating synaptic transmission and in the maintenance of synaptic structure at the frog NMJ.

In contrast to the lack of acute effects, both synaptic structure and function are altered one week after PSC ablation. We have found partial or total retraction of nerve terminals in about 13% of the PSC-ablated NMJs, while AChR clusters remain at the vacated post-synaptic sites. Electron microscopy also shows absence of both nerve terminals and PSCs in approximately 10% of post-synaptic sites, while 'naked' nerve terminals without the overlying PSCs are seen in the remaining sites. The 'naked' nerve terminals appear normal. In addition, we did not observe any sign of Wallerian degeneration in retracting nerve terminals that partially occupy the postjunctional site. Furthermore, there was neither widening of the NMJ cleft between the retracting nerve terminals and muscle surface nor global detachment of 'naked' nerve terminals even after intense muscle contraction in PSC-ablated muscles. Consistent with the morphological changes, NMJ function is also altered one week after PSC ablation. Transmitter release is significantly reduced as shown by the approximately 50% decrease in miniature endplate potential frequency (but not amplitude), endplate potential amplitude and mean quantal contents. The reduction of NMJ transmission also results in the compromise of the overall muscle function as shown by the decrease in muscle twitch-tension in response to nerve stimulation.

Taken together, these PSC ablation studies suggest that PSCs are essential for the long-term, but not short-term, maintenance of synapse structure and function at the adult frog NMJ (Figure 5.4). Is the long-term maintenance by PSCs attributed to the possibility that PSCs mechanically 'glue' the nerve terminals to the postsynaptic site? As discussed above, the absence of global detachment of nerve terminals after PSC removal argues against this suggestion. It has been shown that Schwann cells express an array of trophic factors and extracellular matrix proteins (Scherer and Salzer 1996; Fu and Gordon 1997; Frostick *et al.* 1998). It is possible that PSCs, like other Schwann cells, also express these molecules to provide trophic influence for maintaining nerve terminal structure and function at the frog NMJ. The result is also consistent with the finding that muscle damage alone does not affect nerve terminal maintenance (Dunaevsky and Connor 1998).

Whether PSCs also play a similar role in the maintenance of adult mammalian NMJs is not known, even though, as discussed earlier, PSCs are essential for the growth and maintenance of developing

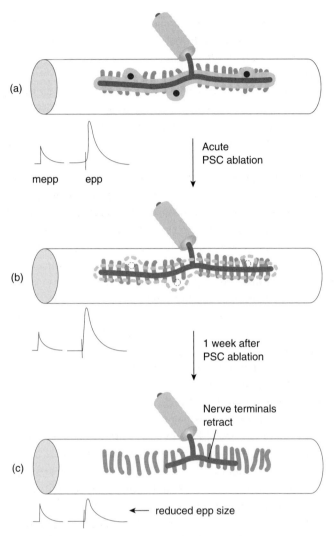

Figure 5.4 A schematic diagram summarising the role of PSCs in the maintenance of frog NMJs. The normal adult NMJ with a sample of miniature endplate potentials (mepp) and endplate potentials (epp) is depicted in (a). PSC ablation does not cause acute effects on the nerve–muscle contact and synaptic transmission (b). However, one week after PSC ablation, retraction of nerve terminals is seen in about 10% of NMJs and the amplitude of endplate potentials, but not miniature endplate potentials, is reduced approximately by half (c). The result suggests that PSCs play an essential role in long-term maintenance of synaptic structure and function. A colour version of this Figure is in the Plate section.

NMJs in both frog and mammalian muscles. Because mAb 2A12 does not recognise mammalian PSCs, the antibody cannot be used for complement-mediated cell lysis as in frog muscles. Recent studies have shown that subsets of auto-antibodies against disialosyl epitopes of gangliosides seen in Miller–Fisher syndrome (MFS) (Willison and Yuki 2002) can bind and damage PSCs via membrane attack complexes at mouse NMJs (Halstead et al. 2004; Halstead et al. 2005). Like mAb 2A12 for frog PSC ablation, these subsets of antibodies may be used to ablate mammalian PSCs in vivo. As shown in a confocal image in Figure 5.5a, PSCs at mouse NMJs are damaged by treatment with an antibody to the ganglioside GD3 epitope, followed by normal human serum containing complement proteins. The death of PSCs is confirmed by the positive staining of ethidium homodimer (in red), a dead cell marker, in PSC somata overlying AChR clusters (in green). The selective damage to PSCs has also been confirmed with electron microscopy (Figure 5.5b). Similar to PSC ablation in frog muscles, mammalian PSCs following complement-mediated lysis also show swollen nuclei (arrowheads) and remnants of lysed PSCs with electron-lucent cytoplasm (arrows). Despite the severe damage of PSCs, the ultrastructure of nerve terminals and muscle fibres is not altered after the treatment. Control muscles treated with Ringer's solution and complement proteins do not show any damage at the NMJ (Figure 5.5c). Intracellular recordings from PSC-ablated muscles show no significant acute changes in transmitter release at mouse NMJs (Halstead et al. 2005). This finding is reminiscent of acute PSC ablation in frog muscles. The long-term effect of PSC ablation at mammalian NMJs remains to be investigated. It would also be interesting to study whether damage to PSCs may contribute to MFS and other peripheral neuropathies. A recent study has shown that Nav1.6 sodium channels are expressed in PSCs (Musarella et al. 2006). However, in a mouse model for the motor endplate disease (med), in which Nav1.6 sodium channels are absent, abundant nerve terminal sprouting and endplates devoid of S100- or GFAP-positive PSCs are observed. The result suggests that defects in PSCs may contribute to the med phenotype (Musarella et al. 2006).

ROLES OF PSCs IN REMODELLING AND REGENERATION

Remodelling

It has been shown that synaptic connections undergo remodelling throughout adult life (Purves and Lichtman 1985; Wernig and

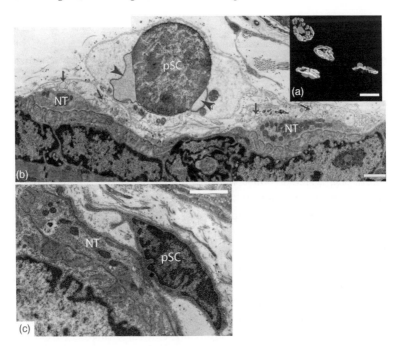

Figure 5.5 Ultrastructural characterisation of perisynaptic Schwann cell injury in the mouse diaphragm. (a) Confocal image of lumbrical NMJs (bungarotoxin, green) exposed to anti-GD3 antibody plus normal human serum (NHS). PSC damage is monitored by ethidium dimmer uptake (red) into PSC nuclei, which is normally excluded by healthy, intact plasma membranes. Scale bar = 20 µm. (b) An NMJ exposed to an anti-GD3 antibody plus NHS. The nerve terminals have a normal appearance whilst the overlying PSC displays severe damage. The electron-lucent PSC cytoplasm is swollen and contains disrupted organelles and vesicle-laden processes (arrows). The nucleus is swollen with a circular appearance and perinuclear bodies (arrowheads), indicative of necrotic cell death. Scale bar =1 µm. (c) Control NMJ profiles exposed to Ringer's solution and NHS as a source of complement. Presynaptic nerve terminals (NT) are of normal appearance, containing densely packed synaptic vesicles and electron-dense mitochondria. The PSC overlies the nerve terminal and contains an electron-dense nucleus, which fills the cell body. The cell cytoplasm is also darkly stained, and is only visible around the perimeter of the cell and as fine processes lying on top of the axonal membrane. (This figure is kindly provided by Dr Sue Halstead and Dr Hugh J. Willison of the University of Glasgow, see Halstead et al. (2005).) A colour version of this Figure is in the Plate section.

Herrera 1986). Extension and retraction of NMJs have also been observed in frog muscles in normal conditions (Wernig et al. 1980; Herrera et al. 1990; Chen et al. 1991). To examine whether PSCs contribute to this synaptic remodelling, we have used PNA as a vital probe for PSCs (Ko 1987) to observe the dynamic behaviour of PSCs in relation to nerve terminals in living frog muscles (Chen et al. 1991; Chen and Ko 1994; Ko and Chen 1996) Results of the repeated in vivo observations have shown that, similar to nerve terminals, PSCs are also undergoing remodelling. Furthermore, PSCs and associated extracellular matrix often extend longer than nerve terminals, which grow along the track of preceding PSC processes. The dynamic relationship between PSCs and nerve terminals has also been confirmed using direct injection of fluorescent dyes into adult toad PSCs and nerve terminals (Macleod et al. 2001; Dickens et al. 2003). These results suggest that PSCs guide nerve terminal sprouting during synaptic remodelling in normal frog muscles. The dynamic behaviour of PSCs may explain the constant remodelling of nerve terminals at frog NMJs. Although mammalian NMJs are relatively stable (Lichtman et al. 1987; Wigston 1989) and do not undergo extensive remodelling as seen at frog NMJs, minor processes of nerve terminal filopodia and lamellipodia are seen adjacent to PSCs (Robbins and Polak 1988). Interestingly, PSCs also extend short and unstable processes beyond AChR clusters at NMJs as seen in transgenic mice expressing GFP in Schwann cells (Zuo et al. 2004).

Regeneration

After nerve injury, peripheral axons are capable of regenerating and reinnervating original synaptic sites in muscles (Letinsky et al. 1976; Nguyen et al. 2002). Axonal Schwann cell basal lamina scaffolding has been shown to provide substrates for axonal regeneration (Kuffler 1986; Son and Thompson 1995b). Does the dynamic behaviour of PSCs seen during NMJ development and synaptic remodelling also contribute to NMJ repair after nerve injury? As shown in the seminal work by Reynolds and Woolf (1992), PSCs sprout profuse numbers of processess after denervation which retract upon reinnervation at the mammalian NMJs (Reynolds and Woolf 1992). Thompson and colleagues have not only confirmed this phenomenon but also demonstrated the significance of PSC sprouting in NMJ reinnervation and nerve terminal sprouting (Son and Thompson 1995a,b).

In mammalian muscles, PSCs extend profuse processes in apparently random directions, beyond the original synaptic sites within days of nerve injury (Reynolds and Woolf 1992; Woolf et al. 1992; Astrow et al. 1994; Son and Thompson 1995b). PSC sprouting can also be induced by paralysis (Brown et al. 1981; Son and Thompson 1995a). Following axonal regeneration (Figure 5.6a–d), axons grow along basal lamina scaffolding lined with activated axonal Schwann cells, and reinnervate the original junctional sites (Son and Thompson 1995b). Upon reinnervation of the junctional sites, PSC sprouts begin to form 'bridges' with other PSC sprouts from adjacent junctional sites, which are not yet reinnervated. Regenerating nerve terminals grow along these PSC 'bridges' and form the so-called 'escaped fibres' to innervate the adjacent endplates. Some of these escaped fibres can continue to grow, but in a retrograde direction, along neighbouring denervated basal lamina scaffolding, and thereby innervate junctional sites in more muscle fibres. These results suggest that PSC processess induced by nerve injury guide regenerating nerve fibres to rapidly restore reinnervation of all NMJ sites (Kang et al. 2003).

The importance of PSC 'bridges' in synaptic repair is further demonstrated in muscles following partial denervation. It has been known that partial damage of nerve fibres innervating the same NMJ target results in nerve terminal sprouting from the remaining intact site to the denervated site (Brown et al. 1981). Son and Thompson (1995) have found that partial denervation also induces PSC sprouting. As shown in Figure 5.6e–g, PSCs sprout from denervated junctional sites to form 'bridges', which make contact with PSC processes at the adjacent innervated junctional sites. Consequently, nerve terminals sprout from the innervated junctional sites along the 'bridges' back to innervate the neighbouring denervated endplates. The results suggest that the formation of these PSC 'bridges' can induce nerve terminal sprouting and restore the NMJ following partial denervation in adult muscles. In contrast to adults, partial denervation in neonates results in PSC apoptosis at denervated sites and lack of nerve terminal sprouting (Lubischer and Thompson 1999), which further demonstrates the importance of PSC 'bridges' in inducing nerve terminal sprouting. Interestingly, in *mdx* mice (a model for Duchenne Muscular Dystrophy), formation of PSC 'bridges' is impaired, which may contribute to the less effective reinnervation and muscle weakness in these mutant muscles (Personius and Sawyer 2005). Poor reinnervation seen during aging may also be attributed to the attenuation of PSC sprouting after nerve injury in aged muscles (Kawabuchi et al. 2001).

The importance of the PSC bridges discussed above, inferred from static observations from fixed tissues, has been confirmed by repeated *in vivo* observations. Using Calcein blue to label PSCs, O'Malley et al. (1999) have shown that regenerating nerve terminals grow along PSC processess in mouse muscles. Recently, Kang et al. (2003) have taken advantage of transgenic mice that express GFP in Schwann cells and CYP in axons to further confirm that PSC processess induced by nerve injury guide regenerating nerve terminals (Kang et al. 2003). The dynamic relationship between PSC and regenerating nerve terminals has also been examined with *in vivo* observations of the same NMJs repeatedly after nerve injury in frog muscles (Koirala et al. 2000). In contrast to mammals, nerve injury alone in frogs does not induce sprouting of PSC processess, as judged by mAb 2A12 and PNA staining for PSCs. Rather, PSC sprouting occurs only after arrival of regenerating nerve terminals. Using PNA as a probe for PSCs, we have shown that PSC sprouts also guide regenerating nerve terminals in frog muscles (Figure 5.7).

What mechanisms induce PSCs to sprout these processes are not known. It has been shown that exogenous application of NRG1 to neonatal muscles or expression of constitutively activated ErbB2 receptors in PSCs induces migration of PSCs away from endplate sites (Trachtenberg and Thompson 1997; Hayworth et al. 2006). These results suggest that NRG1–ErbB signalling is involved in PSC process sprouting. Similar to axonal Schwann cells, PSCs upregulate expressions of GFAP, GAP-43, and p75 NGF receptors after nerve injury or toxin-induced paralysis (Woolf et al. 1992; Georgiou et al. 1994; Hassan et al. 1994). However, whether the upregulation of these molecules plays a role in PSC process formation has not been demonstrated. It has been shown, however, that, in partially denervated muscles, botulinum toxin and α-BTX prevent the formation of PSC 'bridges' that link denervated to adjacent innervated endplates, but without affecting the growth of PSC processess (Love and Thompson 1999). It is interesting to note that direct stimulation or exercise of partially denervated muscles also blocks PSC bridge formation and thereby prevents nerve terminal sprouting (Love et al. 2003; Tam and Gordon 2003). The mechanism by which activity affects PSC process formation is not clear. Although nerve terminal sprouting can be induced by growth factors such as CNTF, IGF-1 (English 2003), whether they mediate PSC-induced nerve terminal sprouting is not known.

In addition to the presynaptic role in guiding regenerating nerve terminals after nerve injury, Schwann cells also express neuronal

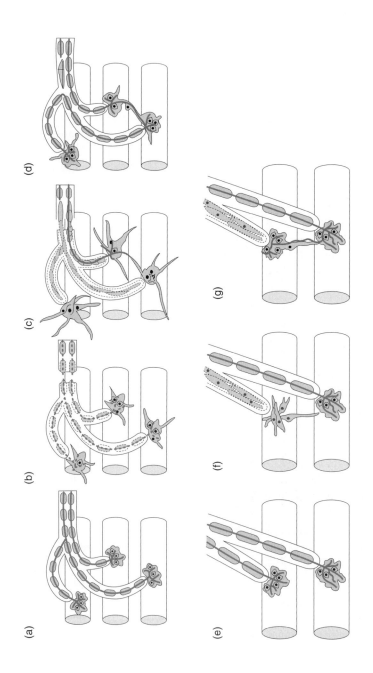

96

Figure 5.6 Schematic diagrams summarising the role of PSCs in reinnervation after nerve injury (a)–(d) and in sprouting after partial denervation (e)–(g) at mammalian NMJs. (a) Three muscle fibres innervated by intramuscular nerves are depicted in a normal muscle. (b) Following nerve injury, PSCs extend processes beyond the synaptic sites. (c) An axon regenerates by growing along the basal lamina scaffolding; and as it reinnervates a synaptic site (the middle muscle fibre), a bridge of PSC process is formed with the adjacent site. (d) The regenerating nerve terminal follows this PSC bridge and innervates the adjacent synaptic site (the lower muscle fibre). The 'escaped fibre' continues to grow in a retrograde direction along another basal lamina scaffolding to innervate the third endplate (the upper muscle fibre). The result suggests that PSCs play a role in the rapid reinnervation of all synaptic sites after nerve injury. (e) Two muscle fibres in the same muscle innervated by two different nerves are depicted. (f) After damage of one, but not the other, nerve, the partial denervation causes PSCs at the denervated endplate site (the upper muscle fibre) to sprout, and one long process forms a bridge with PSC processes at the innervated endplate site (the lower muscle fibre). (g) A nerve terminal branch sprouts from the innervated endplate growing along the PSC bridge to reinnervate the adjacent denervated endplate sites. The result suggests that PSCs induce and guide nerve terminal sprouts following partial denervation and play an important role in synaptic repair. (Modified from Kang *et al.* (2003).) A colour version of this Figure is in the Plate section.

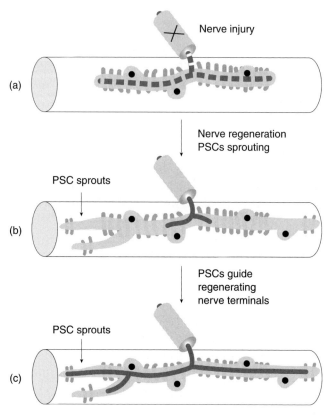

Figure 5.7 A schematic diagram summarising the role of PSCs in the reinnervation of frog NMJs. Unlike mammals, nerve injury *per se* does not induce PSC process sprouting in the frog (a). Large numbers of PSC processess are seen after the arrival of regenerating nerves (b). As reinnervation progresses, regenerating nerve terminals grow along the preceding PSC processess (c). The result suggests that, following nerve injury in frog muscles, regenerating nerve terminals trigger PSC sprouting, and these PSC sprouts in turn guide regenerating nerve terminals. A colour version of this Figure is in the Plate section.

isoforms of agrin and may play a role in inducing clustering of AChRs underneath PSC processess at the frog NMJ (Yang *et al.* 2001). These extrajunctional clusters of AChRs may lay the framework for the subsequent restoration of NMJ synaptic function. A recent study has shown that PSCs also express neuregulin-2, which may promote AChR synthesis and synaptic differentiation at the mammalian NMJ (Rimer *et al.* 2004). Whether PSCs also play a postsynaptic role for synaptic repair at mammalian NMJs remains to be investigated.

CONCLUSIONS

This chapter has highlighted recent advances in understanding of the biology of PSCs. The existence of synapse-associated cells at the NMJ has long been acknowledged since Ranvier's original study in 1878, and confirmed by electron microscopy in the 1950s and 1960s. However, significant advances in our understanding and appreciation of the multiple roles of PSCs have mainly been obtained in the past 10–15 years. During development, while Schwann cells are not required for axon pathfinding and initiation of nerve–muscle contacts, PSCs guide growing nerve terminals and are essential for synaptic growth and maintenance at developing NMJs (Figure 5.3). PSC-derived factors can promote synaptogenesis and potentiate synaptic transmission in tissue culture. Neuregulin released by axons is essential for the survival of PSCs at developing NMJs. At the adult NMJ, nerve stimulation increases the intracellular calcium level in PSCs, indicating that PSCs can detect synaptic activities. In turn, PSCs are capable of modulating synaptic transmission when G-protein pathways or intracellular calcium levels in PSCs are altered pharmacologically. Although removal of PSCs *in vivo* does not cause acute changes, chronic absence of PSCs results in retraction of nerve terminals and reduction in synaptic function, suggesting a long-term maintenance role of PSCs (Figure 5.4). Similar to developing NMJs, PSCs in adult muscles guide nerve terminal growth during synaptic sprouting and repair after nerve injury (Figures 5.6 and 5.7). Like PSCs, astrocytes have also been shown to play multiple roles in the central synapses. Investigations in both CNS and PNS have thus supported the emerging concept that neuroglial cells can make bigger, stronger and more stable synapses. Future challenges would be to unravel the molecular mechanisms of synapse–neuroglia interactions and their likely involvement in neurodegenerative diseases.

6

Cytokine and chemokine interactions with Schwann cells: the neuroimmunology of Schwann cells

ROBERT P. LISAK AND JOYCE A. BENJAMINS

INTRODUCTION

Cytokines, first described as products of the cells of the inflammatory/immune system, are increasingly recognised as acting on non-inflammatory cells as well as being produced by non-inflammatory cells. Nowhere is this more apparent than with cells of the peripheral (PNS) and central (CNS) nervous systems. Studies have clearly indicated that neuroglial cells are targets of cytokines produced by infiltrating immune/inflammatory cells in inflammatory diseases of the PNS and CNS, and are not simply passive targets of lytic destructive processes. In addition there is evidence to show that neuroglial cells, in particular astrocytes, Schwann cells and microglia (cells of the monocyte/macrophage lineage), respond to cytokines by changes in function and phenotype and can themselves produce many of the classically described inflammatory cell cytokines. Perhaps even more interesting are recent studies showing that such cytokines, particularly when produced by cells that are endogenous to the PNS and CNS, are important in PNS and CNS development and perhaps in protection and regeneration of the PNS and CNS in inflammatory, traumatic and even some degenerative diseases.

We and others have been interested in the interactions of Schwann cells and cytokines in the pathogenesis of diseases, modulation and recovery from disease, and regeneration, as well as in normal PNS development and function. Schwann cell–cytokine interactions can be studied in various ways: examining tissue obtained at different stages of development, and at different phases of experimental and naturally occurring diseases, including human disorders of the PNS, and employing different *in vitro* models. The latter include dissociated primary Schwann cell cultures from experimental animals and human biopsy/autopsy material, animal and human Schwann-cell-derived lines, cultures of dorsal root ganglia (DRG), which contain Schwann cells, neurons and support cells, as well as myelinating DRG cultures and myelinating co-cultures of Schwann cells and neurons. All of the *in vitro, in vivo* and *in situ* approaches have their inherent limitations, requiring comparisons of findings of the individual approaches, and ultimately integration of the results of all of these approaches to come to reasonable conclusions and understanding of interactions between cytokines and Schwann cells.

CYTOKINES AND CHEMOKINES

Cytokines are usually defined as proteins secreted by cells, sometimes constitutively but generally in response to a wide variety of stimuli, by which these cells alter the functions of other cells (paracrine or at a distance) or of the secreting cells themselves (autocrine). Originally cytokines were named based on one or more functions (tumor necrosis factor, lymphotoxin, etc.), the cell type first shown to produce or secrete the factor, the affected cell type (interleukins). However, it has become clear that many cell types, including neuroglial cells, can secrete many of these factors, and are affected by these factors. It is also clear that there is tremendous overlap with growth factors, and that classic cytokines, including interleukins, can act directly or indirectly as growth factors in many tissues, including in the PNS and CNS, and that cytokines may even have rapid pharmacologic effects on PNS and CNS cells.

Chemokines consist of a family of small-molecular-weight cytokines rich in cysteine residues, which have as their primary function the ability to attract various inflammatory cells into different tissue compartments via a chemical gradient. As with many of the other cytokines, they were first named by the cell type they attracted or cellular function they seemed to produce. It has become clear that there are a multitude of these chemokines, and while they are relatively specific in the cell types they influence, few cause chemoattraction of a single inflammatory cell type. In addition, they ligate a series of sometimes overlapping receptors. Therefore, it is now recommended that the chemokines be named based on their chemical structure (based on the location and number of the cysteine residues), and the chemokine receptors named on the basis of the type of chemokine they bind; hence the use of names such as CXCL1 or CCL5 for chemokines and CXCR1 or CCR5 for chemokine receptors (Rossi and Zlotnik 2000; Zlotnik and Yoshie 2000; Baggiolini 2001).

CYTOKINES AND SCHWANN CELLS IN DISEASES OF THE PNS

It is important to consider the evidence that cytokines are prominent mediators of inflammatory/immune demyelinating disorders of the PNS. Many studies support a critical role for cytokines in the pathogenesis of PNS diseases as well as in recovery from many of these disorders (Hartung et al. 1996c; Lisak et al. 1997; Kieseier et al. 2002b). There are several clinical disorders of the PNS that are relevant for understanding the interactions of cytokines with Schwann cells. These include acute inflammatory demyelinating polyneuropathy (AIDP) (Lisak and Brown 1987; Griffin et al. 1990; Griffin et al. 1993a; Hartung et al. 1995a,b; Hartung et al. 1998; Ho et al. 1998), the most common form of the Guillain–Barré syndrome (GBS) in the United States and Europe, chronic inflammatory demyelinating polyneuropathy (CIDP) (Lisak and Brown 1987; Griffin et al. 1990; Griffin et al. 1993a; Hartung et al. 1996a,b; Hartung et al. 1998; Steck et al. 1998), trauma to the PNS (Griffin et al. 1992; Griffin et al. 1993a; Stoll et al. 1993b; Stoll and Muller 1999) and certain forms of diabetic neuropathy (Said et al. 1994). There is also evidence that macrophages are important in the axonal form of GBS (Griffin et al. 1996a; Hafer-Macko et al. 1996a). Inflammatory cells and cytokines are probably important in some degenerative and hereditary neuropathies as well. CIDP is a disorder that can relapse and remit, become secondarily progressive, progress from the outset, or progress with superimposed periods of more rapid worsening/relapses (Lisak and Brown 1987; Gorson et al. 1997; Gorson and Chaudhry 1999; Kieseier et al. 2002a). In addition, there are variable degrees of axonal damage and denervation in CIDP often evident early in the disease, and there is often denervation in AIDP. In both AIDP and CIDP much of the long-term disability is determined by the degree of denervation, although axonal regeneration is more rigorous in the PNS than the CNS, for reasons that are not fully understood.

There are other forms of Guillain–Barré syndrome with primary immunologic attack on axons; acute motor axonal neuropathy (AMAN) and acute motor sensory axonal neuropathy (AMSAM), which are almost certainly mediated by antibody directed at components of the axons and complement activation. It has been shown that macrophages are critical in the pathogenesis of these disorders as well (Griffin et al. 1996a; Hafer-Macko et al. 1996a; Ho et al. 1999).

Neurofibromatosis type 1 (NF1) and leprosy are two examples of other diseases of the PNS in which cytokines play a critical role. In NF1, Schwann cells heterozygous for the NF1 mutation (NF1+/−) secrete five times the normal amount of *kit* ligand (stem cell factor), which in turn serves as a chemoattractant for mast cells (Viskochil 2003; Yang et al. 2003). Whether the cytokines and growth factors secreted by the activated mast cells act to maintain the Schwann-cell-derived neurofibromas in a benign state, or influence progression to malignant nerve sheath tumours is not known. In leprosy, cytokines are also postulated to influence disease progression (Modlin 2002; Spierings et al. 2000). The Schwann cell myelin protein, P0 glycoprotein, can serve as a binding protein for *M. leprae*; the structural similarity of P0 glycoprotein to known neuropathogens and to immune-system-related molecules may also serve to drive immune responses in this disease (Vardhini et al. 2004).

Relevant PNS animal models include experimental autoimmune neuritis (EAN), which can be induced by sensitisation with one of several PNS myelin proteins (P2, P0 and PMP 22) (Kadlubowski and Hughes 1979; Kadlubowski and Hughes 1980; Linington et al. 1984; Lisak and Brown 1987; Milner et al. 1987; Rostami et al. 1990; Linington et al. 1992; Rosen et al. 1992; Hartung et al. 1996b; Gabriel et al. 1998), as well as by repeated sensitisation with galactocerebroside (GalC) (Saida et al. 1979), the major galactolipid of PNS and CNS myelin. The former varieties of EAN seem to be initiated by $CD4^+$ T-cells (Linington et al. 1984; Rostami et al. 1985) with an enhancing demyelinating role for antibodies (Linington et al. 1988; Hodgkinson et al. 1994; Pollard et al. 1995; Spies et al. 1995a,b; Taylor and Pollard 2001), whereas the latter are complement-mediated with a secondary role for macrophages, although other antibody-mediated immunopathological mechanisms might also be involved.

Macrophages and lymphocytes infiltrate the PNS in immune-mediated diseases of the PNS including GBS, CIDP, trauma, as well as diabetic neuropathies and other disorders. EAN can be inhibited by blocking the effects of certain cytokines (Tsai et al. 1991; Stoll et al. 1993a; Constantinescu et al. 1996; Redford et al. 1997), while inflammation and demyelination can be induced by intraneural injections of TNF-α (Wagner and Myers 1996a,b) and IL-12 (Pelidou et al. 1999).

There is also deposition of immunoglobulin and complement, suggesting that AIDP is the result of an immunopathologic process comprising cell-mediated immunity as well as humoral (B-cell dependent) processes (Hafer-Macko et al. 1996b). As noted earlier, in the experimental model EAN, changes in the PNS result from a combination of cell-mediated and humoral-mediated immunopathologic mechanisms. T-cells and their products seem to open the blood–nerve barrier (BNB) and probably have deleterious effects on Schwann cells and axons (the innocent bystander effect). Once the BNB is compromised, antibody and complement amplify the lesions, particularly the demyelination. Again, as noted before, there is a different form of EAN induced by repeated sensitisation of animals with galactocerebroside in adjuvant. These animals develop very high titers of antibodies to galactocerebroside (anti-GalC) along with a demyelinating neuropathy which is complement dependent and involves macrophages. Intraneural injection of serum from animals with high titer anti-GalC induces a complement-dependent demyelinating lesion that starts at the paranodal myelin, the most distal part of the Schwann cell, a 'dying back Schwannopathy' (Saida et al. 1978a,b).

Genes and gene products for several cytokines, including interleukin-1 (IL-1), IL-6, TNF-α and TGF-β are upregulated in the PNS in EAN and experimental Wallerian degeneration (Rufer et al. 1994; Kiefer et al. 1995; Kiefer et al. 1996; Wagner and Myers 1996b). Certain matrix metalloproteinases, which can alter the BNB, are also upregulated by and involved in activation and secretion of TNF-α, and are upregulated in EAN and human inflammatory neuropathies (Hughes et al. 1998; Leppert et al. 1999; Mathey et al. 1999). As described, intraneural injection of TNF-α or IL-12, a proinflammatory cytokine produced predominately by cells of the monocyte/macrophage lineage, induces inflammation and demyelination, but in this model one cannot distinguish between direct and indirect effects on the Schwann cell (Redford et al. 1995; Wagner and Myers 1996a; Pelidou et al. 1999; Uncini et al. 1999). In studies of EAN from our laboratories, we found upregulation of the gene for IL-1α as well as IL-1β in infiltrating macrophages, but little IL-1 in Schwann cells (Skundric et al. 2001). Of interest was the finding that there was upregulation of the gene for IL-1 receptor antagonist (IL-1RA) and its protein by Schwann cells. This member of the IL-1 family competes with IL-1α and β for binding to the IL-1 receptors. We had previously demonstrated increased expression of IL-1RA by Schwann cells in vitro (Skundric et al.

1997b). When IL-1RA binds to the receptors it does not cause signalling and thus IL-1RA serves to downregulate the effects of IL-1α and β. Perhaps even more interesting was that the greatest concentration of IL-1RA was in the region of the internode. Telelogically one could speculate that this would be a most important location in which to try to block effects of macrophage secreted IL-1β, since this is where demyelination often starts and is the part of the Schwann cell furthest from the cell body.

Complement-activated macrophages that are incubated with myelin *in vitro* upregulate TNF-α (van der Laan *et al.* 1996), indicating that cytokines are most likely involved in antibody as well as cell-mediated immune damage to the PNS. Increased levels of several cytokines, including TNFα, have been reported in serum of patients with GBS (Sharief *et al.* 1993; Exley *et al.* 1994).

In vitro studies performed by our group and others have demonstrated that the various cytokines present in the PNS in clinical and experimental diseases have the ability to alter several Schwann cell functions, including proliferation, development and maturation (including myelination) and cell viability. By employing our model of Schwann cell cultures treated with cAMP, as an axolemmal-like signal to initiate Schwann cell differentiation, we can examine the effect of cytokines on specific molecular mechanisms involved in differentiation, including expression of myelin-associated galactolipids and proteins (Sobue *et al.* 1986; Shuman *et al.* 1988). Studies employing unstimulated dissociated Schwann cell cultures relate to pre/promyelinating/non-myelinating and myelinating Schwann cells (Scherer 1997) since: (1) such cultures are obtained from sciatic nerve of neonatal rats, a period associated with active myelination in the PNS and in which there are very few Schwann cell precursors or embryonic Schwann cells (Mirsky and Jessen 1999); (2) the majority of Schwann cells in adult PNS in which inflammatory/demyelinating diseases occur are pre/promyelinating/non-myelinating and myelinating Schwann cells; and (3) during different stages of these disorders, as well as in traumatic and metabolic disorders of the PNS, myelinating Schwann cells revert to a pre/promyelinating state, phenotypically identical to non-myelinating Schwann cells *in vitro*. This includes upregulation of phenotypic markers associated with pre/promyelinating Schwann cells, such as NGFRp75 (Morgan *et al.* 1991; Zorick and Lemke 1996) and glial-associated fibrillary protein (GFAP), and downregulation of genes that control myelin proteins such as P0, MBP, etc. (Jessen *et al.* 1990; Mirsky *et al.* 2001).

Schwann cells have been proposed as a source of myelinating cells for transplantation into the CNS of patients with multiple sclerosis (MS) or spinal cord injury. A number of experiments in animal models have given only modest results, suggesting that CNS neuroglial cells interfere with Schwann cell migration, survival, maturation and myelination (reviewed in Franklin 2002; Kocsis et al. 2004; Bunge and Wood in press). We examined whether soluble factors secreted by CNS glia could influence Schwann cell proliferation or differentiation by treating Schwann cells in culture with serum-free conditioned medium from mixed CNS glial cultures (Lisak et al. 2006). The supernatants had no effect on induction of Schwann cell galactolipids, nor did they inhibit cAMP-induced expression of galactoplipids. However, the supernatants increased proliferation and inhibited Schwann cell death induced by TNF-α + TGF-β, indicating that CNS neuroglia under normal conditions secrete factors that can augment CNS repair. The fate of Schwann cells transplanted into injured or demyelinated lesions *in vivo* has been studied immunocytochemically, and more recently using genetic markers. For example, Hill et al. (2006) assessed the interaction of transplanted alkaline-phosphatase-expressing Schwann cell with host tissue in a model of spinal cord injury, and observed poor survival of transplanted Schwann cells, but extensive migration of endogenous p75+ Schwann cells into the lesion site. Survival of transplanted Schwann cells was enhanced by delaying the transplant for one week after injury in combination with immunosuppression by cyclosporin. These results suggest that activation of CNS neuroglial cells along with immune-mediated responses inhibit Schwann cell survival immediately after injury, but subsequently the host environment is more permissive. They further indicate that the CNS neuroglial cells and/or dying transplanted Schwann cells provide factors that stimulate migration of endogenous Schwann cells into the lesion site. The origin of endogenous Schwann cells that remyelinate the CNS following injury may differ according to the nature of the insult. While many studies provide evidence that Schwann cells migrate into areas of CNS damage from the periphery, an increasing number of experiments provide evidence that Schwann cells remyelinating CNS axons can also arise from CNS precursors (reviewed in Blakemore (2005)). The best characterised example is that of X-irradiated ethidium bromide lesions in the CNS. In many other models of Schwann cell transplant following white matter injury, astrocytes are present but not macrophages, whereas after X-irradiation, the lesion site contains macrophages, but is free of astrocytes. This leads to the hypothesis that following demyelination, oligodendrocyte precursor cells differentiate into oligodendrolia in astrocyte-containing areas and into Schwann cells in areas devoid of astrocytes (Blakemore 2005). The role of macrophage-secreted factors in these responses is not known, but in another demyelinating model, acute inflammation stimulated the differentiation of oligodendrocyte precursors into myelinating oligodendroglia in astrocyte-containing areas (Foote and Blakemore 2005).

DEMYELINATION, INHIBITION OF MELIN SYNTHESIS, MYELINATION AND REMYELINATION

Loss of myelin accompanying Wallerian degeneration (secondary demyelination) and primary demyelination in autoimmune/inflammatory disorders are important in producing symptoms and neurophysiologic changes in the PNS as well as in the CNS (Raine and Cross 1989; Bjartmar et al. 2003). Failure to form new myelin is important in preventing recovery and restoration of normal function in myelinated PNS fibres (Bouchard et al. 1999). Even if some PNS diseases are not primarily induced by T-cell-initiated and -mediated immunity, demyelination is produced to some degree by purely cell-mediated processes. Secondary damage to Schwann cell myelin and axons by macrophages and their products probably potentiates antibody-mediated demyelination. As noted, there is in vitro evidence that certain cytokines may mediate harmful reactions (Selmaj and Raine 1988; Selmaj et al. 1991a,b; Soliven et al. 1991; Chao and Hu 1994; Hu et al. 1994; Agresti et al. 1996; Hisahara et al. 1997; Andrews et al. 1998) and over-representation of certain cytokines can induce both inflammation and associated demyelination, albeit most of these experiments have been carried out in studies of the CNS (Probert et al. 1995; Corbin et al. 1996; Akassoglou et al. 1997; Popko et al. 1997; Akassoglou et al. 1998; Akassoglou et al. 1999). Schwann cells are also important in helping maintain normal axonal function and structure. Experimental and clinical genetically determined diseases of Schwann cells result in axonal pathogenesis and pathophysiology, even in these basically non-inflammatory conditions (Krajewski et al. 1999; Krajewski et al. 2000; Sahenk 1999; Sancho et al. 1999; Scherer 1999). Shy et al. discuss this in detail in Chapter 8. The genetic models of Schwann cell disease as well as other studies have shown that adhesion molecules and Schwann cell trophic factors are important in axonal health and function (de Waegh et al. 1992; Scherer and Arroyo 2002; Jones et al. 2003).

SCHWANN CELL DIFFERENTIATION
AND DEDIFFERENTIATION

Schwann cells pass through several stages during development, beginning as cells derived from the neural crest cells. In one scheme the neural crest cells undergo changes to become Schwann cell progenitors and then immature Schwann cells. Some immature Schwann cells then progress to become non-myelinating Schwann cells, while others become pre-myelinating Schwann cells, and after contact with axons and in response to appropriate axonal signals, differentiate to myelin-forming Schwann cells (Arenander and De Vellis 1999). In another slightly different scheme, Schwann cell precursors become immature Schwann cells and then, in contact with axons and axonal signals, become promyelinating and subsequently myelinating Schwann cells. Others establish a different type of relationship to several smaller axons and remain non-myelin-forming Schwann cells (Scherer 1997; Scherer and Arroyo 2002). Promyelinating and non-myelin-forming Schwann cells differ from each other in anatomic appearance and in their relationship to axons *in vivo*, but in culture have a similar pattern of numerous phenotypic markers.

The expression of transcription factors with different patterns of activation and inactivation regulates a sequence of molecular changes associated with development of Schwann cells. As discussed in detail in Chapter 2, Sox 10 is known to be important in establishing the Schwann cell neuroglial lineage in neural crest cells, while Oct-6 (also known as SCIP and Tst-1) and Krox 20 regulate subsequent Schwann cell differentiation and myelination (Wegner 2000a,b). The exact nature of the interactions between these transcription factors and the genes involved in Schwann cell development, differentiation, myelination and myelin maintenance are not fully understood. However, it is clear that different transcription factors seem to provide positive and negative signals for different genes and at different stages (Wegner 2000b). When axonal–Schwann cell interactions are interrupted, myelinating Schwann cells can undergo dedifferentiation back to the pre/promyelinating state, again involving changes in phenotypic markers and Schwann cell function, under control of a series of genes (Taniuchi *et al.* 1986; Taniuchi *et al.* 1988; Shy *et al.* 1996), in turn influenced by transcription factors. The TGF-βs are a family of proteins that have pleotropic and overlapping but not necessarily identical effects on many cell types (Massague *et al.* 1992; Roberts and Sporn 1993; Kingsley 1994; Johnson and Newfeld 2002). During embryogeneis of the rat there are differential changes in levels of expression of each of the isotypes at different stages of development in the PNS (Scherer *et al.* 1993; McLennan and Koishi 2002). In *in vitro* studies (see below) the same group demonstrated various effects of TGF-β1 on mRNA levels for several proteins expressed by denervated and non-myelinating Schwann cells, some of which occur at the transcriptional level (Awatramani *et al.* 2002). As outlined in the Introduction, this is just one example of how *in vitro* experiments can supplement *in vivo* studies by permitting detailed examination of cellular mechanisms.

CYTOKINES, SCHWANN CELLS AND PNS REGENERATION

Trauma to the PNS often results in damage or transection of axons, leading to Wallerian degeneration with associated loss of myelination. Inflammatory cells, primarily macrophages, infiltrate the nerve and phagocytose cellular debris including myelin (Bruck 1997; Be'eri *et al.* 1998; Griffin *et al.* 1992; Hirata *et al.* 1999; Stoll and Muller 1999; Shamash *et al.* 2002). The macrophage infiltration is important in regeneration. Administration of IL-1 receptor antagonist (IL-1RA) also inhibits regeneration, suggesting that IL-1 is an important factor in PNS response to trauma (Guenard *et al.* 1991). Upregulation of IL-1 is seen in experimental nerve trauma (Rutkowski *et al.* 1999) and it has been shown that IL-1 induces NGF production by Schwann cells and fibroblasts (Heumann *et al.* 1987a,b; Lindholm *et al.* 1987). It is likely that the production of NGF is important in regeneration. Increased expression of other growth factors has also been reported in experimental peripheral nerve trauma, probably triggered by cytokines produced by inflammatory cells, such as macrophages, as well as by Schwann cells (Rufer *et al.* 1994; Chen *et al.* 1996; Hirota *et al.* 1996; Reichert *et al.* 1996; Wagner and Myers 1996b; George *et al.* 1999; Shamash *et al.* 2002; Ahmed *et al.* 2005). These other growth factors as well as some of the cytokines themselves may be important in peripheral nerve regeneration (Ahmed *et al.* 2005). As described below, upregulation by Schwann cells of several cytokines besides IL-1 has been demonstrated *in vitro*.

IN VITRO STUDIES OF CYTOKINE AND SCHWANN CELL INTERACTIONS

Effect of major histocompatibility molecules expression

We and others have used various *in vitro* systems to study the roles of cytokines in the pathogenesis of PNS disorders as well as their role in

PNS development. Early studies examined the effects of different cytokines, generally those considered CD4$^+$Th1 or monocyte/macrophage proinflammatory cytokines. Such experiments enhance our ability to understand the findings from *in vivo* and *in situ* studies.

IFN-γ induces upregulation of expression of both major histocompatibility class I (MHC class I) and MHC class II molecules on the surface of Schwann cells (Wekerle et al. 1986; Samuel et al. 1987a,b; Armati et al. 1990; Argall et al. 1992b; Lisak and Bealmear 1992; Gold et al. 1995; Spierings et al. 2001). Upregulation of MHC class I molecules would allow presentation of antigens, most likely endogenous for that cell or from infectious agents in that cell, to the T-cells of the CD8 class. Upregulation of MHC class II molecules is required for a cell to act as an antigen-presenting cell to the CD4 T-cells. There are somewhat controversial studies, demonstrating the ability of Schwann cells to present antigen to T-cells and induce a lymphocyte proliferative response *in vitro* (Wekerle et al. 1986; Argall et al. 1992a,b; Gold et al. 1995). However, most but not all experimental studies of PNS tissue fail to demonstrate significant upregulation of MHC class II molecules on Schwann cells (Olsson et al. 1983; Hughes et al. 1987; Ota et al. 1987; Schmidt et al. 1990; Yu et al. 1990). In addition, for presentation of antigen by an antigen-presenting cell (APC) to induce an immune response, a series of co-stimulatory molecules need to be present on the APC, and its ligand present on the T-cell. Such cells are often referred to as 'professional' APCs; macrophages, dendritic cells and microglia are among such cells. There is little to suggest that Schwann cells are important APCs *in vivo*. Indeed presentation of antigen to a lymphoycte in the context of MHC class II in the absence of co-stimulatory molecules can lead to anergy to that antigen. There is controversy with regard to the presence or absence of MHC class II molecules on Schwann cells in human diseases of the PNS, as well as to their presence in immunopathologically mediated neuropathies versus other non-immune-mediated diseases (Pollard et al. 1986; Pollard et al. 1987; Mancardi et al. 1988; Scarpini et al. 1989; Scarpini et al. 1990).

Effect on expression of adhesion molecules

Certain cytokines have been shown to upregulate expression of adhesion molecules on the surface of Schwann cells (Gold et al. 1995; Armati and Pollard 1996; Lisak and Bealmear 1997; Constantin et al. 1999). In our studies, unfractionated cytokines, obtained from mitogen-stimulated rat spleen cells (inflammatory cells), as well as IFN-γ, TNF-α and IL-1β upregulated expression of intercellular adhesion molecule-1 (ICAM-1; CD54), a member of the immunoglobulin superfamily, but with differering kinetics (Lisak and Bealmear 1997). Incubation of Schwann cells with other cytokines including IL-1β, IL-2 and TGF-β1, all found in the unfractionated cytokines, had no effect on ICAM-1 expression by Schwann cells. In the case of Schwann cell upregulation of ICAM-1, expression by TNF-α is predominately via activation of type I TNF-R (Lisak et al. 1999). Interestingly, there was no effect of IFN-γ, TNF-α or IL-1β on expression of two integrins, which are also adhesion molecules, LFA-1α (CD11a), employing an antibody that binds to the $α_1$ integrin peptide chain, and LFA-1β, employing an antibody that binds to the β2 integrin chain. Thus the effect on ICAM-1 expression is not a non-specific effect on upregulation of all adhesion molecules. Schwann cells express other integrins (Einheber et al. 1993; Feltri et al. 1994; Feltri et al. 1997; Stewart et al. 1997; Frei et al. 1999; Feltri et al. 2002; Previtali et al. 2003b) and at different stages of development other CAMs that are members of the immunoglobulin superfamily, such as neural cell adhesion molecule (NCAM), L1 (Ng-CAM), and N-cadherin (Stewart et al. 1995b; Armati and Pollard 1996; Awatramani et al. 2002), as well as several myelin proteins which act as adhesion molecules, such as myelin-associated glycoprotein (MAG) and P0. TNF-α and TGF-β down-regulate, respectively, the expression of MAG and P0 by Schwann cells (Schneider-Schaulies et al. 1991; Mews and Meyer 1993; Stewart et al. 1995b). Upregulation of ICAM-1 on Schwann cells should enhance Schwann cell–lymphocyte interactions. We have shown that Schwann cells can be induced by cytokines to upregulate MHC class I molecules as well (Lisak and Bealmear 1992), indicating that cytotoxic reactions induced by CD8+ T-cells against Schwann cells presenting an infectious or endogenous Schwann cell antigen would be enhanced. In studies of PNS tissue in experimental models, the majority of the ICAM-1 positive cells are vascular endothelial cells and infiltrating macrophages, with low levels of ICAM-1 seen on Schwann cells in both control and affected animals (Stoll et al. 1993a). ICAM is found predominatly on inflammatory cells in nerve biopsies of patients with CIDP (Van Rhijn et al. 2000a). The reasons for the differences in ICAM-1 expression on Schwann cells between *in vitro* and *in vivo* studies are not clear.

Effect on Schwann cell proliferation

We have shown that unfractionated cytokines stimulate Schwann cell proliferation (Lisak et al. 1985). Specifically, interleukin-1 (mixed IL-1,

IL-1α, IL-1β), IL-2, IL-6 and IFN-γ do not directly induce Schwann cell proliferation (Lisak et al. 1994; Lisak and Bealmear 1991; Lisak and Bealmear 1994), but we confirmed the work of others that TGF-β1 does act as a Schwann cell mitogen (Eccleston et al. 1989; Ridley et al. 1989; Schubert 1992; Einheber et al. 1995; Guenard et al. 1995b; Lisak and Bealmear 1995). Based on the capacity of antibodies to some but not all cytokines to inhibit unfractionated cytokine-induced Schwann cell proliferation and the inhibitory effect of IL-1 receptor antagonist (IL-1RA), we showed that IL-1α and IL-6 are co-mitogens for Schwann cells, and demonstrated the presence of functional receptor for IL-1 on Schwann cells (Lisak and Bealmear 1991; Lisak and Bealmear 1994; Lisak et al. 1994). For reasons that are not clear, IL-1β does not seem to be a co-mitogen for Schwann cells, even though it binds to the same IL-1R as IL-1α, although not necessarily with the same avidity. TGF–β, shown by several groups to induce Schwann cell proliferation at suboptimal concentrations, can be shown to act as a co-mitogen with IL-1 (Lisak and Bealmear 1995; Lisak et al. 1996). These findings are relevant to Schwann cell responses to cytokines *in vivo*, since proliferation is important during development of the PNS as well as during recovery and repair in response to disease.

Schwann cell synthesis of cytokines

We and others have demonstrated that Schwann cells are able to upregulate genes for several cytokines as well as produce proteins *in vitro*. For most of the cytokines this expression of cytokines and specific mRNAs can be shown *in vivo*. Synthesis of IL-1 was shown by Bergsteindottir and colleagues (1991). We have also demonstrated that stimulated Schwann cells upregulate mRNA for IL-1α and β, IL-1RA and IL-1 receptor type I (IL-1R type I), TNF-α and TGF-β, co-express upregulated mRNA for several cytokines, and also synthesise functionally active IL-1, but do not upregulate mRNA for IL-2 (Skundric et al. 1996a,b; Skundric et al. 1997b). Both non-myelinating/premyelinating and myelinating phenotype Schwann cells expressed these cytokines. IL-1, IL-1R and IL-1RA are also expressed in Schwann cells *in vivo* as well (Skundric et al. 2001; Skundric et al. 2002).

Non-myelinating and premyelinating Schwann cells have also been shown to express both TGF-β2 and β3 (Stewart et al. 1995b) *in vitro* and *in vivo* (Scherer et al. 1993; Awatramani et al. 2002). Expression of TNF-α by Schwann cells has been reported *in vivo*

(Mathey et al. 1999; Stoll et al. 1993b; Empl et al. 2001; Shamash et al. 2002) as well as *in vitro* (Skundric et al. 1996a; Skundric et al. 1997a).

Cytokines upregulate mRNA for other cytokines, and interestingly in some instances there is upregulation of mRNA for the stimulating cytokine itself (Skundric et al. 1997a).

As noted elsewhere, Schwann cells also upregulate genes for several cytokines and synthesise cytokines *in vivo* in both inflammatory and traumatic models.

Effect on Schwann cell viability

There is great interest in the effects of various cytokines on Schwann cell viability. The use of *in vitro* systems has been useful in exploring this area of research. Studies of TNF-α failed to demonstrate a cytotoxic effect on Schwann cells in myelinating cultures (Mithen et al. 1990), or in dissociated cultures of human Schwann cells (Bonetti et al. 2000), although cytokines including the related lymphotoxin (TNF-β) had been shown by some investigators to kill oligodendrocytes *in vitro* (Selmaj and Raine 1988; Selmaj et al. 1991a,b).

In our studies of Schwann cell-cytokine interactions, we were unable to demonstrate any killing of rat Schwann cells by TNF-α (Skoff et al. 1998), although we demonstrated that both the p55 and p75 receptors for TNF are present on rat Schwann cells. It is not known whether TNF-α can induce Schwann cell death *in vitro* in other species. Somewhat unexpectedly, we observed that while TGF-β1 alone induced a modest amount of Schwann cell death, when added to TNF-α there was a marked cytotoxic effect. At least some of the cell death was mediated by apoptosis (Skoff et al. 1998). More recently, cooperation between these same two cytokines in inducing death of an oligodendrocyte line has been reported (Schuster et al. 2003). The mechanisms involved in the enhanced cytotoxicity in this paradigm are not known. It should be noted that TNF-α and TGF-β do not share receptor peptides or signalling pathways. Subsequently, it was demonstrated that when Schwann cells are isolated and cultured in the total absence of serum and then incubated with TGF-β, there is marked Schwann cell death (Parkinson et al. 2001). The differences in the results under various experimental conditions suggest that differences in the environment in the PNS could modify the biological effects of different cytokines and mixtures of cytokines. It is also not known if incubation of rat Schwann cells with TNF-α in the presence of other stressors, such as low concentrations of serum, hypoxia, etc. would result in Schwann cell death. One group has reported that TNF-α inhibits the low level of spontaneous Schwann cell proliferation (Chandross et al. 1996).

Effect on Schwann cell development and maturation

There are several studies examining the effects of mixtures of cytokines and purified cytokines on Schwann cell development and maturation. When Schwann cells are incubated in the presence of high concentrations of membrane-permeable analogues of cyclic AMP (cAMP) or high concentrations of forskolin (which, among other effects, increases the intracellular levels of cAMP), they undergo changes in phenotype with regard to cytologic appearance, upregulation of surface expression of galactolipids (Sobue et al. 1986) and myelin-associated proteins (Shuman et al. 1988), and down regulation of expression of NGFRp75, the low-affinity receptor for nerve growth factor (NGF) (Mokuno et al. 1988). Upregulation of galactolipids and myelin-associated proteins are considered indicators of Schwann cell differentiation towards the myelinating phenotype, whereas expression of NGFRp75 is characteristic of promyelinating or non-myelinating phenotype. In this way high concentrations of cAMP can be considered as simulating effects of axons and axolemma. Interestingly, low concentrations of cAMP induce Schwann cell proliferation. Using the experimental paradigm of high concentrations of 8 bromo-cAMP (1 mM), we investigated the effects of supernatants obtained from activated inflammatory cells to examine how cytokines might affect maintenance of myelin and remyelination. Unfractionated cytokines inhibited cAMP-induced upregulation of surface expression of galactoplipids and also reversed ongoing maturation (Lisak et al. 1998). These supernatants induced Schwann cell proliferation, generally associated with lack of Schwann cell differentiation. We inhibited proliferation to determine if the supernatants were still able to inhibit Schwann cell maturation, and found that the inhibitory effect occurred under these conditions as well (Lisak et al. 1998). In an attempt to identify individual cytokines in the supernatants that were capable of inhibiting cAMP-induced Schwann cell maturation, we incubated Schwann cells with IFN-γ, TNF-α, TGF-β1, IL-1α, IL-1β, IL-2 and IL-6. IFN-γ, TNF-α and TGF-β1 inhibited cAMP-induced Schwann cell maturation as assessed by upregulation of galactoplipid expression and downregulation of NGFRp75 expression (Figure 6.1). The other cytokines tested had no effect (Lisak et al. 2001). Blocking TGF-β1-induced Schwann cell proliferation did not interfere

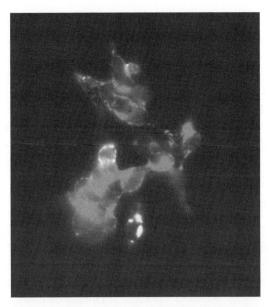

Figure 6.1 Schwann cells were treated with 1 mM 8 Br cAMP + 10 ng/ml TNF-α, and stained after 7 days for GalC (green) with O1 (IgM), and for NGFRp75 (red) with IgGMoAB, and appropriate second AB. The arrow points to where a green (GalC) cell and a red (NGFRp75) cell overlap, producing the yellow colour. Reproduced with permission from Lisak et al. Glia 36: 2001 354–63, Wiley-Liss, Inc. A colour version of this Figure is in the Plate section.

with the ability of TGF-β to inhibit cAMP-induced Schwann cell maturation and differentiation. These findings support the observation of upregulation of NCAM and L1 adhesion molecule, characteristic of premyelinating Schwann cells, with suppression of galactolipids and myelin proteins by TGF-β (Stewart et al. 1995a,b). TGF-β has been shown to inhibit myelin wrapping of axons by Schwann cells in co-culture experiments (Guenard et al. 1995a). There was no inhibition of Schwann cell process formation or initial association of the Schwann cell myelin process with the axon. In another study TGF-β had more profound inhibitory effects on Schwann cell–axonal interactions (Einheber et al. 1995).

TGF-β has been widely regarded as a protective disease-modulating cytokine and suggested as possible therapy in several immunopathologically mediated diseases (Gregorian et al. 1994; Weiner et al. 1994; Kiefer et al. 1996; Calabresi et al. 1998). This is based predominately on the capacity of TGF-β to block the ability of Th1 proinflammatory cytokines such as IFN-γ and TNF-α to activate inflammatory cells and vascular endothelium, which then leads to alteration in the blood–brain and blood–nerve barrier (Dore-Duffy et al. 1994; Gregorian et al. 1994; Fabry et al. 1995; Kiefer et al. 1996). However, these studies on the effects of TGF-β on Schwann cells, as well as studies on the effect of this cytokine on CNS myelination, suggest that increased levels of TGF-β within areas of demyelination may inhibit remyelination (John et al. 2002).

CHEMOKINES AND SCHWANN CELLS IN DISEASES OF THE PNS

Virtually all studies of the role of chemokines in diseases of the PNS as well as in normal PNS function have been *in vivo* or *in situ*. There have been several studies demonstrating several chemokines of both the CXC (alpha chemokines) and CC (beta chemokines) chemokines in the PNS and specifically in Schwann cells. Macrophage chemoattractant protein-1 (MCP-1; CCL2), macrophage inflammatory protein-1alpha (MIP-1α; CCL3) and RANTES (regulated on activation, normal T cell expressed and secreted; CCL5) are upregulated in the PNS after peripheral nerve transection (Taskinen and Roytta 2000; Perrin et al. 2005). The earliest increase of MCP-1 was seen in Schwann cells, peaking in those cells at 24 h, and observed in inflammatory cells and endothelial cells later in the evolution of the lesion. MIP-1α-positive cells were seen by 24 h after trauma, but the expression in Schwann cells was seen at 5 days. RANTES was basically seen exclusively in inflammatory cells. MIP-1α was also upregulated in Schwann cells present in PNS tumours, suggesting that the chemokine contributes to the recruitment of macrophages into Schwann-cell-derived tumours (Mori et al. 2004).

In Guillain–Barré syndrome (GBS) there was upregulation of MCP-1 in Schwann cells as well as inflammatory cells and in the endothelial extracellular matrix. CCR2, the receptor for MCP-1, was seen, but at low levels (Orlikowski et al. 2003). In EAN there was much less expression of MCP-1 and there was infiltration of CCR2-positive inflammatory cells during the plateau of the disease, not present in the PNS in animals who had recovered from clinical disease (Fujioka et al. 1999a,b). In two other studies there were minor differences in the kinetics of expression of the chemokines and their receptors (Zou et al. 1999; Kieseier et al. 2000; Orlikowski et al. 2003). In all of these studies in EAN, histologic studies have focused on localisation to cells other than Schwann cells, including endothelial cells and infiltrating inflammatory cells, although MCP-1 was detected in Schwann cells in at least one study of GBS.

Stromal-cell-derived factor-1 (SDF-1, also called CXCL12; several isotypes exist) which binds to CXCR4 (also an accessory receptor for HIV) has been demonstrated in the PNS during embryonic and postnatal development. SDF-1β predominates in embryonic and early postnatal stages of development, and SDF-1γ is more highly expressed in adult PNS (Gleichmann et al. 2000). Neurons and Schwann cells were the predominant SDF-1 expressing cells. After trauma SDF-1β is transiently increased, suggesting that SDF-1β is important in PNS development and in repair from trauma. SDF-1 and CXCR4 are important in cerebellar granule cell migration, axonal pathfinding and other developmental processes in the CNS, and are found in adult brain as well (McGrath et al. 1999; Tham et al. 2001; Ragozzino et al. 2002; Chalasani et al. 2003a,b; Lazarini et al. 2003; Stumm et al. 2003; Belmadani et al. 2005; Li et al. 2005b).

We can reasonably hypothesise that different chemokines, including those produced by Schwann cells, are important in attracting inflammatory cells into peripheral nerve in trauma and inflammatory disorders. It is not known if SDF-1 and CXCR4 plays any role in PNS development or repair and regeneration of axons in diseases of the PNS. In Schwann cell cultures, cAMP, which simulates the effect of axolemma on Schwann cells *in vitro*, upregulates CXCR4 on Schwann cells, and TNF-α downregulates CXCR4 expression. Interestingly, ligation of CXCR4 by SDF-1 results in increased Schwann cell death (Kury et al. 2003). Additional characterisation of the effects of chemokines on Schwann cell function and identification of the cellular and molecular signals that regulate Schwann cell expression of chemokines await further *in vitro* studies.

CONCLUSION

Cytokine and chemokine interactions with Schwann cells, as well as expression and production of cytokines and chemokines by Schwann cells, are important in the pathogenesis of several different types of disease of the PNS in normal development and PNS repair. Additional *in vitro* studies are clearly needed to further elucidate the role of these cytokines and chemokines in health and disease. These in turn will also need to be supplemented by in *vitro* functional experiments as well as additional *in vivo* and *in situ* studies.

7

Schwann cells as immunomodulatory cells

BERND C. KIESEIER, WEI HU AND HANS-PETER HARTUNG

INTRODUCTION

The nervous system has long been considered an immunologically privileged site. This concept was based on the premises that: (1) there is a more or less strict anatomic separation between the systemic immune compartment (blood) and the neural tissue; (2) molecules required for antigen presentation are absent under normal circumstances; (3) there is no lymphatic drainage; and (4) immune surveillance by T cells is lacking. It is now obvious that most of these assumptions are not tenable. The blood−nerve barrier (BNB) does restrict access of immune cells and soluble mediators to a certain degree; however, this restriction is not complete, either anatomically (e.g. the BNB is absent or relatively deficient at the roots, in the ganglia and the motor terminals) or functionally. Activated T lymphocytes can penetrate intact barriers irrespective of their antigen specificity, and, under certain circumstances, release cytokines that upregulate the expression of major histocompatibility complex (MHC) class II molecules, key molecules required for antigen presentation. In the central nervous system (CNS) tissue-resident neuroglial cells are present that actively participate in the regulation of immune responses within the tissue. In recent years, several lines of evidence have pointed to Schwann cells as immunocompetent cells within the peripheral nervous system (PNS), which, in addition to their physiological roles, exhibit a broad spectrum of immune-related functions and might be involved in the local immune response in the PNS. In this chapter we will elaborate on the expanding recognition of Schwann cells as immunocompetent cells that form part of the local immune circuitry within the PNS. Interestingly, present data suggest that the entire spectrum of an immune response can be displayed by Schwann cells; recognition of antigens, presentation of antigens, mounting an immune response, and, finally, terminating an immune response within the inflamed peripheral nerve.

SETTING THE STAGE: IMMUNE-MEDIATED INFLAMMATORY DISORDERS OF THE PNS

Immune-mediated inflammatory disorders of the PNS are characterised by cellular infiltration, demyelination and axonal loss in the affected part of the nerve. Most of these changes can be reproduced in the model disorder, experimental autoimmune neuritis (EAN), an animal model of the human Guillain–Barré syndrome (GBS). EAN can be elicited in susceptible animals by active immunisation with whole peripheral nerve homogenate, myelin, myelin proteins P0 and P2 or peptides thereof, and galactocerebroside. It can also be produced by adoptive transfer of P2, P2-peptide-specific, P0, and P0-peptide-specific T cell lines (Gold et al. 2000).

The pathological hallmark of EAN is the infiltration of the PNS by lymphocytes and macrophages, which results in multifocal demyelination of axons predominantly around venules. Macrophages actively strip off myelin lamellae from axons, induce vesicular disruption of the myelin lamellae, and phagocytose both intact and damaged myelin, as shown by electron microscopy (Griffin et al. 1993a; Kiefer et al. 2001). Macrophages, which are numerous as resident cells in the endoneurium, represent the predominant cell population in the inflamed PNS, and they reside in spinal roots as well as in more distal segments of the affected nerves (Hartung et al. 1998).

Crucial to the pathogenesis of inflammatory demyelination is the early invasion of the PNS by T cells. Circulating autoreactive T cells need to be activated in the periphery in order to cross the BNB and to incite a local immuno-inflammatory response. Breakdown of the BNB is one of the earliest morphologically demonstrable events in lesion development in EAN (Gold et al. 1999).

At present, how the cascade of autoimmune responses targeting PNS structures is ignited remains elusive. One pathogenic mechanism of special relevance to autoimmune neuropathies is molecular mimicry, in a proportion of patients with GBS, epitopes shared between the enteropathogen *Campylobacter jejuni, cytomegalovirus* (CMV) or *Haemophilus influenzae* and nerve fibres have been identified as targets for aberrant cross-reactive B cell responses (Kieseier et al. 2004). In a recent study, antibody responses to the ganglioside GM1 were linked to axonal and motor injury, as seen in one clinical variant of GBS, in an experimental model induced in rabbits by active immunisation with this glycoconjugate (Yuki et al. 2001). These observations cannot explain the entire clinical spectrum and laboratory findings of this disorder. Therefore, GBS should be defined as an organ-specific immune-mediated disorder emerging from a synergistic interaction of cell-mediated and humoral immune responses to still incompletely characterised peripheral nerve antigens (Hahn 1998; Sheikh et al. 1998).

ANTIGEN RECOGNITION BY SCHWANN CELLS

Two types of responses to invading organisms can take place in the human immune system: an acute response launched within hours, and a delayed response occurring within days. The immediately responding system is called innate immune system, and it evolves stereotypically and at the same magnitude regardless of how often the infectious agent is encountered. The strategy of the innate immune response may not be to recognise every possible antigen, but rather to focus on a few highly conserved structures present in large groups of microorganisms. These structures are referred to as 'pathogen-associated molecular patterns', and the corresponding receptors of the innate immune system are called 'pattern-recognition receptors' (Janeway 1992). Examples of pathogen-associated molecular patterns are bacterial lipopolysaccharide (LPS), peptidoglycan and bacterial DNA. Although chemically quite distinct, these molecules display common features: (1) they are only produced by microbial pathogens, and not by their host; (2) they represent in general invariant structures shared by large classes of pathogens; and (3) they are usually relevant for the survival or pathogenicity of microorganisms (Medzhitov and Janeway 2000).

The family of toll-like receptors (TLRs) belong to the group of pattern-recognition receptors that are key to recognising specific conserved components of microbes (Takeda et al. 2003), such as LPS, and that have been shown to play a critical role in various inflammatory disorders (Cook et al. 2004; Hoebe et al. 2004). To date, 11 TLRs have been identified in humans and seven in rats (Iwasaki and Medzhitov 2004). LPS, a major component of the outer membrane of gram-negative bacteria, that appears to be a relevant antigen-triggering immune-mediated demyelination of the PNS (Kieseier et al. 2004; Koller et al. 2005), is recognised by TLR-4. TLRs are usually found on antigen–presenting cells, such as dendritic cells. TLR-2 has been shown to be constitutively expressed on primary human and rat Schwann cells, and has been invoked as a target receptor for *M. leprae* (Oliveira et al. 2003; Harboe et al. 2005). Under inflammatory conditions, expression of various TLRs, especially TLR-4, is inducible on rat Schwann cells *in vitro*. Selective stimulation of TLR-4 on Schwann cells with LPS elicited various inflammatory mediators, including chemokines, protease inhibitors and growth factors (Hu et al. 2004). Thus, in aggregate these findings suggest that Schwann cells can detect LPS fragments and might act as a link between innate and acquired immunity via TLR-4 activation in the inflamed PNS.

SCHWANN CELLS AS ANTIGEN-PRESENTING CELLS

The task of displaying the antigens of cell-associated microbes for recognition by T lymphocytes is performed by specific molecules that are encoded by genes comprising the major histocompatibility complex (MHC). The physiologic function of MHC molecules is the presentation of peptides to T cells. Two types of MHC gene products can be distinguished, MHC class I molecules and MHC class II molecules, which differ functionally. Each MHC molecule consists of an extracellular peptide-binding cleft, or groove, and a pair of immunoglobulin-like domains, containing binding sites for the T cell surface markers CD4 and CD8. MHC molecules are anchored to the cell surface by transmembrane domains. MHC molecules exhibit a broad specificity for peptide binding, whereas the fine specificity of antigen recognition resides in large parts in the T cell receptor. Peptides from the cytosol are expressed by MHC class I molecules and presented to $CD8^+$ T cells, whereas peptides generated in vesicles are bound to MHC class II molecules and recognised by $CD4^+$ T lymphocytes. The two classes of MHC molecules are expressed differentially on cells. All nucleated cells exhibit MHC class I molecules, although hematopoietic cells express them at highest densities. MHC class II molecules, in contrast, are normally only expressed on 'professional' antigen-presenting cells, such as macrophages, dendritic cells, and B lymphocytes. The levels of both class I and II molecules can be markedly upregulated by cytokines, in particular interferon-γ (IFN-γ) and tumor necrosis factor-α (TNF-α).

As one would predict, human and rat Schwann cells *in vitro* constitutively express low levels of MHC class I but not MHC class II (Samuel et al. 1987a,b; Armati et al. 1990; Bergsteinsdottir et al. 1992; Gold et al. 1995; Lilje and Armati 1997). Significant numbers of MHC class II molecules can be detected in the presence of activated T lymphocytes upon stimulation with the pro-inflammatory cytokine IFN-γ, which can be synergistically increased by the addition of TNF-α (Kingston et al. 1989; Armati et al. 1990; Tsai et al. 1991; Gold et al. 1995; Lilje and Armati 1997). Moreover, a number of molecules that are key in the intracellular signalling cascade of MHC class can be visualised in Schwann cells *in vitro* (Kieseier et al., unpublished observation). In human nerve biopsies from patients with GBS and its chronic variant, chronic inflammatory demyelinating polyradiculoneuropathy (CIDP), Schwann cells stained positive for MHC class II. This suggests that these cells may indeed act as antigen-presenting cells (APC) in immune-mediated disorders of the PNS (Pollard et al. 1986; Pollard et al. 1987). APCs are characterised by their ability to phagocytose exogenous antigen and its degradation to antigenic peptides, which can be expressed in the cleft of the MHC class 11 molecule (Cambier et al. 2001). Schwann cells *in vitro* have been shown to immunogenically present foreign and exogenous autoantigen, such as myelin basic protein (MBP), to antigen-specific syngeneic T line cells (Wekerle et al. 1986; Bigbee et al. 1987). Moreover, the presentation of endogenous antigen, such as the myelin component P2, by Schwann cells via MHC class II molecules can restimulate resting antigen-specific CD4 T^+ cell lines (Lilje 2002).

For optimal T cell activation and differentiation to occur, at least two distinct signals delivered during the interaction with an APC are required. These include antigen-specific signalling via MHC and signalling through co-stimulatory molecules. If the T cell does not receive adequate co-stimulation, it is rendered anergic or undergoes apoptosis. Thus, co-stimulation signal is central to T cell activation and survival (Coyle and Gutierrez-Ramos 2001; Sharpe and Freeman 2002). Recent data suggest that BB-1, a member of the family of co-stimulatory molecules, can be detected on unmyelinated Schwann cells and appears upregulated on myelinating Schwann cells in nerve biopsies from CIDP patients. This further indicates that Schwann cells possess the necessary markers enabling them to act as an APC in the inflamed PNS (Murata and Dalakas 2000; Pollard 2002). The extent to which these cells may propagate a T cell response via antigen presentation *in situ* needs to be elucidated in greater detail. Similarly, whether Schwann cells can suppress such a response *in vivo* needs to be investigated, as it has been reported *in vitro* (Lisak et al. 1997). At any rate, the concept that Schwann cells actively control the local T cell response within the peripheral nerve by acting as an antigen-presenting cell appears very appealing.

SCHWANN CELLS AS REGULATORS OF AN IMMUNE RESPONSE

For a long time cytokines as mediators of an immune response within the peripheral nerve were considered to be the exclusive product of inflammatory cells. There is now a large body of evidence implying that Schwann cells are capable of producing and secreting a large variety of cytokines, which could act as immunomodulators (Lisak et al. 1997). Interleukin (IL)-1, a cytokine relevant to the initiation of an immune response, has been demonstrated to be produced by cultured Schwann cells (Bergsteinsdottir et al. 1991). In addition, other proinflammatory cytokines, such as IL-6 (Bergsteinsdottir et al. 1991; Bourde et al. 1996; Gadient and Otten 1996; Murwani et al. 1996), TNF-α (Bergsteinsdottir et al. 1991; Bourde et al. 1996; Gadient and Otten 1996; Murwani et al. 1996; Wagner and Myers 1996b), and TGF-β (Scherer et al. 1993; Stewart et al. 1995b), are generated and released by Schwann cells under certain conditions, some in vitro, others also in vivo (Kiefer et al. 1996; Mathey et al. 1999; Skundric et al. 2001). Schwann cells are able to regulate the production of proinflammatory cytokines, at least in part, in a specific autocrine manner, as was shown for IL-1 (Skundric et al. 1997b; Skundric et al. 2001). The specific receptors for some of the cytokines, such as the TNF receptor, are constitutively expressed on Schwann cells, rendering these cells susceptible, for example, to a TNF-α response (Bonetti et al. 2000; Oliveira et al. 2005). Other proinflammatory and immunoregulatory mediators, such as prostaglandin E_2, thromboxane A_2 and leukotriene C_4, are synthesised in large amounts by Schwann cells, and may regulate the immune cascade within the inflamed PNS (Constable et al. 1994; Constable et al. 1999). Further mediator-produced Schwann cells include osteopontin, which is constitutively expressed and can be easily induced in Schwann cells (Ahn et al. 2004); glial-cell-line-derived neurotrophic factor (GDNF), a known survival factor for neurons; and GDNF family receptor α-1 (GFRα-1), displayed on the Schwann cell surface (Hase et al. 2005; Iwase et al. 2005).

Nuclear transcription factor-κB (NF-κB) plays a pivotal role in the regulation of the host innate antimicrobial response. It governs the expression of many immunological mediators, including cytokines, their receptors and components of their signal transduction. Recent studies suggest that two NF-κB complexes, p65/p50 and p50/p50, can be activated and regulated in human Schwann cells under certain conditions (Pereira et al. 2005). Interestingly, the natural inhibitor of

NF-κB, IκB, can also be detected in large amounts in Schwann cells (Andorfer et al. 2001). These observations further point to an active rather than a passive role of Schwann cells in the inflamed PNS. Apart from engaging such cellular factors, Schwann cells can also modulate local immune reactivity by humoral mechanisms. One pathway may operate through the production and release of nitric oxide (NO), a multipotential mediator with neurotoxic and immunosuppressive properties. Schwann cells are endowed with the inducible nitrite oxide synthase (iNOS), which is upregulated after stimulation with proinflammatory cytokines (Wohlleben et al. 1999). Taken together, a large number of pro- and antiinflammatory mediators can be induced and released by Schwann cells. To what extent these mediators have a significant impact on the clinical course of an immune-mediated disorder in the PNS clearly warrants further investigation.

SCHWANN CELLS AS TERMINATORS OF THE IMMUNE RESPONSE

In order to control the massive expansion of cellular and soluble immune mediators within the target tissue, certain mechanisms must operate with high fidelity to regulate the immune response. Once the target antigen has been eliminated or infection abated, the activated effector cells are no longer needed. When the antigenic stimulus is no longer present, the cells will succumb to programmed cell death or apoptosis (Gold et al. 1999). The survival of lymphocytes depends on a delicate balance between death-promoting and death-inhibiting factors. Various mechanisms can induce apoptosis. It can for example be effected through the interaction of the cell surface receptor Fas on T cells with its ligand FasL, a member of the TNF family (Krammer 2000). Schwann cells reveal surface expression of FasL after stimulation with proinflammatory cytokines *in vitro*. Functional analysis indicates that the interaction between Fas on T cells and FasL on Schwann cells promotes apoptosis of T lymphocytes. This raises the possibility that Schwann cells are important in terminating the immune response in the inflamed PNS (Wohlleben et al. 2000) (Figure 7.1).

SCHWANN CELLS AS IMMUNOCOMPETENT CELLS

In summary, our current knowledge of the potential immunocompetence of Schwann cells suggests that Schwann cells can induce an immune response within the peripheral nerve via pattern-recognition receptors, but also trigger a T cell response via the presentation of

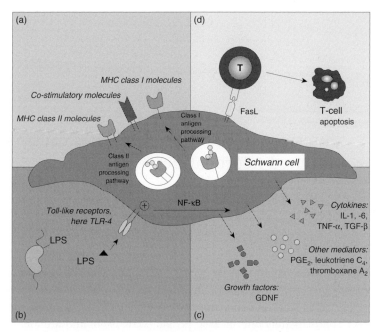

Figure 7.1 Immunocompetence of Schwann cells. (a) Antigen presentation: Schwann cells are able to process and present endogenous antigens to immune cells. Schwann cells are able to express the major histocompatibility (MHC) class I molecules as well as MHC class II molecules; in addition co-stimulatory molecules are expressed on the cellular surface. (b) Antigen recognition: via toll-like receptors Schwann cells can recognise antigens, such as LPS, activating the nuclear transcription factor NF-κB. (c) Regulation of an immune response: Schwann cells, after stimulation, secrete various soluble factors, such as cytokines, growth factors and other immune mediators. (d) Termination of an immune response: via the interaction of Fas and FasL Schwann cells can drive inflammatory T cells into apoptosis. LPS: lipopolysaccharide; GDNF: glial-cell-line-derived neurotrophic factor; PGE$_2$: prostaglandin E$_2$; IL: interleukin; TNF: tumor necrosis factor; TGF: transforming growth factor. A colour version of this Figure is in the Plate section.

antigen fragments in the cleft of MHC class II molecules in the context of co-stimulatory molecules. Through the release of immunomodulators, Schwann cells could regulate the immune reaction *in situ* and, by inducing apoptosis, appear even to be able to terminate an ongoing immune response. This evidence collected over recent years indicates that the multifaceted functions of Schwann cells go far beyond forming the myelin lamellae.

8

Mutations in Schwann cell genes causing inherited neuropathies

MICHAEL E. SHY, JOHN KAMHOLZ AND JUN LI

INTRODUCTION

Charcot–Marie–Tooth disease (CMT) refers to inherited peripheral neuropathies named for three investigators who described them in the late 1800s (Charcot and Marie 1886; Tooth 1886). CMT neuropathies affect approximately one in 2500 people (Skre 1974), and are among the most common inherited neurological disorders. The majority of CMT patients have autosomal dominant inheritance, although X-linked dominant and autosomal recessive forms also exist. Apparent sporadic cases occur, since dominantly inherited disorders may begin as a new mutation in a given patient. The majority of CMT neuropathies are demyelinating, although up to one third appear to be primary axonal disorders. Most patients have a 'typical' CMT phenotype characterised by onset in childhood or early adulthood, distal weakness, sensory loss, foot deformities (pes cavus and hammer toes) and absent reflexes. However, some patients develop severe disability in infancy (Dejerine Sottas disease or congenital hypomyelination), while others develop few, if any, symptoms of disease. Thus far at least 30 genes are known to cause inherited neuropathies, and more than 45 distinct loci have been identified. A summary of the genes associated with CMT neuropathies can be found online at http://molgen-www.uia.ac.be/CMTMutations/. Genetic testing for several forms of CMT is now available, which, in addition to providing accurate diagnosis, also provides for genotypic–phenotypic correlations. Progress has been made toward understanding how particular mutations cause disease, but pathogenic mechanisms remain largely unknown.

In landmark studies, Dyck and Lambert (1968a,b) as well as Harding and Thomas (1980a,b) subdivided hereditary motor and sensory neuropathies into dominantly inherited demyelinating CMT1 (HMSN) and dominantly inherited axonal CMT2 (HMSNII) forms, based on electrophysiological and neuropathological criteria. Other types were then classified as HMSN (CMT) III–VII, depending on inheritance pattern and accompanying features. In general, this classification has remained intact, although we now add CMTX if the patient has an X-linked neuropathy, and CMT4 if the neuropathy is recessive (reviewed in Shy et al. (2005)).

As might be expected, mutated genes responsible for causing demyelinating CMT1 have proven to be expressed by myelinating Schwann cells. The gap junction beta 1 (*GJB1*) responsible for CMTX1 is also expressed in myelinating Schwann cells (Bergoffen et al. 1993), as are several of the genes responsible for CMT4 (reviewed in (Shy et al. 2005)). A list of the known CMT genes expressed in Schwann cells is given in Table 8.1. The identification of precise genetic causes of many of the Schwann cell forms of CMT have revolutionised the diagnosis of these disorders and have also clarified several confusing issues about CMT1. For example, the Roussy–Levy syndrome, a demyelinating peripheral neuropathy associated with tremor, is now known to be caused by mutations in the myelin protein zero (MPZ) gene (Plante-Bordeneuve et al. 1999), as well as by the duplication on chromosome 17 that causes the most common form of demyelinating CMT disease, CMT1A (Thomas et al. 1997). In addition, affected individuals within the same family may or may not have tremor. The Roussy–Levy syndrome is thus a phenotypic variant with variable expression, of several forms of CMT1, rather than a separate disease entity. Also, Dejerine–Sottas neuropathy, a severe form of CMT with early onset, is known to be caused by mutations in one of several genes expressed by Schwann cells, including *PMP22, MPZ* and *EGR2* that are also mutated in less severe forms of neuropathy, with later onset. The severity of the clinical neuropathy is thus due to the nature of the specific mutation, and is not the result of a separate disease process. However, the identification and investigations of CMT have also provided significant lessons about Schwann cell biology, the subject of this book. Some of these lessons will be discussed below. However, we will first briefly review some of the basics of the biology of myelinating Schwann cells.

BIOLOGICAL BACKGROUND

Most peripheral nerves are mixed, containing both motor and sensory axons that are ensheathed along their length by Schwann cells.

Table 8.1. *Inherited neuropathies: Schwann cell aetiology*

Gene chromosomal localisation	Phenotype	Cellular localisation
PMP22 17p11.2	CMT1A HNPP DSS	Compact myelin
MPZ 1q22	CMT1B DSS CH 'CMT2'	Compact myelin
CX32 (GJB1) Xq13.1	CMTX	Non-compact myelin
L-Periaxin 19q13.1–13.2	CMT4F	Ab/Ad-axonal membrane
EGR2 10q21.1	CMT1 DSS CH	Nucleus
SOX10 22q13	CMT1+CNS+Waardenburg +Hirschprung	Nucleus
MTMR2 11q22	CMT4B1	Cytoplasm
MTMR13 11p15	CMT4B2	Cytoplasm
GDAP1 8q13	CMT4A	Unknown
PLP1 Xq22	CMTX+CNS	?

During development, Schwann cell precursors from the neural crest migrate out and contact the developing peripheral axons (Harrison 1924; Le Douarin and Dupin 1993). These 'immature' Schwann cells then ensheath bundles of developing axons, a process called 'radial sorting', and further differentiate into myelinating or non-myelinating Schwann cells (Webster 1993). Schwann cells that establish a one-to-one association with an axon, called the 'promyelinating stage' of Schwann cell development, initiate the programme of myelination and become myelinating Schwann cells (Webster 1993; Scherer 1997). In contrast, Schwann cells that do not establish this relationship with an axon do not activate the programme of myelin gene expression, and become non-myelinating Schwann cells (Webster 1993; Mirsky and Jessen 1996). Interestingly, this decision process is directed by axons, so that all immature Schwann cells have the potential to become either myelinating or non-myelinating cells.

The primary function of myelin is to increase axonal conduction velocity without a significant increase in axonal diameter. This is accomplished by the process of saltatory conduction, in which nerve impulses jump between electrically excitable regions of the axon, called nodes of Ranvier, located between the electrically insulated areas ensheathed by myelinating Schwann cells. To determine nerve conduction velocity in patients, nerve responses are usually detected by surface recording. The onset points of the responses are routinely used to calculate the conduction velocity that is determined mainly by the speed of the largest diameter myelinated fibres, even though peripheral nerves are mixed, with bundles of both large- and small-diameter myelinated axons. This has proved to be important since many myelin-related disorders preferentially affect the largest diameter myelinated fibres.

Recent investigations of myelinated axons and their nodes of Ranvier have demonstrated a surprising structural complexity. As can be seen in the cartoon of a myelinated axon and its node of Ranvier in Figure 8.1, the myelin lamellae has two regions, compact and noncompact, each of which contains a unique non-overlapping set of protein constituents. The compact region contains the myelin structural proteins, myelin protein zero (MPZ), peripheral myelin protein-22 (PMP22), and myelin basic protein (MBP), which participate in forming the highly organised myelin lamellae and in electrically insulating axons. The noncompact region is composed of two subdomains, the paranode and the juxtaparanode. The paranodal region, the loops of Schwann cell membrane and interacting axonal membrane adjacent to the node of Ranvier, contains the Schwann cell proteins myelin-associated glycoprotein (MAG), Connexin-32, Neurofascin-155, and the axolemmal proteins Caspr and Contactin. These proteins participate in Schwann cell–axonal or Schwann cell–Schwann cell interactions, and act to electrically isolate the nodal region. The juxtaparanodal region, the portion of Schwann cell and interacting axonal membrane adjacent to the paranode, contains potassium channels and Caspr-2, both expressed by axons (reviewed in Arroyo and Scherer (2000) and Salzer (2003)). The complex cellular structures formed by myelinating

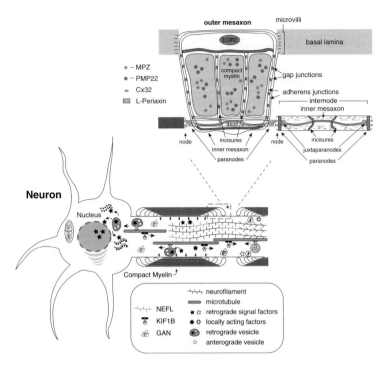

Figure 8.1 Schematic view of myelinated axon and myelinating Schwann cell in which the localisation of proteins associated with some of the causes of inherited neuropathies are illustrated at the site of their localisation in the cell. The region enclosed in the rectangle on the bottom panel is shown in detail on the top. MPZ is localised to compact myelin whereas Cx32 is localised to the paranodal loops, incisures, and inner mesaxon composed of non-compact myelin. The figure was taken from Shy et al. 2005. It was originally modified from two figures by Arroyo and Scherer (2000) and Brady et al. (1999), respectively. A colour version of this Figure is in the Plate section.

Schwann cells and their axons are thus analogous in many respects to the neuromuscular junction formed between motor axons and muscle cells; both are highly ordered multicomponent systems formed by the interaction of two distinct cell types in order to carry out a specific biological function related to nerve transmission.

The process of nerve development and associated Schwann cell differentiation, myelination and establishment of an electrically insulated node of Ranvier capable of salutatory conduction, provides the biological framework for understanding the pathogenesis of all types of peripheral neuropathy, including CMT. In fact, one might anticipate that inherited peripheral neuropathies would be caused by mutations that alter crucial aspects of this biological process, such as the critical interactions between Schwann cells and their axons at the paranodal region, or the process of myelin compaction. For the purpose of the following discussion, then, we have grouped the known Schwann cell mutations causing CMT according to their presumed cell biological functions.

PERIPHERAL MYELIN PROTEIN-22 KD

PMP22 mutations causing CMT1A and HNPP demonstrate the importance of gene dosage for normal myelination

The most common form of CMT, CMT1A, is caused by a 1.4 Mb duplication on chromosome 17 in the region carrying the gene encoding *PMP22* (Lupski *et al.* 1991; Raeymaekers *et al.* 1991). Duplication of the *PMP22* gene within the region is the likely cause of the disease, since mice and rats with extra transgenic copies of the *Pmp22* gene develop a similar demyelinating neuropathy (Huxley *et al.* 1996; Magyar *et al.* 1996; Sereda *et al.* 1996), as do some patients with *PMP22* point mutations (Roa *et al.* 1993b,c; Vallat *et al.* 1996). Although demyelination is the pathological and physiological hallmark of CMT1A, the clinical signs and symptoms of this disease, progressive weakness and sensory loss, are produced by axonal degeneration, not slowed conduction velocities from demyelination (Krajewski *et al.* 2000). The molecular basis for the progressive axonal degeneration is unknown, but a number of studies have demonstrated that they are likely to be the result of abnormalities in Schwann cell–axonal interactions. Brady and co-workers, for example, have demonstrated that Trembler mice, which have a dysmyelinating peripheral neuropathy due to a point mutation in *pmp22* similar to CMT1A, have significant changes in both axonal structure and function, including alterations in neurofilament phosphorylation, increased neurofilament density and decreased axonal transport (de Waegh and Brady 1990), and similar changes have been found in patients with CMT1A (Sahenk 1999). Transplantation of a segment of Trembler and CMT1A nerve into normal nerve also produces similar changes in axons that have regenerated through the nerve graft, but not in the surrounding nerve (de Waegh and Brady 1990), demonstrating that this effect is induced by contact with the abnormal Schwann cells. Finally, neurofilament packing density is increased and axonal caliber is decreased at the node of Ranvier of normal nerve, where there is no axonal contact with Schwann cells, compared to the adjacent myelinated internode (de Waegh and Brady 1990). These data thus convincingly demonstrate that axonal contact with myelinating Schwann cells has a significant effect on underlying axonal physiology in both normal and abnormal nerves.

In contrast, deletion of the same 1.4 Mb region on chromosome 17 that is duplicated in CMT1A does not primarily cause a progressive length-dependent neuropathy but rather an asymmetric multifocal neuropathy, called hereditary neuropathy, with a liability to pressure palsies (HNPP) (Li et al. 2002a; Li et al. 2004). Decreased expression of PMP22 is the cause of this syndrome (Chance et al. 1993), since it has been demonstrated that a family with HNPP has a frame shift 'null' mutation in the proximal portion of the *PMP22* gene, that results in a stop codon at the 7th amino acids from the start of translation (Nicholson et al. 1994). Clinically, HNPP is characterised by transient reversible episodes of focal weakness and sensory loss. Between episodes, patients are often completely asymptomatic and demonstrate little evidence of axonal loss except for very slow progressive axonal loss in elderly patients with HNPP (Li et al. 2002a; Li et al. 2004). Thus, the episodes cannot be adequately explained by axonal degeneration. Rather, it is hypothesised that nerves in HNPP are predisposed to develop transient blockage of saltatory conduction at sites of mechanical compression, the conduction block (Li et al. 2002a; Li et al. 2004). Alternatively, there might be an impairment of myelin repairing after compression injury on myelin takes place.

In summary, evaluation of patients expressing too much (CMT1A) or too little (HNPP) PMP22 develop neuropathy but by distinct mechanisms, information that was only gleaned through the investigation of human patients. How an increase in PMP22 expression causes a demyelinating neuropathy with secondary axonal damage is not known, but must involve additional mechanisms beyond the normal function of PMP22, since the neuropathy is distinct from that caused by a deficiency of PMP22. Alternatively, HNPP demonstrates that PMP22 has a functional role in maintaining myelin stability, since there is a predisposition to develop conduction block and focal neuropathies at sites of compression when its levels are reduced.

Over 80 different mutations in the *PMP22* gene also cause dominantly inherited neuropathies, although these are much less common than *PMP22* duplications. Several of these mutations are shown in Figure 8.2. The clinical phenotype of patients with *PMP22* point mutations is determined, at least in part, by the nature of the mutation: some individuals develop a neuropathy similar to that in patients with CMT1A and gene duplications (Roa *et al.* 1993b,c); others, however, develop the more severe Dejerine–Sottas/congenital hypomyelination phenoytpe with onset of neuropathy in the early neonatal period

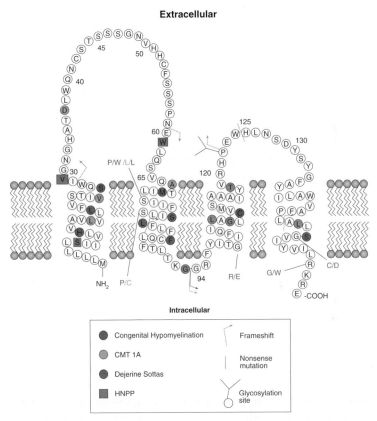

Figure 8.2 Mutations in the open reading frame of PMP22 associated with inherited neuropathies. Different mutations have caused different neuropathy severity (see text). PMP22 is an intrinsic membrane protein with four transmembrane domains, one intracellular and two extracellular loops and an amino and carboxyterminal cytoplasmic tail. Most inherited neuropathies caused by PMP22 abnormalities are caused by the chromosome 17p11.2 duplication (CMT1A) or deletion (HNPP) that contains the entire PMP22 gene. The figure was taken from Shy *et al.* (2002). A colour version of this Figure is in the Plate section.

(Hoogendijk et al. 1993; Roa et al. 1993a; Ionasescu et al. 1995; Tyson et al. 1997; Marques et al. 1998). Some patients even develop an HNPP phenotype if the mutation causes a truncated protein (Nicholson et al. 1994).

Certain of these mutations have provided clues into potential disease mechanisms involved in demyelinating disease. For example, the Leu16Pro (Valentijn et al. 1992) and Leu147Arg (Navon et al. 1996) missense mutations each cause a demyelinating neuropathy in humans and the naturally occurring demyelinating trembler J (Tr^J) (Suter et al. 1992a) and trembler (Tr) (Suter et al. 1992b) mouse mutants. Since both of these mutations are more severe in humans than HNPP (also more severe than CMT1A caused by *PMP22* duplication) they are likely to cause disability by causing an abnormal gain of function by the mutant PMP22, rather than by a simple loss of PMP22 function. Normally, only a small proportion of the total PMP22 acquires complex glycosylation and accumulates in the Golgi compartment. This material is translocated to the Schwann cell membrane in detectable amounts only when axonal contact and myelination occur (Pareek et al. 1997). However, when epitope-tagged Tr, Tr^J and wild type Pmp22 were microinjected into sciatic nerves of rats and analysed by immunohistochemistry, wild type Pmp22 was transported to compact myelin, but both Tr and Tr^J Pmp22 were retained in a cytoplasmic compartment that co-localised with the endoplasmic reticulum (Colby et al. 2000). Other studies have also shown that mutant Tr and Tr^J proteins aggregate abnormally in transfected cells (Tobler et al. 1999). In fact, aggresome-like structures have been identified in sciatic nerves of Tr^J mice, surrounded by chaperones and lysosomes, suggesting that abnormalities in the clearance of mutant PMP22 were contributing to the pathogenesis of the neuropathy (Fortun et al. 2003). More recent studies have shown that there are impairments of proteosome activity resulting in the accumulation of ubiquitinated substrates in the Tr^J model (Fortun et al. 2005). Transfection studies have demonstrated that other PMP22 as well as MPZ mutations result in mutant proteins being retained in intracellular compartments (Shames et al. 2003). Whether these also disrupt proteosome activity or cause disease by other mechanisms, such as activating the unfolded protein response (UPR) (Southwood et al. 2002), are areas of active investigation.

PMP22 and axonal degeneration

As discussed above, evaluation of patients and animal models with *PMP22* mutations have the role of Schwann cell–axonal interactions in causing disability in patients with CMT1. It is now generally recognised that axonal degeneration, secondary to disruptions in Schwann cell–axonal interactions, is a major cause of disability in CMT1 (Dyck 1975). Children with CMT1A, for example, have slow nerve conduction velocities before the onset of symptoms, and the velocities do not change appreciably as the disease progresses; thus, demyelination *per se* does not seem to be sufficient to cause the neurological signs and symptoms (Berciano et al. 2000). In addition, Krajewski and coworkers have shown that the amplitudes of compound motor action potentials, not nerve conduction velocities, are most closely related to weakness in patients with CMT1A (Krajewski et al. 2000), again suggesting that axonal loss is the cause of weakness. Finally, there is anatomical evidence of progressive length-dependent axonal loss in mouse and rat models of CMT1A (Suter and Nave 1999).

Although the nature of the Schwann-cell-derived signaling that modulates neurofilament packing density, neurofilament phosphorylation and axonal transport is not known, it must be altered or modified in dysmyelinating Schwann cells. In addition, it may be expressed as a part of the coordinated programme of myelin-specific gene expression, since this signal is also produced by normal Schwann cells. An attractive location for the origin of this signalling pathway is the paranodal region of the myelinating Schwann cell, where specialisations occur in both the Schwann cell and its underlying axolemma (reviewed in Salzer (2003)). Further definition of the molecular architecture of the paranode and surrounding region of axolemma should delineate the molecular pathways by which Schwann cells and axons communicate and contribute to understanding of axonal degeneration in demyelinating neuropathies.

It is also important to note that disruption of particular Schwann cell–axonal signalling pathways is not the only mechanism to produce axonal degeneration. Increased energy demands on the neuron to propagate action potentials, and decreased trophic factor support from denervated Schwann cells or muscle are other potential mechanisms that may also contribute to axonal degeneration in CMT1 (reviewed in Massicotte and Scherer (2004)).

Skin biopsies to evaluate PMP22 levels in CMT1A and HNPP

Most pathological material from CMT patients has been obtained from sural nerve biopsies. Sural nerves are biopsied just above the ankle. Following biopsy, patients lose sensation over the region on the lateral foot that is innervated by the sural nerve. Because sural nerve biopsies are invasive and because they are not needed to diagnose CMT, it has become increasingly difficult to obtain pathological material to investigate disease mechanisms in CMT patients.

We have recently begun to analyse myelinated nerves in skin biopsies from controls and CMT patients. Skin biopsies are minimally invasive procedures in which 2–3 mm punch biopsies are taken from the patient's skin. Unlike sural nerve biopsies, skin biopsies can be repeated many times (Polydefkis et al. 2002), and performed at various sites along the body permitting evaluation of both proximal and distal nerves. Previously, skin biopsies have mainly been used to investigate non-myelinated nerves (Kennedy et al. 1996; Holland et al. 1998; Polydefkis et al. 2002). There has been evidence to support skin biopsy in evaluating myelinated nerves. Ceuterick-de Groote and colleagues used EM from a skin biopsy to demonstrate the same hypomyelination and onion bulb formation that had been identified in a sural nerve biopsy of a patient with a Ser72Leu mutation in PMP22 (Ceuterick-de Groote et al. 2001). A recent study by Nolano and colleagues demonstrated that myelinated nerve fibres innervating Meissner's receptors in skin were readily identifiable (Nolano et al. 2003).

We have characterised the morphological and molecular phenotype of myelinating Schwann cells from skin biopsies. We found that light microscopy, EM and immunohistochemistry routinely identified myelinated nerve fibres in glabrous skin that appeared similar to myelinated fibres in sural and sciatic nerve (Li et al. 2005a) (Figure 8.3). We also identified myelin abnormalities in a group of patients with CMT, including the predicted increased levels of PMP22 in samples from patients with CMT1A and decreased levels in patients with HNPP (Figure 8.4). We have also performed preliminary studies that suggest that immunohistochemistry, semi-thin and EM studies from patients with various forms of CMT1 may prove useful in identifying morphological abnormalities in these diseases (Li et al. 2005a). However, these findings still need to be quantitatively confirmed in larger numbers of controls and CMT patients.

MYELIN PROTEIN ZERO (MPZ)

MPZ and CMT1B

CMT1B is caused by mutations in the major myelin protein zero (MPZ), which comprises approximately 50% of myelin protein, and is necessary

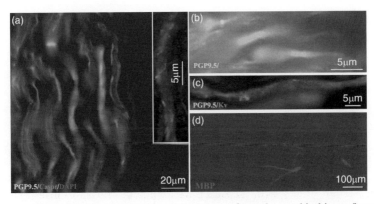

Figure 8.3 Immunohistochemistry was performed on a skin biopsy from the forearm of a normal subject. (a) Double-staining with PGP9.5 (green) and Caspr (red) antibodies shows that Caspr discrete bands in the paranodal regions flank the node of Ranvier bilaterally. Spiralling distribution of Caspr on the axolemma opposing inner mesaxon can also be visualised (insert). (b) Voltage-gated sodium channels are restricted to the nodes of Ranvier (arrows). (c) Potassium channels are located in the juxtaparanodes (arrowheads) and spiral along the axolemma opposing the inner mesaxon of the internode (arrows). (d) MBP (blue) is also stained in these nerves (arrows). These data suggest that skin myelinated nerves have a molecular architecture that is similar to that of other PNS-myelinated axons. Figure taken from Li *et al.* (2005a). A colour version of this Figure is in the Plate section.

for both normal myelin structure and function (Greenfield *et al.* 1973; Eylar *et al.* 1979). MPZ is a transmembrane protein of 219 amino acids and is a member of the immunoglobulin supergene family. It has a single immunoglobulin-like extracellular domain of 124 amino acids, a single transmembrane domain of 25 amino acids, and a single cytoplasmic domain of 69 amino acids (Lemke and Axel 1985; Uyemura *et al.* 1995). The first 29 amino acids of MPZ serve as a leader peptide that is cleaved prior to insertion of MPZ into the myelin lamellae. MPZ is also posttranslationally modified by the addition of an N-linked oligosaccharide at a single asparagine residue in the extracellular domain, as well by the addition of sulphate, acyl and phosphate groups (D'Urso *et al.* 1990; Eichberg and Iyer 1996). MPZ, like other members of the immunoglobulin superfamily, is a homophilic adhesion molecule (Filbin *et al.* 1990). Heterologous cells expressing MPZ adhere to each other in an *in vitro* cell interaction assay (Xu *et al.* 2001), while absence of MPZ expression *in vivo* in *Mpz* knockout mice produces poorly compacted myelin lamellae (Giese *et al.* 1992).

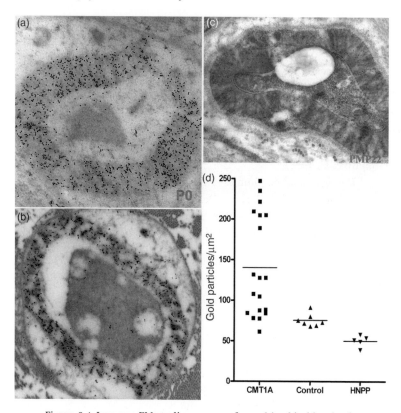

Figure 8.4 Immuno-EM studies were performed in skin biopsies from normal controls in panels (a), (b), and (c). Myelin protein zero (MPZ, panel a), myelin basic protein (MBP, panel b), and PMP22 (panel c) were labelled with antibodies conjugated with gold particles. In panel d, studies were subsequently performed in patients with CMT1A and HNPP. PMP22 levels were increased in patients with cmt1A (PMP22 duplication) and decreased in patients with HNPP (PMP22 deletion), as has been previously described in sural nerve biopsies (Vallat et al. 1996). Gold particles were quantified with ImagePro Plus software. Figure taken from Li et al. (2005a).

Interestingly, overexpression of MPZ disrupts myelination as well, by inhibiting Schwann cell wrapping of axons, also consistent with an adhesive function for the protein (Wrabetz et al. 2000). Taken together, these data demonstrate that MPZ plays an essential role in myelination, probably by holding together adjacent wraps of myelin membrane through MPZ-mediated homotypic interactions.

Crystallographic analysis of the extracellular domain of the rat MPZ demonstrates that it forms a compact sandwich of beta-sheets held together by a disulphide bridge, similar to that of other members of the Ig-superfamily. In addition, each Ig-domain monomer interacts by way of a 'four-fold' interface to form a homotetramer, a doughnut-like structure with a large central hole, as well as by way of a separate 'adhesive' interface (Shapiro et al. 1996). These data suggest that MPZ monomers interact within the plane of the membrane forming a lattice of homotetramers that interact with similar structures on the opposing membrane surface to mediate myelin compaction (Shapiro et al. 1996).

The cytoplasmic domain of MPZ is also necessary for MPZ-mediated homotypic adhesion. Deletion of 28 amino acids from the carboxy-terminus of the protein abolishes MPZ-mediated adhesion *in vitro* (Filbin et al. 1999), and non-sense mutations, such as Q186X, within the cytoplasmic domain cause particularly severe forms of demyelinating peripheral neuropathy in patients (Mandich et al. 1999). A PKC substrate motif, RSTK, located between amino acids 198 and 201 of the cytoplasmic domain, is also necessary for MPZ-mediated adhesion *in vitro*, and mutation of this residue in a patient causes peripheral neuropathy (Xu et al. 2001). Recently, we have also found that both RACK1, an activated PKC binding protein, and a 65 kDa protein identified through a yeast two-hybrid screen (p65), interact with the cytoplasmic domain of MPZ (Gaboreanu et al. 2004). Although the mechanisms by which the cytoplasmic domain is involved in homotypic adhesion are not known, MPZ probably participates in an adhesion-mediated signal transduction cascade, similar to that of other adhesion molecules such as the cadherins and the integrins, by interacting with the cell cytoskeleton by way of its cytoplasmic domain. Consistent with this notion, we have previously shown that the absence of MPZ expression leads to both the dysregulation of myelin-specific gene expression and abnormalities of myelin protein localisation in Schwann cells (Xu et al. 2000; Menichella et al. 2001). Taken together the above data suggest that MPZ plays at least two separate roles in myelination in the peripheral nervous system. The first is a predominantly structural role to hold together adjacent wraps of myelin membrane through MPZ-mediated homotypic interactions. The second is a 'regulatory' role, a consequence of the MPZ-mediated signal transduction cascade. The clinical phenotype of patients with mutations in MPZ might thus be expected to reflect these two separate roles of the protein.

To date there are over 88 different mutations in MPZ known to cause CMT1B in patients. In a recent study (Shy et al. 2004), we personally evaluated 13 individuals with CMT1B from 12 families with eight unique MPZ mutations, and reviewed the data from 64 cases from the literature in which there was sufficient information to determine the patient's clinical phenotype. These data show that 90% of these patients fall into two distinct phenotypic groups: one with extremely slow nerve conduction velocities and onset of symptoms during the period of motor development; and a second with essentially normal nerve conduction velocities and the onset of symptoms as adults. Interestingly, there is little overlap between the early-onset and late-onset groups. In addition, both groups of patients are also clinically and physiologically distinct from patients with CMT1A caused by a duplication of the PMP22 gene region. Although the precise molecular mechanisms are not known, these data suggest that MPZ mutations causing CMT1B also fall into two functional groups: one which effects both the processes of myelin development, causing delayed motor development and markedly slowed nerve conductions; and a second which allows developmental myelination, but eventually leads to axonal degeneration and weakness in later life, with minimal evidence of demyelination.

MPZ mutations that disrupt protein structure or truncate the cytoplasmic domain are associated with early onset phenotype

In order to understand further how the MPZ mutations cause these two phenotypes, we mapped the known mutations onto a model of the secondary structure of the protein. These results are shown in Figure 8.5. Our numbering system does not include the 29 amino acid leader sequence that is cleaved prior to insertion into the myelin membrane. Mutations causing the early-onset phenotype are in red, while those causing the late-onset phenotype are in blue. Mutations that introduced a charged amino acid, removed or added a cysteine residue, or altered an evolutionarily conserved amino acid were more likely to cause early-onset neuropathy. All the five mutations in this series, for example, which introduced a charge change within the extracellular domain of MPZ, resulted in early onset disease as did 8 out of 9 mutations that disrupted or added a cysteine residue, as well as 18 of 23 mutations which altered an amino acid conserved throughout vertebrate evolution (Shy et al. 2004; Shy 2005). These data suggest that mutations disrupting the tertiary structure of MPZ are more likely to cause severe, early-onset disease. In addition,

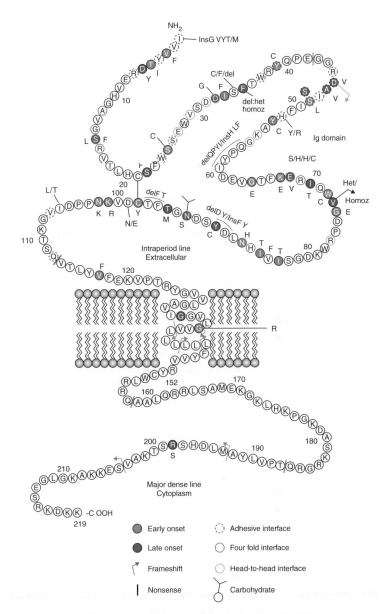

Figure 8.5 Mutations in the open reading frame of MPZ that are associated with early-onset or late-onset symptoms of neuropathy. Early-onset cases are defined as those in which there was a delayed onset of walking or other early milestones. Late-onset cases are defined as those in which the first symptoms of neuropathy occurred after the age of 21 years. MPZ is an intrinsic membrane protein with a single extracellular domain, transmembrane domain and cytoplasmic domain. Figure taken from Shy *et al.* (2004). A colour version of this Figure is in the Plate section.

5 of 6 mutations that truncate the cytoplasmic domain of MPZ, eliminating amino acids 198–201 (RSTK) have caused severe early onset disease, as has a point mutation altering Ser204, which has been hypothesised to be phosphorylated via PKC binding to the RSTK domain (Table 8.2).

Dominant negative and gain of function models for early-onset CMT1B

Patients with the Q186X mutation have presented with a very severe, early-onset demyelinating neuropathy and very slow nerve-conduction velocities, that have been characterised as congenital hypomyelination (CH) (Warner et al. 1996; Mandich et al. 1999). The Q186X protein has a truncated cytoplasmic domain, and is thus missing the PKC α interaction site, RSTK, described above. In addition, the protein is not able to mediate homotypic adhesion in an *in vitro* assay. Transfection of the mutant protein in Chinese hamster ovary (CHO) cells demonstrates that Q186X protein fused to GFP can be transported through the secretory pathway and inserted into the plasma membrane. Co-expression of the wild type MPZ along with a similarly truncated protein inhibits adhesion of the wild type protein (Kamholz, unpublished observations). Interestingly, there are several other known mutations that cause a truncation of the MPZ cytoplasmic domain, and all of which produce severe, early-onset disease, similar to that caused by Q186X (reviewed in Shy et al. (2004)). These data suggest, at least in part, that the Q186X protein acts as a dominant negative inhibitor of MPZ-mediated adhesion.

The R69C mutation in the extracellular domain of MPZ also causes an early onset, severe, de-/dysmyelinating neuropathy as determined by clinical and neurophysiological analysis of affected patients. However, since this mutation does not truncate the cytoplasmic domain of the protein but rather introduces an additional cystein into its extracellular domain, R69C is likely to cause early-onset CMT1B by a toxic gain of function rather than dominant negative effect. We had the opportunity to perform a sural nerve biopsy on a 53-year-old patient with an R69C mutation and compared it with results from her previous sural nerve biopsy on the other ankle 20 years previously. Morphometric analysis of the biopsies revealed a loss of large diameter myelinated fibres that were similar in both sural nerves, suggesting that there had been little increase in axonal loss over the past 20 years. In the biopsy we performed, segmental demyelination and numerous 'onion bulbs' were conspicuous

Table 8.2a. *Features of mutations affecting amino acids critical for homotypic interactions as identified by Shapiro et al. (1996).*

Interface	Early-onset neuropathy	Late-onset neuropathy
Four-fold	R69C, R69S, R69H, N87H	S15F, R69H
Adhesive	I1M, C21del, Y39C, S49L, H52R	D46V, S49L, H52Y
Head to head	C21del	D32G
Remaining amino acids	T5I, I32F, Del35F (homo), Y53C, Del56QPYI/InsHLF, D61E, G64E, W72C, V73fs (homo), G74E, I85T, Del89DY/InsFY, G94C, Del96FT, C98Y, K101R, I106T/L, N102K, V117F, G138R	D6Y, H10P, S22F, F35del, I70T, V73fs (hetero), Y90C, N93S, T95M, S111N, G134R, R198S, Del207K

Table 8.2b. *Features of mutations that affect phenotypes.*

Early onset		Introduction of new ± charges	G64E, G74E, N87H, N102K, G138R
		Taking away ± charges	R6S, R69C, D99N
	Extracellular	Deletion truncating the EC domain	C21del, Q56del, D89del, F96del
		Disruption involving Cys	C21del, S34C, Y39C, Y52C, R69C, W72C, G94C, C98Y
		Changes in polarity and hydrophobicity	T5I, S34F, S49L, T85I, I106T
	Transmembrane	Frame shift that truncates TM and IC domain	G138fs, L145fs
	Cytoplasmic	Frame shift that truncates IC domain	A159fs, A192fs, Q186stop
Late onset		Taking away ± charge	D6Y, H10P, D32G, D42V, H52Y
		Change at/near glycosylation site	N93S, Y90C, T95M
	Extracellular	Introduction of Cys	Y90C
		Changes in polarity and hydrophobicity	D6Y, H10P, S15F, S22F, D32G, D46V, H52Y, I69T, T95M
	Transmembrane		G134R
	Cytoplasmic	Change at RSTK site (PKC)	R198S
		Deletion	K207del

EC = extracellular; TM = transmembrane; IC = intracellular; PKC = protein kinase C.

although axonal loss of large diameter nerve fibres was also severe (no fibres >7 μm). Macrophages were found in areas of segmental demyelination but only in a few nerve fibres (<2%). Teased fibre immunohistochemistry shows voltage-gated sodium channel subtype 1.8 (Nav1.8) expressed at the nodes of Ranvier around the areas of segmental demyelination. In some nerve fibres, Nav1.8-formed hemi-nodes are reminiscent of the expression pattern of Nav1.2 during myelin development (Boiko et al. 2001). Interestingly, many myelinated fibres had 5–8 uniform short internodes sequentially (all of them <150 μm) with no segmental demyelination, no nodal Nav 1.8 expression and symmetric paranodes. These features are inconsistent with segmental demyelination and remyelination. We believe that these shorter internodes never achieved their normal developmental length, consistent with the developmental abnormalities caused by this mutation (Bai et al. in press). Interestingly, nodal Nav1.8 expression was also found in two animal models of inherited neuropathies, Trembler-J mice (Devaux and Scherer 2005) and P0 knockout mice (Ulzheimer et al. 2004). The same expression was also detected in the cerebellar tissue of the multiple sclerosis model (Waxman 2005). However, the significance of this subtype expression is still unclear at this time.

H10P causes a 'dying back gliopathy' and accumulation of proteinaceous material within the intralaminar space but not demyelination

How might MPZ mutations causing late-onset neuropathy interfere with axonal function? One attractive possibility is that these mutations produce subtly abnormal myelin lamellae that causes alteration in Schwann cell–axonal interactions, and thereby leads to axonal degeneration. Most patients with late-onset neuropathy, for example, have relatively normal or only slightly slowed nerve conductions, and show predominant axonal loss with mild or no demyelination on nerve biopsy. Clues to how this might occur were provided in an autopsy on a 73-year-old woman with a late-onset form of CMT1B caused by a H10P mutation. We found that unlike R69C, there was little evidence of segmental demyelination. Rather, length-dependent axonal loss dominated the pathology. Additionally, we identified large amounts of amorphous material that had accumulated in the inner intralaminar spaces of proximal, large-diameter myelinated nerves, particularly in dorsal roots. At high EM magnification, this material could be seen in the space between two split leaflets of major dense-line, which suggest an intracellular accumulation of Schwann cell. This preferential involvement of adaxonal myelin wraps was frequent, consistent with a dying back pathology of the distal Schwann cell. Finally, we identified abnormalities in the molecular organisation of the paranodal and juxtaparanodal regions of axolemma, including the spreading of voltage-gated potassium channels and Caspr (see below). Taken together, we hypothesise that a dying back of adaxonal myelin results in a breakdown of myelin–axonal interactions that ultimately lead to axonal degeneration in our patient (Li et al. 2006). Whether these abnormalities are widespread in patients with late-onset CMT1B is not yet known. Recently we have had the opportunity to re-examine an autopsy of a patient with a T95M mutation who also had a late-onset neuropathy. Unfortunately, this autopsy did not include either ventral or dorsal roots. In general the findings were similar to those observed in the H10P autopsy. We did identify two myelinated fibres from the T95M sciatic nerve that contained a large amount of vesicular material in the periaxonal space (Bai et al. in press),

CONNEXIN 32 (CX32)

CMT is caused by mutation in the gene encoding Cx32, a gap junction protein localised to the paranodal region and incisures of the myelinating Schwann cell

Mutations in the *GJB1* gene encoding connexin 32 (Cx32), a gap junction protein localised to the paranodal region and incisures of myelinating Schwann cells, occurs in 10–15% of patients with CMT (Nelis et al. 1996). Because the Cx32 gene is located on the X-chromosome, this form of the CMT is called CMTX1. Over 280 different mutations in the Cx32 gene have been identified (http://www.molgen.ua.ac.be/CMTMutations/default.cfm), which produce a clinical phenotype characterised by varying amounts of weakness, muscle atrophy and sensory loss. Many of these mutations are illustrated in Figure 8.6 (Wrabetz et al. 2004). Since Cx32 forms gap junctions between adjacent loops of Schwann cell membranes at the paranodal region, mutations in the *Cx32* gene should only minimally disrupt the structure of compact myelin. Consistent with this hypothesis, many individuals with CMTX1 have near normal nerve conduction velocities with markedly decreased amplitudes (Nicholson and Nash 1993; Lewis and Shy 1999), suggesting that axonal loss rather than demyelination is a prominent feature of the neuropathy (Hahn et al. 2000; Hahn et al. 2001). Analysis of

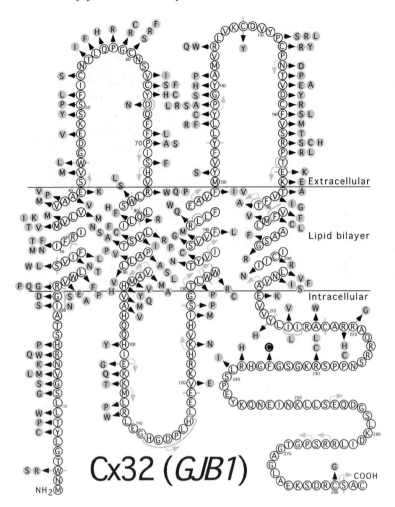

Figure 8.6 Mutations associated with CMTX1. Connexin 32 (Cx32) is an intrinsic membrane protein with four transmembrane domains, one intracellular and two extracellular loops, and an amino and carboxyterminal cytoplasmic tail. The positions of the amino acids affected by the known mutations are indicated. Taken with permission from Scherer and Kleopa (2005).

sural nerve biopsies from patients with CMTX also demonstrates that axonal loss is prominent in CMTX. Interestingly, *Gjb1/cx32*-null mice develop a neuropathy similar to that in patients with CMTX1 (Anzini *et al.* 1997; Scherer *et al.* 1998), demonstrating that absence of functional paranodal gap junctions in Schwann cells is sufficient to cause axonal loss. How loss of these gap junctions causes prominent axonal degeneration, however, is not known, but probably is due to alterations in cell–cell interactions between paranodal loops of Schwann cell membrane and their associated axons.

How might mutations in *GJB1* cause CMTX?

Cx32 belongs to a family of proteins, all of which have a similar structure (Bruzzone et al. 1996; Bruzzone and Ressot 1997). Each connexin protein has four membrane-spanning domains connected by one intracellular and two extracellular loops. Six connexin molecules assemble to form a connexon or hemichannel, with the third transmembrane domain lining the central pore. When hemichannels meet at opposed cell membranes, gap junction channels can form through which ions and small molecules can pass. In theory, mutations which limit the ability of ions and small molecules to pass through Cx32 channels might cause neuropathy by a loss of function mechanism. Multiple *GJB1* mutations have been evaluated in functional assays in paired *Xenopus* oocytes (reviewed extensively in Wrabetz et al. (2004)). G12S, R22G, R22P, C53S, C60F, C64S, L90H, V95M, R142W, P172S, 175 frameshift, E186K, E208K, Y211stop, R215W were not able to form functional channels. In contrast, S26L, I30N, M34T, V35M, V38M, L56F, T86A, T86S, T86N, P87A, E102G, 111–116 deletion, C217stop, R220stop, F235C, R238H, C280G, S281stop mutant proteins formed gap junctions (25–31) (Dahl et al. 1992; Bruzzone et al. 1994; Rabadan-Diehl et al. 1994; Oh et al. 1997; Ballesteros et al. 1999; Castro et al. 1999; Lin et al. 1999). Most of these gap junctions had abnormal electrophysiological characteristics, including diminished level of junctional conductance (C217stop, R220stop, R265stop, and S281stop), but some (R238H and C280G) were indistinguishable from wild type Cx32, raising the question of how these mutations cause CMTX. Thus, whereas the oocyte system may be optimal for examining the biophysical properties of mutant channels, this analysis has not identified potential pathological mechanisms of all mutants.

Since Cx32, like PMP22 and MPZ, is transported through the endoplasmic reticulum prior to being inserted in the myelin lamellae, researchers have investigated whether particular mutations are retained within the ER or other cytoplasmic compartments, as is the case with these other proteins. In addition to 16 *GJB1* mutations that have been expressed in mammalian cells (Omori et al. 1996; Deschenes et al. 1997; Oh et al. 1997; Liu et al. 1999), Scherer and colleagues have also expressed at least two of these mutations in cultured Schwann cells and obtained similar results. The mutant protein encoded by the 175 frameshift as well as the R142Q mutation could not be detected, although their cognate mRNAs were present (Deschenes *et al.* 1997). G12S, M34T, M34K, V38M, A39V, A39P, A40V, T55I, C53S, R75W, R75P, R75Q, M93V, R142W, R164Q, R164W, P172R, R183C, E186K, S205I, E208K, Y211stop, and C217stop mutants do not reach the cell surface, being retained in the ER and/or Golgi, and hence cannot form functional gap junctions. R15Q, S26L, M34V, M34I, V35M, V37M, L56F, C60F, V63I, V139M, R183H, R183S, N205S, I213V, R215W, R219C, R219H, R220G, R220stop, R230G, R230C, R230L, F235C, R238H, L239I, C280G, and S281stop all reached the cell surface, although the bulk of Cx32 in the M34V, M34I, V37M, M93V, R183H, R183S, N205S mutant protein was intracellular (reviewed in Wrabetz *et al.* (2004)). Whether the mutant connexins that reached the cell surface formed functional gap junctions has not been well studied – S26L and R220stop makes functional gap junctions, whereas R215stop does not. Scherer and colleagues also demonstrated that at least two mutations that were retained in the ER of mammalian cells were also retained in the ER of cultured Schwann cells.

Do most CMTX1 patients have loss of function phenotypes?

Although different mutations appear to disrupt intracellular trafficking or function of Cx32, these mutations do not appear to cause abnormal gain of function deficits as frequently occurs with MPZ or PMP22 mutations, as described above. In fact, while a detailed analysis has not yet been done, patients in whom the Cx32 protein has been entirely deleted have a similar phenotype to that of at least most other patients with Cx32 mutations (Ainsworth *et al.* 1998; Nakagawa *et al.* 2001b). For example, Hahn and her colleagues recently performed genotype–phenotype correlations on 68 patients with 25 distinct *GJB1* mutations. These investigators found that the CMT Neuropathy Score (CMTNS) and related measurements worsened with age in all patients, including those in which Cx32 was deleted. Scores did not differ between mutations, with the exception of Lys187Glu and Glu208Lys, which appeared slightly more severe than the other mutations. Preliminary conclusions by the authors suggested that virtually all mutations resulted in a 'loss of function' phenotype (Hahn *et al.* 2005).

Transient CNS abnormalities in CMTX1

Cx32 is expressed by oligodendrocytes as well as by myelinating Schwann cells. However, while mutations in Cx32 typically cause demyelinating neuropathies, they spare CNS myelin clinically, although some patients may have prolonged brainstem auditory evoked responses (BAERS) (Nicholson and Corbett 1996). Why this occurs is not known but may be because Cx32 is localised differently in the PNS and CNS. For example, PNS Cx32 is localised in non-compact regions of myelin including Schmidt−Lanterman incisures and paranodal loops, whereas Cx32 appears localised to the oligodendroycte soma. Recent studies over the past few years have demonstrated that some *GJB1* mutations cause transient clinical CNS dysfunction accompanied by diffuse and symmetric white matter abnormalities. In some cases these events followed potential metabolic stresses such as a rapid return from skiing at high altitudes (Paulson *et al.* 2002) (Figure 8.7). In other cases the episodes occurred within multiple family members (Panas *et al.* 2001; Hanemann *et al.* 2003).

Why these events occur is unknown but likely involves other connexins besides Cx32. The nature of the Cx32 mutations in at least two of the above patients makes it unlikely that they can utilise Cx32 to form gap junctions. The Arg142Trp mutation is unable to form gap junctions in *in vitro* assays (Bruzzone *et al.* 1994). The Cys168Tyr mutation in the second patient is also unlikely to form functional junctions since it disrupts one of the six conserved cysteines of the extracellular

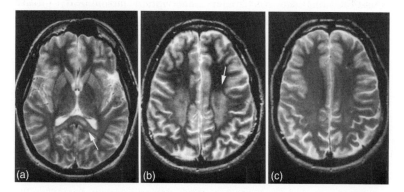

Figure 8.7 Symmetric white matter abnormalities on T2-weighted MR images of a CMTX patient with transient CNS dysfunction (a and b). Note involvement of the splenium of the corpus callosum in (b). The signal changes, which persisted in an MRI two months later, had resolved one year later (c). Figure taken with permission from Paulson *et al.* (2002).

loops of Cx32 that are necessary to dock apposing hemichannels into a gap junction, through the formation of disulphide bonds (Dahl et al. 1992). In the CNS, oligodendrocytes, which express Cx47, Cx45 as well as Cx32 form heterotypic gap junctions with astrocytes which express Cx30 and Cx43 hemi-channels. Therefore, gap junctions between oligodendrocytes and astrocytes in both patients are probably limited to those not involving Cx32 on the oligodendrocyte. The Cx32 mutations in our patient probably cause CNS dysfunction by reducing the number of functioning gap junctions between oligodendrocytes and astrocytes, making both cells more susceptible to abnormalities of intercellular exchange of ions and small molecules in situations of metabolic stress. Interestingly, the CNS of mice that lack either Cx32 or Cx37 is normal. However, when both are knocked out the mice die by 6 weeks of age from a severe leukodystrophy (Menichella et al. 2003). Moreover, mutations in Cx47 are known to cause an X-linked leukodystrophy in humans similar to Pelizaeius–Merzbacher disease (Uhlenberg et al. 2004). Taken together, these results suggest that a lack of heterotopic connexin formation by Cx32 and Cx47 may contribute to the transient CNS episodes in these patients.

CMT1 is caused by mutation in the gene encoding EGR-2, a transcriptional regulator of myelin gene expression

Several investigators have recently identified mutations in the gene encoding the zinc finger transcription factor, early growth response 2 (EGR2; also called Krox 20), as the cause of a novel form of dominantly inherited demyelinating neuropathy (Warner et al. 1998; Bellone et al. 1999; Timmerman et al. 1999; Pareyson et al. 2000; Rogers et al. 2000; Yoshihara et al. 2001). EGR-2 expression is upregulated in Schwann cells at the promyelinating stage when they have established a one-to-one relationship with axons, and is expressed at the highest levels in myelinating Schwann cells. EGR-2 and its mRNA also accumulate in parallel with the other major myelin gene products. Like other myelin genes, EGR-2 expression requires Schwann cell–axonal interactions, and is abolished after nerve transection and reestablished during reinervation (Topilko et al. 1994; Zorick et al. 1996). Mice lacking EGR-2 expression have a dysmyelinating peripheral neuropathy in which Schwann cells are arrested at the promyelinating stage of development, suggesting that EGR-2 expression is necessary for progression past this stage (Topilko et al. 1994). Data from Millbrandt and colleagues demonstrates further that EGR-2 can upregulate a number of myelin-specific genes in Schwann cells, and that mutations in EGR-2 probably causes neuropathy by an alteration of the expression of these genes (Nagarajan et al. 2001).

Inactivation of the gene encoding the POU-domain transcription factor, Oct 6, in mice, also produces an arrest of Schwann cell development at the promyelinating stage, similar to that seen in egr-2 knockout. Developmental arrest of Schwann cell development in the Oct 6 knockout, however, is only transient, and peripheral nerves are eventually myelinated normally (Bermingham et al. 1996). Interestingly, the onset of Oct 6 expression begins in Schwann cells before that of EGR-2, peaks at the promyelinating stage, and is turned off in myelinating Schwann cells (Scherer et al. 1994). Expression of EGR-2 does not require expression of Oct 6, however, since EGR-2 is expressed in Schwann cells in the Oct 6 knockout mice when they myelinate. Taken together, these data suggest that Oct 6 acts upstream of EGR-2 during Schwann cell development, and that Oct 6 may play a role in the timely onset of EGR-2 expression and myelination. Mutations in the Oct 6 gene causing CMT have not been identified, probably because of the transient nature of the peripheral myelination defect.

LITAF/SIMPLE mutations cause CMT1C

Street et al. reported that mutations in the putative protein-degradation protein LITAF/SIMPLE cause the demyelinating autosomal dominant disorder CMT1C (Street et al. 2003). These initial patients with CMT1C have had phenotypes that strongly resemble those with CMT1A with uniformly slow nerve conduction velocities (NCV), in the range of 20–25 m/s, along with mild weakness and sensory loss that first presents within the first two decades of life. Although the precise function of SIMPLE is unknown its murine orthologue has been shown to interact with Nedd4, an E3 ubiquiton ligase. Mono-ubiquitination of plasma proteins by Nedd4 family members serve as internalisation signals that are recognised by protein TSG101, which facilitate the sorting of membrane proteins to the lysosome for degradation (Bennett et al. 2004). Although SIMPLE is expressed in many cell types when mutated it seems to cause only a demyelinating neuropathy, which suggests that the disease specificity may come from impaired degradation of specific Schwann cell proteins such as PMP22.

CMT4F is caused by mutation in the gene encoding periaxin, a protein involved in Schwann cell–basal lamina interactions

Mis-sense mutations in *L-periaxin* have been shown to cause a severe autosomal recessive demyelinating inherited neuropathy, called CMT4F (Boerkoel et al. 2001; Guilbot et al. 2001). L-periaxin is a Schwann-cell-specific membrane-associated protein initially identified in a screen for proteins associated with the Schwann cell cytoskeleton (Gillespie et al. 1994). Like its alternatively spliced isoform S-periaxin, L-periaxin has a known protein interaction motif at its carboxy terminus, called a PDZ domain, suggesting that both isoforms are involved in protein–protein interactions (Dytrych et al. 1998). L-periaxin is expressed on the abaxonal surface of myelinated Schwann cells, the region of the membrane not next to the axon but apposed to the extracellular matrix (Scherer et al. 1995). Brophy and co-workers have shown that L-periaxin forms a protein complex with dystrophin-related protein 2 (DRP2), a cytoplasmic protein with homology to the muscle protein, dystrophin (Sherman et al. 2001). The periaxin–DRP2 complex also interacts with the cytoplasmic domain of dystroglycan, a transmembrane protein known to bind to laminin-2 (merosin), a component of the basal lamina or extracellular matrix (Yamada et al. 1994) and L-periaxin is required for normal clustering of DRP2-dystroglycan complexes in sciatic nerve (Sherman et al. 2001). These data thus demonstrate that L-periaxin participates in a dystrophin–dystroglycan complex similar to that found in skeletal muscle linking the basal lamina, outside the cell, to the actin cytoskeleton within the cell. Like its role in muscle, the dystrophin–dystroglycan complex probably stabilises the outer Schwann cell membrane. In addition, Schwann cell–basal lamina interactions probably also participate in a signal transduction pathway, since Schwann cell contact with the basal lamina is known to modulate myelin gene expression. Mutation of the *L-periaxin* gene thus causes demyelinating peripheral neuropathy by altering these Schwann cell–basal lamina interactions.

Although L-periaxin is localised to the abaxonal membrane of myelinating Schwann cells, its subcellular localisation changes during development. In embryonic Schwann cells, for example, L-periaxin is localised to the nucleus (Sherman and Brophy 2000), while in perinatal Schwann cells, it is found in the adaxonal or periaxonal cytoplasm (Scherer et al. 1995). Interestingly, pathological evaluation of peripheral nerve from additional patients with CMT4F demonstrates structural abnormalities at the paranodal region with disruption of interactions between paranodal loops and the adjacent axon (Schroder et al. 2001), suggesting that these changes are caused by alteration of the function of periaxin at its periaxonal localisation. Further work, however, will clearly be required to elaborate the developmental function of L-periaxin.

Recent studies have demonstrated that during development periaxin-null mice (Prx $-/-$) were unable to generate longitudinal and transverse bands of cytoplasm (Cajal bands) and the Schwann cells were unable to elongate normally along elongated PNS axons. Resultant internodes were uniformly short and nerve conduction velocities were significantly slow at approximately 10 m/s, because of the shortened internodes (Court et al. 2004). Whether similar changes occur in patients with CMT4F and whether shortened internodes contribute to slow nerve conduction in other demyelinating neuropathies will likely be the subject of many future investigations.

CMT4B 1 and 2 are caused by mutations in genes encoding myotubularin-related phosphatases

Mutation in the gene encoding myotubularin-related phosphatase 2 (MTMR2) has been shown to cause CMT4B, a recessively inherited demyelinating neuropathy, also referred to as hereditary motor and sensory neuropathy with focally folded myelin lamellae (Figure 8.8) (Gambardella et al. 1999; Bolino et al. 2000; Houlden et al. 2001). MTMR2 is a member of a family of proteins characterised by a 10-amino acid sequence with similarity to the active site of both tyrosine and serine phosphatases, which are thus referred to as dual-specific phosphatases (Laporte et al. 2001). Mutation in myotubularin, the founding member of this group, has been found to cause myotubular myopathy. Interestingly, recent studies of myotubularin suggest that it cleaves the phosphate from phosphatidylinositol 3-monophosphate, and is thus involved in the regulation of the phosphatidylinositol 3-kinase (PI 3 kinase) signal transduction cascade (Laporte et al. 2001). Although the function of MTMR2 in the PNS is not known, teased fibres from sural nerve biopsies demonstrate segmental demyelination associated with redundant loops of myelin (Gambardella et al. 1999) suggesting that MTMR2 may play a role in the regulation of myelin homeostasis

Four children in a Turkish inbred family were subsequently identified with neuropathies characterised by focally folded myelin,

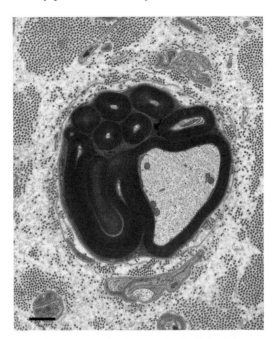

Figure 8.8 Transverse electron micrograph through a myelinated nerve fibre from a patient with CMT4B1, caused by a mutation in MTMR2 (Houlden et al. 2001) (see text). Note the multiple myelin outpouchings characteristic of the disorder. Scale bar = 1 μm. The figure is a generous gift from Dr Rosalind King.

a characteristic of CMT 4B1 as cited above. However homozygosity mapping demonstrated linkage to a distinct locus on chromosome 11p15, leading to the designation of the locus as **CMT4B2**. Recently, mutations have been identified in a large, novel gene myotubularin-related phosphatase 13 (MTMR13), also named *SET binding factor 2 (SBF2)*, that lies within the interval on 11p15 and segregates with the neuropathy in affected families. MTMR13 (SBF2) is a member of the pseudo-phosphatase branch of myotubularins with striking homology to myotubularin-related protein 2 (MTMR2) (Senderek et al. 2003). MTMR13 lacks a functional phosphatase domain so that it cannot cleave phosphate moieties by itself. Rather, it is likely that it interacts directly with MTMR2 in a common pathway. MTM1 and MTMR2, for example, have been shown to interact with pseudophosphatases MTMR12 and MTMR5 for example (Kim et al. 2003). Unfortunately, the subcellular localisation and tissue distribution of many of these myotubulin-related proteins remains unknown, limiting our ability to determine the biological relevance of these potential interactions. Some mutations in MTMR13 appear to cause glaucoma instead of neuropathy for unclear reasons.

CMT4A is caused by mutation in the gene encoding the ganglioside-induced-differentiation-associated-protein-1 (GDAP1)

Recently, mutations in a gene encoding a novel protein of unknown function, ganglioside-induced-differentiation-associated-protein-1 (GDAP1), have been shown to cause the recessively inherited CMT4A. Depending on the patients, the mutations appear to cause either demyelinating (Baxter *et al.* 2001) or axonal (Cuesta *et al.* 2001) neuropathies. RT-PCR studies suggest that the *GDAP1* gene is expressed ubiquitously, including neurons and Schwann cells (Baxter *et al.* 2001; Cuesta *et al.* 2001), although the gene was originally identified in a neuronal cell line (Liu *et al.* 1999). Thus, whether mutations in GDAP1 cause neuropathy by disrupting Schwann cell, or neuronal function, or a combination of the two is not yet known. A recent study has demonstrated that mRNA levels are expressed much more robustly in neurons. Interestingly, this same study demonstrates that GDAP1 associates with mitochondria in cells (Pedrola *et al.* 2005). The role of GDAP1, however, in neurons, Schwann cells or mitochondria remains to be discovered.

SOX10 AND PLP1 MUTATIONS CAUSE DISEASE IN SCHWANN CELLS AND OLIGODENDROCYTES

Sox10 mutations

Sox proteins are a family of transcription factors expressed at the neural plate border in response to neural-crest-inducing signals. Sox10 is thought to play an important role in the survival of neural-crest precursor cells and for the proper differentiation of neural-crest-derived glia and melanocytes. Sox10 mutations disrupt neural crest development in Dominant megacolon (Dom) mice and the Waardenburg–Hirschsprung syndrome type IV. Both these disorders consist of hypopigmentation, cochlear neurosensory deafness, and enteric aganglionosis (reviewed in Mollaaghababa and Pavan (2003) and Hong and Saint-Jeannet (2005)). Sox10 is also expressed in mature myelin-forming cells of both the PNS and CNS. Innoue and colleagues reported a patient with a Waardenburg–Hirschsprung syndrome who also had a leukodystrophy resembling Pelizaeus–Merzbacher disease (PMD) and a demyelinating peripheral neuropathy consistent with CMT1. The mutation did not disrupt the coding region of Sox10 but extended the length of the peptide and was hypothesised to act as a dominant-negative allele (Inoue et al. 1999). Histopathological studies showed an absence of PNS myelin despite normal numbers of Schwann cells (Inoue et al. 2002). Similarly, a young girl has been identified with a demyelinating neuropathy characterised by hypomyelination, deafness and chronic intestinal pseudo-obstruction with a frameshift mutation in Sox10 (Pingault et al. 2000). The timing and mechanism by which Sox10 mediates these neuropathies in Schwann cells is not yet known. Innoue and colleagues also noted that more severe patient phenotypes occurred when mutant mRNAs escaped the non-sense-mediated decay pathway, in which mRNAs bearing truncating nonsense mutations are degraded before they can be translated into abnormal proteins (Inoue et al. 2004). Similar results were also identified in several patients with *MPZ* truncating mutations, suggesting that triggering non-sense mediated decay may be a widespread process in the pathogenesis of CMT (Inoue et al. 2004).

PLP1 mutations

Proteolipid protein (PLP1) and its alternatively spliced isoform, DM20, are the major proteins in compact myelin in the CNS, but are also expressed by Schwann cells in the PNS. Both proteins have an identical sequence except for 35 amino acids in PLP1 (the PLP1-specific domain) not present in DM20. Mutations of *PLP1/DM20* cause Pelizaeus–Merzbacher disease (PMD), an X-linked leukodystrophy. Garbern and colleagues evaluated a PMP family in which a guanine deletion in the initiating ATG codon resulted in a null allele for PLP1 and DM20. Male patients in this family were found to have a mild, asymmetric demyelinating peripheral neuropathy (Garbern et al. 1997). Subsequently, we identified several other families with null alleles, in which no PLP1 or DM20 protein were made, in which affected males had demyelinating neuropathies. In addition, we identified four new *PLP1* point mutations that cause both PMD and peripheral neuropathy, three of which truncated PLP1 expression within the PLP1-specific domain, but do not alter DM20. The fourth mutation altered both PLP1 and DM20 due a change in mRNA splicing, and is probably a null mutation. Six other *PLP1* point mutations predicted to produce proteins with an intact PLP1-specific domain, however, did not cause peripheral neuropathy, even though the CNS manifestations of these later point mutations were more severe than those in the 'null' patients. Sixty-one individuals with *PLP1* duplications, the most frequent cause of PMD, also had normal peripheral nerve function. Taken together, these data demonstrate that Schwann cell expression of PLP1 but not DM20 function is necessary to prevent abnormal demyelinating peripheral neuropathy, suggesting that the 35 amino acid PLP1-specific domain plays an important role in normal peripheral nerve function (Shy *et al.* 2003).

9
Guillain–Barré syndrome and the Schwann cell

RICHARD A. C. HUGHES

THE DEFINITION AND THEN SUBDIVISION OF GUILLAIN–BARRÉ SYNDROME

Guillain–Barré syndrome (GBS) emerged from the confusion of nineteenth century descriptions of paralytic disorders with the descriptions of two case histories by Guillain, Barré and Strohl in 1916 (Guillain et al. 1916; Pritchard and Hughes 2004). Their patients had an acute paralysing disorder characterised by absent tendon reflexes and an increased cerebrospinal fluid protein with a normal cell count. They deduced that the disease involved the spinal nerve roots and Guillain insisted that it had a benign prognosis. This statement is sadly not true but does reflect the regenerative capacity of the Schwann cell, which turned out to be the principal focus of the pathology in the common form of the disease.

During the 1970s and 1980s the clinical picture was pinned down by a working definition which has continued in use (Asbury et al. 1978; Asbury and Cornblath 1990). The definition has inclusion criteria of progressive weakness of the limbs reaching its worst typically within four weeks, associated with loss or substantial loss of the tendon reflexes. It also has criteria of features which cast doubt on the diagnosis, such as marked asymmetry or bladder and bowel dysfunction at the onset, and exclusion criteria of alternative diagnoses, such as porphyria or toxin exposure. The characteristic features of GBS are that it occurs after, not during, an infection and that the pathology is largely confined, like Schwann cells, to the peripheral nervous system (PNS).

Since the mid 1980s it has been appreciated that the clinical picture portrayed by this definition may be produced by more than one pathological entity. The common underlying pathology in Europe and North America is acute inflammatory demyelinating polyradiculoneuropathy (AIDP) (Asbury et al. 1969). This long-winded name accurately defines the temporal evolution, spatial distribution of lesions and principal pathology. It disguises the fact that axonal damage also occurs. In some patients the pathology is predominantly axonal as first described by Feasby and colleagues (Feasby et al. 1986; Feasby et al. 1988). Among these patients, some have acute motor and sensory axonal neuropathy (AMSAN) and others acute motor axonal neuropathy (AMAN). This last form of the disease was described especially in summer epidemics in northern China, and is sometimes so severe as to cause death and yet leave the sensory axons spared (McKhann et al. 1991; McKhann et al. 1993; Ho et al. 1995; McKhann et al. 1999).

Additional disorders share features with GBS but have more focal distributions of lesions. Of these the commonest and best known is a combination of ophthalmoplegia, ataxia and loss of tendon reflexes called the Fisher syndrome, after the person who first described it (Fisher 1956). In the usual form of this condition, the deduction from the clinical signs and clinical neurophysiology is that the pathology lies in the ocular motor nerves and in the sensory nerves or posterior spinal roots or ganglia. The clinical picture may be confused with a benign form of brain stem encephalitis called Bickerstaff's encephalitis (Al-Din et al. 1982). There are also formes frustes of Fisher syndrome with various combinations of ophthalmoplegia, ataxia, facial or bulbar palsy, and areflexia.

Other disorders resemble GBS closely but have a different time course. By arbitrary definition, GBS reaches its nadir within four weeks. There is however a continuum and some patients reach their nadir in five, six or seven weeks and others have a much more chronic course. Conventional definitions separate patients with a progressive phase lasting more than eight weeks as having chronic inflammatory demyelinating polyradiculoneuropathy (CIDP) (Hughes et al. 2005b). To establish the continuum, there is also a subgroup of patients with an intermediate course, which has been termed subacute inflammatory demyelinating polyradiculoneuropathy (Hughes et al. 1992; Oh et al. 2003).

PATHOLOGY OF GBS

Acute inflammatory demyelinating polyradiculoneuropathy

In the common AIDP form of GBS post-mortem studies identified multifocal endoneurial lymphocytic infiltration as the earliest pathological event (Asbury et al. 1969). This was associated with macrophage invasion and segmental demyelination, which resembled that reported in experimental autoimmune neuritis (EAN) (Waksman and Adams 1956). At the electron microscope level, this macrophage invasion involved the penetration of the Schwann cell basal lamina and the myelin lamellae by the macrophage processes (Figures 9.1 and 9.2). The crucial observations were that the invaded myelin lamellae were morphologically normal and did not show any sign of previous disruption by soluble factors (Prineas 1971; Prineas 1981). Furthermore, the Schwann cells appeared intact and did not show signs of pre-existing damage. All these observations were mirrored by the lesions of EAN in rats (Lampert 1969), which is now known to be mediated predominantly by T helper cells directed against myelin proteins (Linington et al. 1984). These images are consistent with a T helper cell mediated mechanism for the pathogenesis of AIDP. However, it remains unclear how the macrophages are targeted so specifically by the release of cytokines by the T helper cell. In experimental situations it has not been possible to reproduce primary demyelination by inflammatory lesions in nerve tissue in the absence of autoimmunity (Powell et al. 1984; Pollard et al. 1995). The explanation might be that there is some unknown mechanism by which T helper cells having reactivity against myelin or Schwann cells antigens confer specificity on nearby macrophages directing it onto the myelin lamellae. Alternatively or additionally, there may be antibodies directed against surface epitopes on the myelinated Schwann cell, which act as opsonins and permit the attachment of macrophages via Fc receptors.

According to an alternative hypothesis, the primary mechanism driving the inflammatory process is a complement-mediated antibody-driven attack on the outer Schwann cell surface. This is a difficult area to investigate because of the ubiquity of immunoglobulin and the difficulty of identifying significant specific staining. Earlier studies found either no immunoglobulin (Ammoumi et al. 1980; Brechenmacher et al. 1987; Tsukada et al. 1987), complement and IgM but not IgG (Luitjen and Faille-Kuyper 1972), or IgG but not IgM (Nyland et al. 1981) in some but not all cases. Complement components were identified on the myelin lamellae in two biopsies and one autopsy in other studies (Koski et al. 1987; Hays et al. 1988). In a more recent and influential post-mortem study, Hafer-Macko and colleagues demonstrated the deposition of complement component C3d on the myelinated nerve fibres in the spinal roots from a patient with very early AIDP (Hafer-Macko et al. 1996b). C3d is a breakdown product of C3b which binds stably to the

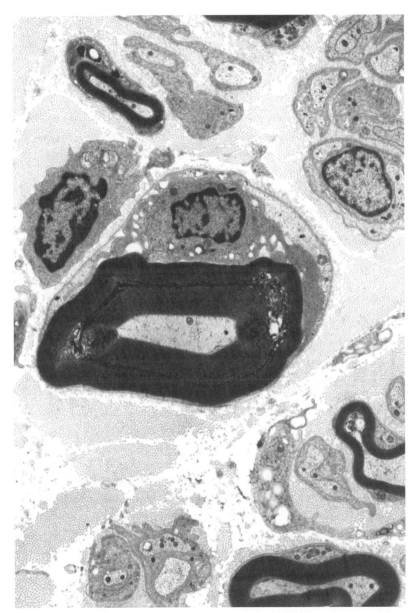

Figure 9.1 Sural nerve biopsy from a patient with GBS. Electron micrograph transverse sections showing stages in the invasion of the myelin lamellae by macrophage processes. From Hughes (1990) with permission.

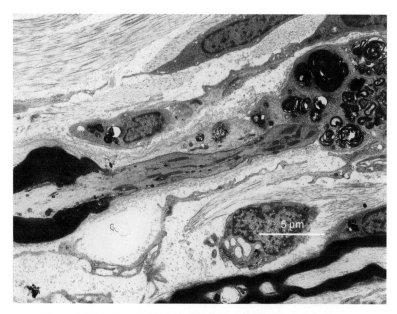

Figure 9.2 Sural nerve biopsy from a patient with GBS. Electron micrograph longitudinal section showing a demyelinated hemi-node on the right in which a phagocyte within the basement membrane contains myelin debris, as does a macrophage in the endoneurium above. The left hemi-node is intact. From Hughes (1990) with permission.

surface of membranes, which have been the site of complement fixation for which it acts as a marker. Most previous studies had not used this as a marker and none used high definition imaging techniques. Hafer-Macko et al. used resin sections and then etched out the plastic and immunostained them. In one 3-day and another 8-day autopsy they showed C3d deposited not on the myelin lamellae but on the surface of the Schwann cells, separated from the myelin by the abaxonal Schwann cell cytoplasm and, at times, by the Schwann cell nucleus. In both cases they also illustrated the presence of the complement membrane attack complex C5b−9. In a 9-day case, deposition of complement was much less frequent although there was punctuate staining of C3d in macrophages. This leads to the hypothesis that a very early event is the binding of antibody to an antigen, as yet undefined, on the surface of myelinating Schwann cells, followed by the fixation of complement with activation of the complement cascade.

In support of the idea that complement-mediated demyelination is important, Hafer-Macko et al. also showed at the electron microscope level vesicular dissolution of the myelin lamellae similar to that which has been produced experimentally with antibody to galactocerebroside (Saida et al. 1981; Sumner et al. 1982b). Similar vesicular changes do occur sporadically in the myelin lamellae of the spinal roots in autopsies. However, the material studied by Hafer-Macko et al. was collected very early and exceptionally well-preserved. The interpretation is that the complement membrane attack complexes punched holes in the abaxonal Schwann cell membrane, which allowed the ingress of calcium ions. Similar vesicular dissolution of myelin was shown to be the earliest sign of damage in experimental studies in which a calcium ionophore was injected into the rat nerve (Smith and Hall 1988). This antibody-mediated mechanism is certainly plausible in AIDP. However, similar changes have not been prominent in biopsy studies of GBS nerves in which vesicular dissolution of myelin is usually observed in the presence of macrophage infiltration (Prineas 1972; Brechenmacher et al. 1987). It is then uncertain whether the myelin damage is the consequence of earlier antibody- and complement-mediated attack that has targeted the macrophages, or of enzymes released by the invading macrophages. Both mechanisms might well co-exist.

Conventional wisdom is that following demyelination the myelin debris must be cleared, Schwann cells proliferate and one of the daughter Schwann cells assumes the responsibility for re-myelinating a length of axon (Griffin and Sheikh 2005). This length is often shorter than that of the original internode, so that where one internode existed before two or more are present after. The supernumerary Schwann cells may initially produce an extra spiral encircling the newly myelinated axon, but these Schwann cells eventually disappear and the only evidence of previous demyelination is the presence of lengths of relatively thinly remyelinated axons with short internodes. Whether the excess Schwann cells migrate away or undergo apoptosis is not known.

Axonal forms of GBS

In both AMSAN and AMAN, the pathological process is different. Hafer-Macko et al. (Hafer-Macko et al. 1996a) examined early autopsies with immunohistochemistry on plastic sections and showed the deposition of C3d and IgG on the nodal axolemma in the anterior roots of large myelinated nerve fibres from patients with AMAN. Probably targeted by antibody and complement, the macrophages invaded the nodes of Ranvier and then inserted between the axon and the surrounding Schwann cell axolemma, leaving the myelin lamellae initially intact.

In the severest cases, the axons were so damaged that Wallerian degeneration resulted (Griffin et al. 1995; Griffin et al. 1996b; Hafer-Macko et al. 1996a). The difference between AMSAN and AMAN lies in the distribution of lesions which are confined to the anterior roots in AMAN but also affect the posterior roots in AMSAN (Griffin et al. 1996b). Not all AMAN is severe. The average rate of recovery is about the same as that of AIDP (Ho et al. 1997a). This suggests that in many cases there is a reversible process, which may be due to the location of the lesions near the motor nerve terminals where the blood–nerve barrier is relatively leaky and the axons may be accessible to circulating antibody (Ho et al. 1997b).

Fisher syndrome and experimental models

The pathology of Fisher syndrome is not known from direct observations because patients with pure forms of the condition do not develop respiratory failure and make good recoveries. The presumption from the clinical features and neurophysiological observations of loss of sensory nerve action potentials is that there are demyelinating lesions in the ocular motor nerves and terminal segments of sensory nerve fibres (Guiloff 1977).

NATURAL HISTORY

In typical cases, GBS is a monophasic disease with an onset phase of one, two or three weeks, a variable plateau phase and then gradual recovery with return of proximal followed by distal strength over weeks or months (Figure 9.3). The first symptoms are pain, numbness, paraesthesiae or weakness in the limbs. The weakness may initially be proximal, distal, or a combination of both. In severe cases, muscles become wasted after about two weeks. The numbness and paraesthesiae usually affect the extremities and spread proximally. Between 4% and 15% of patients die, and up to 20% are left with persistent disability (Winer et al. 1988; Van et al. 2000). Even the non-disabled survivors may complain of persistent fatigue, probably due to the loss of nerve fibres leaving reduced reserves. The facial nerves are commonly affected, and less often the bulbar and ocular motor nerves. In 25% of cases, weakness of the respiratory muscles requires artificial ventilation. Retention of urine, ileus, sinus tachycardia, hypertension, cardiac arrhythmia and postural hypotension, due to autonomic nervous system involvement are all common (Hughes 1990).

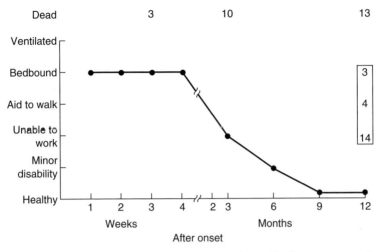

Figure 9.3 Diagram of the clinical course of GBS. The line represents the median course of the disease from an epidemiological study of 100 patients in South-East England. After Winer et al. (1988).

TREATMENT

Since the common form of GBS, AIDP, is an inflammatory disease, it would have been expected that corticosteroids would be beneficial. This has been extraordinarily difficult to show, and the balance of evidence from randomised trials is that they do not have a significantly beneficial effect (Hughes and van der Meche 2000; Van et al. 2004; Hughes 2004). This lack of effect has not been satisfactorily explained. Corticosteroids might interfere with the normal clearance of myelin debris from the demyelinated internodes. Corticosteroids are known to exacerbate muscle necrosis in denervated muscle in animal models (Rich et al. 1998; Rich and Pinter 2001). Either mechanism might counterbalance any possible beneficial effect from reducing the amount of inflammation. Fortunately, other treatments are available that are effective. Plasma exchange accelerates recovery, significantly reducing the time to recover the ability to walk unaided and the time on the ventilator for ventilated patients. In a North American trial (The Guillain–Barré Syndrome Study Group 1985), the median time to recover walking unaided was 53 days in the plasma exchange group ($p<0.001$) compared to 85 days in the control group. In a French trial (French Cooperative Group on Plasma Exchange in Guillain–Barré Syndrome 1987), this time was 70 days in the plasma exchange and 111 days in the control group ($p<0.001$). In a meta-analysis of randomised trials, there were 35/321 patients with severe sequelae after a year in the treated group, compared to 55/328 in the control group. Thus, the relative risk of having severe sequelae after a year was 0.65 (95% CI 0.44 to 0.96, $p = 0.03$) in favour of treatment with plasma exchange (Raphael et al. 2002). Because of these results, plasma exchange was adopted as the gold standard treatment for severe GBS. A trial in The Netherlands then showed that intravenous immunoglobulin has a similar effect to plasma exchange (van der Meche and Schmitz 1992), a conclusion which has been confirmed in subsequent trials and endorsed by a Cochrane review (Hughes et al. 2001).

Unfortunately, neither intravenous immunoglobulin nor plasma exchange prevent the development of severe weakness, and ventilatory support is still needed in about 25% of patients. Caring for these patients requires the resources of intensive care and multidisciplinary rehabilitation teams (Hughes et al. 2005b).

PATHOGENESIS

Antecedent events

About two thirds of patients with GBS develop symptoms of peripheral neuropathy between one and six weeks after symptoms of an antecedent infection, most commonly an upper respiratory infection but also after gastrointestinal infection. The organisms responsible are not commonly identified because the infection has cleared by the time the patient presents to their neurologist. Using serological techniques, large series have established the recent occurrence of *Campylobacter jejuni* infection in about 23–32% of patients, cytomegalovirus in 8–18%, Epstein–Barr virus in 2–7% and *Mycoplasma pneumoniae* in 9% (Van et al. 2000; Hadden et al. 2001). Other agents encountered less frequently include Herpes zoster and *Haemophilus influenzae*, but many other organisms have been implicated in occasional cases and series.

The occurrence of so many different infections preceding the onset raises the possibility that the infection only acts as a non-specific stimulus, releasing an autoimmune inflammatory process in the endoneurium as a result of failure of the normal immunoregulatory mechanisms. Transgenic mice which over-express CD86 co-stimulatory molecule required for T-cell stimulation develop spontaneous inflammatory demyelinating lesions in their CNS and spinal roots (Zehntner et al. 2003). CD86 engages with CD28 to deliver a positive stimulatory signal to effector T-cells, which might deregulate a balance between the activating and inhibiting autoreactive T-cells infiltrating the peripheral nervous system. In a NOD mouse model, depletion of the regulatory CD25+CD4+ T cells by injection of anti-IL-2 monoclonal antibody induced autoimmune neuropathy as well as other T-cell-mediated diseases (Setoguchi et al. 2005). In our own series of patients with GBS, circulating CD25+CD4+ T cells were reduced in the acute stage compared with convalescent samples and compared with healthy controls (Pritchard et al. 2004).

Molecular mimicry

According to an alternative, but not mutually exclusive, hypothesis, GBS arises because the immune response to the antecedent infection induces an adaptive immune response against antigens on the infective organism, which coincidentally resemble neural antigens. This has been convincingly shown in AMAN and Fisher syndrome but not in AIDP. Although in Western series *C. jejuni* may be responsible for any form of GBS including Fisher syndrome, detailed neurophysiological studies of a recent Japanese series identified all of 21 *C. jejuni* associated cases as having AMAN (Kuwabara et al. 2004). The strains of *C. jejuni* that induce GBS uniformly have the gene which codes for sialic acid synthase and leads to the expression of ganglioside mimicking epitopes in the bacterial lipooligosaccharide (Godschalk et al. 2004). Injection of the lipooligosaccharide into rabbits induced antibodies to ganglioside GM1 and a peripheral neuropathy with the histological hallmarks of AMAN, including macrophage invasion of the nodes of Ranvier and periaxonal space (Yuki et al. 2004). Patients with AMAN commonly do have antibodies to ganglioside GM1 and this hypothesis is very convincing. It leaves unanswered the question why some individuals with *C. jejuni* infection get AMAN and others just enteritis. There must be a host susceptibility factor, presumably immunogenetic, that is responsible.

Fisher syndrome

Most patients with Fisher syndrome have IgG antibodies to ganglioside GQ1b (Chiba et al. 1992). Willison and colleagues have applied antibodies to ganglioside GQ1b *ex vivo* to the mouse phrenic nerve–diaphragm model in the presence of complement. Some monoclonal antibodies induced a massive discharge of the muscle fibre due to an α-latrotoxin-like effect of the antibody followed by conduction block. This was associated with complement and immunoglobulin localisation on the motor nerve terminal or its perisynaptic Schwann cell or both, depending on the fine specificity of the antibody (Plomp et al. 1999; O'Hanlon et al. 2001; Willison 2005). The fixation of these factors led to complement-dependent destruction The perisynaptic Schwann cells swelled, their cytoplasm became translucent, the nuclear membrane blebbed and perinuclear organelles developed, suggesting necrotic cell death (Figure 9.4). These observations raise the novel hypothesis that Fisher syndrome is due to an autoimmune attack on the perisynaptic Schwann cell and the terminal motor axon. It is possible that the perinuclear Schwann cell is also targeted by antibodies to ganglioside GM1 in AMAN (Willison 2005).

The molecular mimicry hypothesis extends to Fisher syndrome. Some patients with Fisher syndrome have preceding *C. jejuni* infection. Some *C. jejuni* strains have ganglioside GQ1b-like epitopes in their

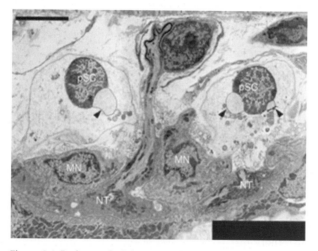

Figure 9.4 Perisynaptic Schwann cell necrosis following passive exposure of mouse neuromuscular junctions to monoclonal antibody against disialosyl antibody in the presence of complement. A motor axon is wrapped by the last myelinating Schwann cell just proximally to the nerve terminals (NTs) that form synaptic contact with a muscle fibre. The NTs have a normal morphology, with electron-dense mitochondria and tightly packed synaptic vesicles. Two Schwann cells sitting on either side of the terminal axon appear severely damaged with swollen and electron lucent cytoplasm, damaged organelles, nuclear membrane blebbing, and perinuclear bodies (arrowheads), all indicative of necrotic cell death. Two myonuclei (MN) beneath the post-synaptic membrane look normal. Scale bar = 5 μm. From Halstead et al. (2005) with permission.

lipooligosaccharide and antibodies from patients with Fisher syndrome cross-react with them (Yuki et al. 1994; Jacobs et al. 1995; Salloway et al. 1996). These antibodies are the ones which have the α-latrotoxin-like effect on the neuromuscular junction (Goodyear et al. 1999).

Immune responses in AIDP

There are many early reports of antibodies directed against a variety of crude nerve antigen preparations (Hughes et al. 1999). In particular, Koski used an especially sensitive C1 fixation and transfer assay to detect circulating IgM antibodies to peripheral nerve myelin in >90% of sera from patients with acute GBS, which decreased following plasma exchange and increased during relapses (Koski et al. 1985; Vriesendorp et al. 1993). The target antigen of these antibodies has defied subsequent attempts at identification, although subsequent studies have identified antibodies to a variety of different lipid antigens, including cardiolipin, sulphoglucuronyl paragloboside (Frampton et al. 1988; Ilyas et al. 1991), and galactocerebroside (Kusunoki et al. 1995). Most studies of antibodies to myelin proteins have not found them or, when they have been present, they have only been detected in low titre. Of particular interest would be reports of immune responses to the myelin proteins, P2, P0 and PMP22, which are known to produce experimental autoimmune neuritis. There is only limited evidence of T-cell responses to such antigens (Dahle et al. 1997; Khalili-Shirazi et al. 1992; Hughes et al. 1999). There are some reports of antibodies to P0 or PMP22 but also contradictory reports so that more research is needed (Gabriel et al. 2000; Kwa et al. 2001).

There have been many attempts to identify antibodies in the serum of GBS patients that will induce demyelination. In myelinated (Morris et al. 1999) rodent dorsal root ganglion cell cultures most investigators obtained demyelination by adding human GBS serum or IgM (Sawant-Mane et al. 1991; Mithen et al. 1992; Sawant-Mane et al. 1994) but one study was negative (Lisak et al. 1984). Intraneural injection of GBS serum or immunoglobulin into rat sciatic nerve has given variable results. Most studies obtained significant conduction block and demyelination greater than with control human serum, though not as marked as with EAN serum (Feasby et al. 1982; Saida et al. 1982; Sumner et al. 1982a,b; Harrison et al. 1984; Metral et al. 1989). Some investigators found GBS serum did not cause significantly more demyelination or conduction block than control serum (Oomes et al. 1991; Harvey et al. 1995; Loeb et al. 1999). The results from this rather unphysiological model have not provided consistent evidence of relevant demyelinating antibodies in GBS. In the related condition, CIDP, there is more consistent evidence of antibodies to P0 in 20% of patients and these antibodies do induce demyelination (Yan et al. 2001; Allen et al. 2005).

Since complement components are located on the surface of the abaxonal Schwann cell and not on the myelin lamellae in the earliest stage of AIDP (Hafer-Macko et al. 1996b), attempts to find antibodies to myelin have been misguided and the search should be redirected to Schwann cell surface antigens. Kwa et al. (2003) identified IgG antibodies against cultured proliferating, non-myelinating human Schwann cells in 24% of 233 GBS sera and rarely in sera from normal subjects or patients with non-inflammatory neurological disorders. The antigenic target was located at the distal tips of the Schwann cell processes and also neurites of non-myelinated neurons. Kwa et al. therefore proposed that part of the immune response in GBS is directed against non-myelin proteins and epitopes possibly involved in Schwann cell–axon interaction. The cells used in their experiments were non-myelinating. Future research needs to be directed at identifying the antigens expressed at the surface of myelinating Schwann cells and discovering which are the targets in GBS.

10

Chronic idiopathic demyelinating polyneuropathy and Schwann cells

JOHN D. POLLARD

Chronic inflammatory demyelinating polyneuropathy (CIDP) is the most common chronic treatable neuropathy in the Western world with a prevalence of 1−2 per 100,000 (Lunn et al. 1999; McLeod et al. 1999). It is the chronic variety of the inflammatory demyelinating neuropathies, of which the Guillain−Barré syndrome (GBS) is the acute form. Because it is treatable, CIDP needs to be considered in the differential diagnosis whenever a patient presents with an acquired chronic or progressive neuropathy. The inflammatory demyelinating neuropathies are characterised pathologically by multifocal areas throughout the peripheral nervous system (PNS) of inflammatory infiltrates associated with demyelination. When demyelination is assessed by the study of teased nerve fibres it is seen to be segmental, i.e. by and large restricted to the territory of individual Schwann cells (Figure 10.1). This characteristic pathological description suggests that the myelin loss, which occurs in these neuropathies, results from some insult which targets Schwann cells rather than the myelin itself.

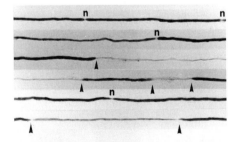

Figure 10.1 Contiguous sections of a teased nerve fibre from a patient with CIDP showing segmental demyelination (between arrowheads). n = node of Ranvier. Magnification × 260.

Despite considerable advances in treatment strategies, which have developed effective therapies for a majority of patients, the pathogenic mechanisms which disrupt Schwann cell function resulting in demyelination remain unknown in most patients. It is likely moreover that CIDP, like GBS, is comprised of a number of different subtypes, and the study of patients within these homogenous subgroups may permit the recognition of meaningful patterns of pathogenicity. To date, research into CIDP and indeed its acute counterpart GBS has centred largely around the animal model experimental autoimmune neuritis (EAN). This model, similarly to its analogous model for central nervous system (CNS) demyelination, experimental autoimmune encephalomyelitis (EAE), has focussed the attention of researchers on myelin proteins and their peptides, such as P2 and P0 protein, since these molecules, in addition to crude nerve or myelin extracts, may induce this autoimmune neuropathy. In more recent years neural gangliosides have also become the focus of considerable research, since they have been shown to be intimately related to precipitating infections and pathogenic antibodies in several subtypes of GBS. However, to date, this enormous research effort has not explained the mechanism of demyelination in patients with CIDP, or in the majority of those with GBS who have the demyelinating variety of that disease – acute inflammatory demyelinating polyneuropathy (AIDP) (Willison 2005). There is clearly a need to focus attention back on the cell of Schwann rather than a limited number of particular proteins it produces, to gain an understanding of the demyelinating process.

CLINICAL FEATURES

The typical case of CIDP presents with a chronically progressive or relapsing and remitting symmetrical mainly motor neuropathy with proximal and distal weakness which develops for more than 2 months. Cranial nerves may be affected, but less commonly than in GBS, and the deep tendon reflexes are absent or reduced (Dyck et al. 1975; McCombe et al. 1987b; Barohn et al. 1989; EFN and PNS Task Force 2005). For the diagnosis of CIDP to be confirmed, electrodiagnostic studies should show changes consistent with demyelination and the protein level be elevated in the cerebrospinal fluid (CSF). Electrophysiological criteria for demyelination have been much debated in the literature (Koller et al. 1995; EFN and PNS Task Force 2005), since, if too restrictive, patients who may respond to therapy could be denied a therapeutic trial (Latov 2002; Magda et al. 2003).

CIDP variants have been recognised and describe differing clinical phenotypes which may be related to differing pathogenic mechanisms and/or host genetic factors.

These variants include patients with:

1. Predominantly distal weakness – DADS (distal acquired demyelinating symmetric) phenotype.
2. Asymmetric presentation – MADSAM (multifocal asymmetric demyelinating sensory and motor), or the Lewis–Sumner syndrome
3. Pure motor CIDP
4. Pure sensory CIDP
5. Focal presentation, i.e. one or more nerves or a focal group of nerves
6. CIDP with acute onset
7. CIDP with CNS involvement (EFN and PNS Task Force 2005; van Doorn 2005).

A diagnosis of CIDP may be made when a patient with appropriate clinical features is shown to have electrodiagnostic criteria consistent with demyelination (see above). The diagnosis is supported by the finding of elevated CSF protein levels without a rise in leukocyte count, nerve biopsy findings of primary demyelination, MRI evidence of gadolinium enhancement and/or hypertrophy of nerve roots or plexuses or improvement following immunotherapy (EFN and PNS Task Force 2005).

The prevalence of CIDP is between one and two per hundred thousand of population (Lunn et al. 1999; McLeod et al. 1999). Whereas GBS commonly follows a preceding infective illness, such a precipitating event is less commonly identified in CIDP suggesting differing pathogenic mechanisms for these disorders (Prineas and McLeod 1976; McCombe et al. 1987b). Occasional cases have been reported to follow specific infections or vaccinations (McCombe et al. 1987b; Donaghy et al. 1989) and a significantly increased risk of relapse in pregnancy has been identified (McCombe et al. 1987a).

PATHOLOGY

Post-mortem examination has shown that the most common sites of involvement are the spinal roots, major plexuses and proximal nerve trunks (Hyland and Russell 1930; Krucke 1955; Dyck et al. 1975). These are also the sites in which MRI examination may show gadolinium enhancement or nerve hypertrophy in approximately 50% of patients when followed for prolonged periods (Duggins et al. 1999). However, lesions may occur at any level within PNS including the distal ends of motor nerves and the autonomic nerves. The lesions consist of patchy regions of demyelination and oedema with variable inflammatory infiltrates (Prineas and McLeod 1976; Prineas 1981).

The pathological changes seen within nerve biopsy material (usually the sural nerve) may be minimal but on occasions can strongly support a diagnosis of CIDP. Such changes may include: endoneurial or subperineurial oedema; demyelinated axons – large fibres devoid of myelin; macrophage phagocytosis of myelin (Figure 10.2); myelin stripping or vesicular dissolution in the presence of mononuclear cells; and thinly myelinated fibres and inflammatory infiltrates within the endoneurium (Prineas 1971; Dyck et al. 1975; Prineas and McLeod 1976; Pollard et al. 1983) (Figure 10.3). Since inflammatory demyelination is maximal in proximal regions, the main abnormality seen in distal regions may be axonal degeneration.

The presence of inflammatory cells within the endoneurium is associated with upregulation of MHC Class I and II on Schwann cells (Pollard et al. 1986) (Figure 10.4). Onion bulb formations in which a central axon, myelinated or demyelinated, is surrounded by concentric rings of Schwann cell processes (Figure 10.5) indicate the chronic nature of the pathological process, since they result from repeated episodes of demyelination and remyelination. Likewise, cluster formations may be seen due to axonal degeneration and regeneration (Dyck et al. 1975; Prineas and McLeod 1976; Pollard et al. 1983; Barohn et al. 1989). The ultrastructural finding of macrophage stripping of myelin lamellae or vesicular dissolution of myelin surrounding a normal axon, is an important one, since in clinical practice it is virtually diagnostic of inflammatory demyelinating polyneuropathy (Prineas 1971; Prineas 1972).

IMMUNOPATHOLOGY

Immunogenetics

Significant HLA associations have been established for CIDP, particularly for the class I antigen HLA–B8 (Stewart et al. 1978; Adams et al. 1979; Vaughan et al. 1990). HLA–CW7, another class I antigen, has also

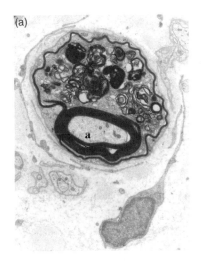

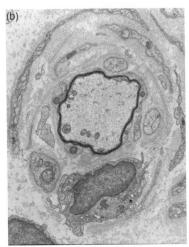

Figure 10.2 Electron micrographs of sural nerve biopsy from a patient with CIDP. (a) Macrophage-mediated demyelination. Myelin is being stripped from around the axon (a) by a process of a macrophage which contains myelin debris.
(b) A thinly myelinated axon is central to an onion bulb which is associated with several mononuclear cells (lacking basal lamina).

been reported with increased frequency in CIDP patients (Vaughan et al. 1990). The abnormal M3 allele of α-1, antitrypsin, the major proteinase inhibitor has also been reported with increased frequency in CIDP (McCombe et al. 1985).

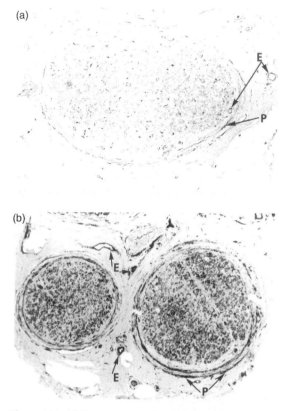

Figure 10.3 (a) Transverse section of normal human sural nerve immunostained with (i) mouse monoclonal antibody to human common specificity Ia and (ii) peroxidase-conjugated goat antimouse IgG. Note reaction product (black) is present sparsely within the nerve fascicle and perineurium (P). Capillary endothelium (E) accounts for the majority of positive staining. (b) Sural nerve from CIDP patient immunostained as above. Reaction product has stained densely within perineurium (P), endothelium (E) and within nerve fascicle. × 170.

Cellular immune mechanisms

There is general agreement that CIDP is an autoimmune disorder in which self-tolerance is lost and autoreactive T and B cells, part of the normal immune repertoire, become activated with subsequent organ specific damage. The basis of this belief includes the following:

1. The pathology, including nerve invasion by mononuclear cells, antibody and complement (Pollard 1994).

Chronic idiopathic demyelinating polyneuropathy and Schwann cells

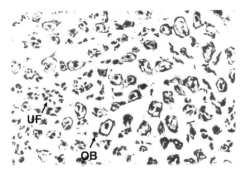

Figure 10.4 Transverse section of sural nerve from a CIDP patient immunostained with (i) mouse monoclonal antibody to human Class I MHC, (ii) peroxidase-conjugated goat anti-mouse IgG. Note reaction product labels of Schwann cells both within onion bulb formations (OB) and associated with unmyelinated fibres (UF).

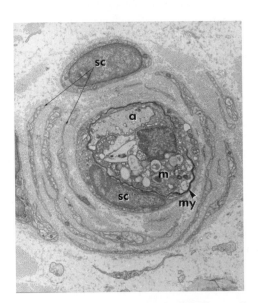

Figure 10.5 Electron micrograph of sural nerve from a CIDP patient. This figure shows macrophage-mediated demyelination within an established onion bulb thus illustrating the formation of these structures. The newly demyelinated axon will be surrounded by another Schwann cell displacing the former cell (sc) to add further lamellae to this arrangement. a = axon, m = macrophages, my = myelin, sc = Schwann cell.

2. The response of patients to immune therapy including corticosteroids, immune-suppressive drugs, plasma exchange and intravenous immunoglobulin (Koller et al. 2005).
3. The clinical, electrophysiological and pathological similarities between CIDP and the animal model chronic experimental autoimmune neuritis (CREAN) (Harvey et al. 1987) (Figure 10.6).
4. The association, albeit rare, between CIDP and melanoma, since both Schwann cells and melanocytes derive from neural crest tissues and share antigens (Fuller et al. 1994; Weiss et al. 1998).

CELLULAR AND IMMUNE RESPONSES

T and B lymphocytes both occur within the nerve lesions of CIDP (Pollard et al. 1986; Schmidt et al. 1996). Upregulation of class I and II molecules on Schwann cells within nerve (Pollard et al. 1986; Mancardi et al. 1988) provides further evidence of T cell activity within the endoneurium and provides the milieu for possible cytotoxic T cell (CD8+) damage to Schwann cells or a role for Schwann cells in antigen presentation. T cells also play a crucial role in the breakdown of the

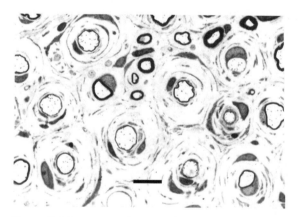

Figure 10.6 Low power electron micrograph of a nerve root from a rabbit with chronic relapsing EAN (CREAN). The numerous well-developed onion bulbs are highly reminiscent of those that occur in long-standing CIDP. It is pertinent that it is well-established that these onion bulbs occur almost exclusively in the nerve roots in blood–nerve barrier leakiness. Scale bar = 5 μm.

blood–nerve barrier (BNB), by virtue of adhesion molecules such as lymphocyte function-associated antigen-1 (LFA-1) and very late antigen-4 (VLA-4), which are expressed on the surface of activated T cells. These can bind and adhere to their counter receptor molecules, intracellular cell adhesion molecule-1 (ICAM-1) and vascular cell adhesion molecule-1 (VCAM-1), on endothelial cells. The process of transmigration by which T cells pass through endothelial and basal lamina barriers into the nerve is facilitated by the production of matrix metalloproteinases and various chemokines and cytokines such as tumor necrosis factor alpha (TNFα), interferon gamma (IFNγ) and interleukin-2 (IL-2) (Leppert et al. 1999; Mathey et al. 1999; Kieseier et al. 2002b). This process produces a leaky BNB which also allows access of soluble factors, such as antibody, into the endoneurium (Pollard et al. 1995; Spies et al. 1995a,b) (Figure 10.7).

Within the endoneurium, activated T cells may undergo clonal expansion if they encounter antigen presented by the appropriate MHC molecules and costimulatory signals. Such T cells developing along the Th1 pathway may produce proinflammatory cytokines such as TNFα, IFNγ and IL-2, which may activate resident macrophages to produce toxic molecules such as proteases, toxic oxygen species and nitric oxide metabolites, which in turn may damage Schwann cells or

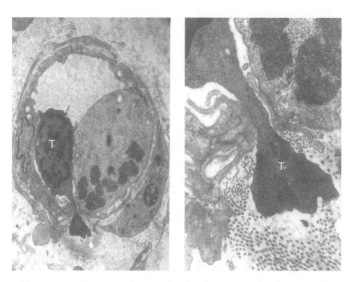

Figure 10.7 Electron micrograph of a blood vessel from a rat with EAN showing transmigration of a T cell from the blood vessel into nerve. T = T cell.

myelin membranes (Gold et al. 1999). Macrophages or perivascular dendritic cells may also act as antigen-presenting cells and have been shown to express MHC class II molecules and the class I-like CD1a molecules in CIDP nerve biopsy specimens (Pollard et al. 1986; Murata and Dalakas 2000; Van Rhijn et al. 2000a). The costimulatory molecules B7−1 and B7−2 have also been demonstrated on macrophages in CIDP and may determine whether T cells develop down the Th1 (proinflammatory) or Th2 (anti-inflammatory) pathways, thus regulating the local immune response (Van Rhijn et al. 2000a; Murata and Dalakas 2000).

Schwann cells in inflammatory neuropathy, including CIDP, express both class I and class II molecules, which may enable them to participate in immune regulation or become targets for a cytotoxic CD8+ T cell attack (Pollard et al. 1986; Mancardi et al. 1988; Armati et al. 1990). The costimulatory molecule BB-1 is expressed on Schwann cells in CIDP, and since the counter receptors CTLA4 and CD28 are deployed on adjoining lymphocytes, these Schwann cells may act as facilitative antigen-presenting cells (Murata and Dalakas 2000). In vitro studies have shown that both CD8+ and CD4+ T cells can be cytotoxic to Schwann cells. CD8+ T cells sensitised to *M. leprae* have been shown to lyse Schwann cells expressing *M. leprae*, whilst T cells raised against the 65 kDa mycobacterium heat shock protein (HSP) antigen lysed IFNγ-stimulated Schwann cells (Steinhoff and Kaufmann 1988; Steinhoff et al. 1990). Argall et al. (1992a,b) have shown that CD4+ P2 T cells capable of producing EAN are cytotoxic to Schwann cells *in vitro* (Argall et al. 1992a). Since both CD4+ and CD8+ T cells have been shown within CIDP nerve, cytotoxicity of class I and II positive Schwann cells by such mechanisms remains a distinctly possible mechanism for demyelination.

HUMORAL RESPONSES

The efficacy of plasma exchange therapy in many patients with CIDP has suggested an important pathogenic role for humoral factors such as antibody. Dalakas and Engel (1981) first reported immunoglobulin bound to Schwann cells or myelin membranes in nine of twelve cases of CIDP (Dalakas and Engel 1981) but subsequent studies have failed to confirm such a high positivity rate (Yan et al. 2000; Yan et al. 2001; Allen et al. 2005). Recent studies have reported the presence of antibodies to various Schwann cell or myelin protein antigens, such as protein zero (P0), peripheral myelin protein-22 (PMP22) and connexin, in small percentages of CIDP patients by ELISA, immunoblot, immunofluorescence or western blot (Khalili-Shirazi et al. 1993; Melendez-Vasquez et al. 1997; Gabriel et al. 2000; Ritz et al. 2000; Kwa et al. 2001). However, two groups have shown antibodies to P0 in a significant minority of patients (33% – Yan et al. (2001); 25% – Allen et al. (2005)) and the study of Yan et al. (2001) also showed that the antibodies were pathogenic by passive transfer into rats. These anti-P0 antibodies were of the IgG subclass and bound complement (Yan et al. 2001). Antibodies to gangliosides, which have been shown to play an important pathogenic role in subtypes of GBS (AMAN and MFS) (see Chapter 9), do not appear to play a major role in CIDP. Antibodies to GM-1, LM-1, SGPG and various acidic glycolipid have been reported in small percentages of patients but their pathogenicity has not been shown (Ilyas et al. 1992; Melendez-Vasquez et al. 1997; Yuki et al. 1996) nor has the site of reactivity (Schwann cell, myelin or axon) been demonstrated for these antibodies. Kwa and colleagues (2003) have recently shown serum reactivity against non-myelinating Schwann cells in 25% of CIDP patients studied (Kwa et al. 2003), reawakening the question as to whether Schwann cells or myelin may contain the primary antigenic target.

It may be pertinent that the best studied and most powerfully demyelinating antibody relevant to inflammatory demyelinating neuropathies is not an antibody to a myelin protein but rather antigalactocerebroside antibody (anti-GalC) – the major pathogenic antibody found in rabbits with EAN (Saida et al. 1979; Saida et al. 1981; Stoll et al. 1986). Anti-GalC binds to Schwann cells and is highly toxic to them (Armati and Pollard 1987) (Figure 10.8). Saida et al. (1981) described extensive Schwann cell damage in their original description of galactocerebroside-induced EAN. Hence the demyelination seen in rabbit EAN is probably due to complement-mediated Schwann cell lysis rather than anti-myelin activity. Since Halstead et al. (2004) have shown that certain anti-disialosyl antibodies that bind epitopes on gangliosides are cytotoxic to the perisynaptic Schwann cell (Halstead et al. 2004), it is clearly important to assess whether other antiglycolipid or antiganglioside antibodies, associated with CIDP, cause Schwann cell cytotoxicity.

Antibodies to myelin-associated glycoprotein (MAG), seen in approximately 50% of patients with IgM-associated paraproteinaemic neuropathy, provide another example of an antibody associated with demyelination that does not display simple antimyelin activity. These antibodies bind principally in regions of uncompacted myelin in

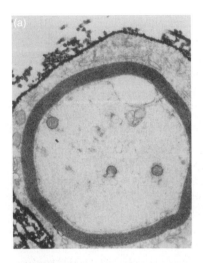

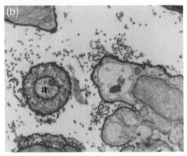

Figure 10.8 Electron micrograph from sciatic nerve of rabbit with CREAN. Peroxidase-conjugated anti-IgG labels galactocerebroside antibodies found on Schwann cell plasmalemma of both myelinated (a) and unmyelinated (b) fibres.

the periaxonal membranes and membranes of the paranodal loops, Schmidt–Lantermann incisures and mesaxons (Griffin et al. 1993b) where MAG is located (Figure 10.9).

Given the relatively low frequency of specific antibodies reported in CIDP patients, it is likely that different pathogenic mechanisms are involved in different patients. It will be important to study homogenous subgroups of CIDP patients to see whether particular mechanisms underlie these various groups as has been shown in GBS. Moreover, these low frequencies of specific responses indicate the need to test cellular and humoral responses to the cell that produces and maintains myelin, rather than only to specific epitopes, which these Schwann cells express.

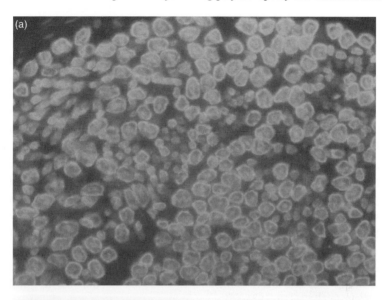
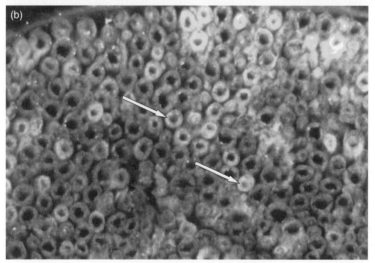

Figure 10.9 Immunofluorescence of normal human nerve labeled with antibodies to (a) MAG and (b) P0 protein. Note that in (a), anti-MAG antibodies bind mainly to regions of uncompacted myelin as described in the text, whereas in (b), anti-P0 antibodies highlight compacted myelin. A colour version of this Figure is in the Plate section.

TREATMENT

Despite our limited understanding of the pathogenesis of CIDP and the mechanism by which Schwann cell function is perturbed, effective treatment is available for a majority of patients. Controlled trials have shown that corticosteroids, intravenous immunoglobulin (IVIg) and plasma exchange are each effective; however, for a majority of patients long-term therapy is required (van Doorn 2005). Some patients require treatment with immunosuppressive agents reflecting the pathogenic heterogeneity of the disease, but there have been no adequate controlled trials of these agents to prove their efficacy (Hughes et al. 2003; van Doorn 2005). When treatment is initiated early with proven therapies, improvement occurs in a high percentage of patients (70–90%). Immunosuppressive agents are usually reserved for patients who are non-responsive to first-line therapies, such as IVIg, steroids or plasma exchange, or in patients who need to be given these treatments too frequently or in too high a dose. If treatment is delayed, patients may suffer axonal loss that is irreversible, and presumably reflects the important relationship between the Schwann cell and axon (Bouchard et al. 1999). Axonal loss in sural nerve biopsy correlates strongly with long-term disability (Bouchard et al. 1999).

References

Adam, A. M., Atkinson, P. F., Hall, S. M., Hughes, R. A. and Taylor, W. A. (1989). Chronic experimental allergic neuritis in Lewis rats. *Neuropathol Appl Neurobiol* **15**, 249–64.

Adams, D., Festenstein, H., Gibson, J. D. *et al.* (1979). HLA antigens in chronic relapsing idiopathic inflammatory polyneuropathy. *J Neurol Neurosurg Psychiatry* **42**, 184–6.

Agresti, C., D'Urso, D. and Levi, G. (1996). Reversible inhibitory effects of interferon-gamma and tumour necrosis factor-alpha on oligodendroglial lineage cell proliferation and differentiation in vitro. *Eur J Neurosci* **8**, 1106–16.

Ahmed, M. R., Basha, S. H., Gopinath, D., Muthusamy, R. and Jayakumar, R. (2005). Initial upregulation of growth factors and inflammatory mediators during nerve regeneration in the presence of cell adhesive peptide-incorporated collagen tubes. *J Peripher Nerv Syst* **10**, 17–30.

Ahn, M., Lee, Y., Moon, C. *et al.* (2004). Upregulation of osteopontin in Schwann cells of the sciatic nerves of Lewis rats with experimental autoimmune neuritis. *Neurosci Lett* **372**, 137–41.

Ainsworth, P. J., Bolton, C. F., Murphy, B. C., Stuart, J. A. and Hahn, A. F. (1998). Genotype/phenotype correlation in affected individuals of a family with a deletion of the entire coding sequence of the connexin 32 gene. *Hum Genet* **103**, 242–4.

Akassoglou, K., Probert, L., Kontogeorgos, G. and Kollias, G. (1997). Astrocyte-specific but not neuron-specific transmembrane TNF triggers inflammation and degeneration in the central nervous system of transgenic mice. *J Immunol* **158**, 438–45.

Akassoglou, K., Bauer, J., Kassiotis, G. *et al.* (1998). Oligodendrocyte apoptosis and primary demyelination induced by local TNF/p55TNF receptor signaling in the central nervous system of transgenic mice: models for multiple sclerosis with primary oligodendrogliopathy. *Am J Pathol* **153**, 801–13.

Akassoglou, K., Bauer, J., Kassiotis, G. *et al.* (1999). Transgenic models of TNF induced demyelination. *Adv Exp Med Biol* **468**, 245–59.

Akassoglou, K., Kombrinck, K. W., Degen, J. L. and Strickland, S. (2000). Tissue plasminogen activator-mediated fibrinolysis protects against axonal degeneration and demyelination after sciatic nerve injury. *J Cell Biol* **149**, 1157–66.

Akassoglou, K., Yu, W.M., Akpinar, P. and Strickland, S. (2002). Fibrin inhibits peripheral nerve remyelination by regulating Schwann cell differentiation. *Neuron* **33**, 861–75.

Al-Din, A.N., Anderson, M., Bickerstaff, E.R. and Harvey, I. (1982). Brainstem encephalitis and the syndrome of Miller Fisher: a clinical study. *Brain* **105** (Pt 3), 481–95.

Albrecht, D.E. and Froehner, S.C. (2004). *DRP2 and Dp116 Form Spatially Distinct Dystrophin-like Complexes in the Schwann Cells of Peripheral Nerves.* American Society for Cell Biology, Washington DC, Abstract.

Allen, D., Giannopoulos, K., Gray, I. *et al.* (2005). Antibodies to peripheral nerve myelin proteins in chronic inflammatory demyelinating polyradiculoneuropathy. *J Peripher Nerv Syst* **10**, 174–80.

Allen, N.J., Barres, B.A. (2005). Signaling between glia and neurons: focus on synaptic plasticity. *Curr Opin Neurobiol* **15**, 542–8.

Altevogt, B.M., Kleopa, K.A., Postma, F.R., Scherer, S.S. and Paul, D.L. (2002). Cx29 is uniquely distributed within myelinating glial cells of the central and peripheral nervous systems. *J Neurosci* **22**, 6458–70.

Alvarez-Buylla, A. and Lim, D.A. (2004). For the long run: maintaining germinal niches in the adult brain. *Neuron* **41**, 683–6.

American Academy of Neurology (1991). Research criteria for diagnosis of chronic inflammatory demyelinating polyneuropathy (CIDP). Report from an Ad Hoc Subcommittee of the American Academy of Neurology AIDS Task Force. *Neurology* **41**, 617–18.

Ammoumi, A.A., Pertschuk, L., Daras, M. and Rosen, A.D. (1980). Guillain–Barré syndrome; results of direct immunofluorescent study. *NY State J Med* **80**, 1434–5.

Andorfer, B., Kieseier, B.C., Mathey, E. *et al.* (2001). Expression and distribution of transcription factor NF-kappaB and inhibitor IkappaB in the inflamed peripheral nervous system. *J Neuroimmunol* **116**, 226–32.

Andrews, T., Zhang, P. and Bhat, N.R. (1998). TNFalpha potentiates IFNgamma-induced cell death in oligodendrocyte progenitors. *J Neurosci Res* **54**, 574–83.

Anzini, P., Neuberg, D.H., Schachner, M. *et al.* (1997). Structural abnormalities and deficient maintenance of peripheral nerve myelin in mice lacking the gap junction protein connexin 32. *J Neurosci* **17**, 4545–51.

Aoki, E., Semba, R. and Kashiwamata, S. (1991). Evidence for the presence of L-arginine in the glial components of the peripheral nervous system. *Brain Res* **559**, 159–62.

Apostolski, S., Sadiq, S.A., Hays, A. *et al.* (1994). Identification of Gal(beta 1–3)GalNAc bearing glycoproteins at the nodes of Ranvier in peripheral nerve. *J Neurosci Res* **38**, 134–41.

Araque, A., Parpura, V., Sanzgiri, R.P. and Haydon, P.G. (1999). Tripartite synapses: glia, the unacknowledged partner. *Trends Neurosci* **22**, 208–15.

Archelos, J.J., Roggenbuck, K., Schneider-Schaulies, J., Linington, C., Toyka, K.V. and Hartung, H.P. (1993). Production and characterization of monoclonal antibodies to the extracellular domain of P0. *J Neurosci Res* **35**, 46–53.

Arenander, A. and De Vellis, J. (1999). Development of the nervous system. Siegel, G. (Ed.) *Basic Neurochemistry: Molecular, Cellular, and Medical Aspects.* Raven Press, New York, pp. 573–606.

Argall, K.G., Armati, P.J., Pollard, J.D. and Bonner, J. (1992a). Interactions between CD4+ T-cells and rat Schwann cells in vitro. 2. Cytotoxic effects of P2-specific CD4+ T-cell lines on Lewis rat Schwann cells. *J Neuroimmunol* **40**, 19–29.

Argall, K.G., Armati, P.J., Pollard, J.D., Watson, E. and Bonner, J. (1992b). Interactions between CD4+ T-cells and rat Schwann cells in vitro. 1. Antigen presentation by Lewis rat Schwann cells to P2-specific CD4+ T-cell lines. *J Neuroimmunol* **40**, 1–18.

Armati, P.J. and Pollard, J.D. (1987). Cytotoxic response of serum from patients with chronic inflammatory demyelinating polyradiculoneuropathy (CIDP). *Acta Neurol Scand* **76**, 24–7.

Armati, P.J. and Pollard, J.D. (1996). Immunology of the Schwann cell. *Baillieres Clin Neurol* **5**, 47–64.

Armati, P.J., Pollard, J.D. and Gatenby, P. (1990). Rat and human Schwann cells in vitro can synthesize and express MHC molecules. *Muscle Nerve* **13**, 106–16.

Arroyo, E.J. and Scherer, S.S. (2000). On the molecular architecture of myelinated fibers. *Histochem Cell Biol* **113**, 1–18.

Arroyo, E.J., Sirkowski, E.E., Chitale, R. and Scherer, S.S. (2004). Acute demyelination disrupts the molecular organization of PNS nodes. *J Comp Neurol* **479**, 424–34.

Arteaga, M.F., Gutierrez, R., Avila, J., Mobasheri, A., Diaz-Flores, L. and Martin-Vasallo, P. (2004). Regeneration influences expression of the Na+, K+-atpase subunit isoforms in the rat peripheral nervous system. *Neuroscience* **129**, 691–702.

Asbury, A.K. and Cornblath, D.R. (1990). Assessment of current diagnostic criteria for Guillain–Barré syndrome. *Ann Neurol* **27** Suppl, S21–4.

Asbury, A.K., Arnason, B.G. and Adams, R.D. (1969). The inflammatory lesion in idiopathic polyneuritis. Its role in pathogenesis. *Medicine (Baltimore)* **48**, 173–215.

Asbury, A.K., Arnason, B.G.W., Karp, H.R. and McFarlin, D.F. (1978). Criteria for diagnosis of Guillain–Barré syndrome. *Ann Neurol* **3**, 565–6.

Astrow, S.H., Son, Y.J. and Thompson, W.J. (1994). Differential neural regulation of a neuromuscular junction-associated antigen in muscle fibers and Schwann cells. *J Neurobiol* **25**, 937–52.

Astrow, S.H., Tyner, T.R., Nguyen, M.T. and Ko, C.P. (1997). A Schwann cell matrix component of neuromuscular junctions and peripheral nerves. *J Neurocytol* **26**, 63–75.

Astrow, S.H., Qiang, H. and Ko, C.P. (1998). Perisynaptic Schwann cells at neuromuscular junctions revealed by a novel monoclonal antibody. *J Neurocytol* **27**, 667–81.

Asundi, V.K., Erdman, R., Stahl, R.C. and Carey, D.J. (2003). Matrix metalloproteinase-dependent shedding of syndecan-3, a transmembrane heparan sulfate proteoglycan, in Schwann cells. *J Neurosci Res* **73**, 593–602.

Atanasoski, S., Notterpek, L., Lee, H.Y. et al. (2004). The protooncogene Ski controls Schwann cell proliferation and myelination. *Neuron* **43**, 499–511.

Augustien, G.J., Burns, M.E., DeBello, W.M. et al. (1999). Proteins involved in synaptic vesicle trafficking. *J Physiol* **520**, 33–41.

Auld, D.S. and Robitaille, R. (2003a). Glial cells and neurotransmission: an inclusive view of synaptic function. *Neuron* **40**, 389–400.

Auld, D.S. and Robitaille, R. (2003b). Perisynaptic Schwann cells at the neuromuscular junction: nerve- and activity-dependent contributions to synaptic efficacy, plasticity, and reinnervation. *Neuroscientist* **9**, 144–57.

Auld, D.S., Colomar, A., Belair, E.L. et al. (2003). Modulation of neurotransmission by reciprocal synapse–glial interactions at the neuromuscular junction. *J Neurocytol* **32**, 1003–15.

Awasaki, T. and Ito, K. (2004). Engulfing action of glial cells is required for programmed axon pruning during Drosophila metamorphosis. *Curr Biol* **14**, 668–77.

Awatramani, R., Shumas, S., Kamholz, J. and Scherer, S. S. (2002). TGFbeta1 modulates the phenotype of Schwann cells at the transcriptional level. *Mol Cell Neurosci* **19**, 307–19.

Bachelin, C., Lachapelle, F., Girard, C. et al. (2005). Efficient myelin repair in the macaque spinal cord by autologous grafts of Schwann cells. *Brain* **128**, 540–9.

Baggiolini, M. (2001). Chemokines in pathology and medicine. *J Intern Med* **250**, 91–104.

Bai, Y., Ianokova, E., Pu, Q. et al. (in press). R69C mutation in myelin protein zero causes early onset CMT1B with demyelination, dysmyelination and axonal loss. *Arch Neurol*.

Balarezo, F. S., Muller, R. C., Weiss, R. G. et al. (2003). Soft tissue perineuriomas in children: report of three cases and review of the literature corrected. *Pediatr Dev Pathol* **6**, 137–41.

Balice-Gordon, R. J., Bone, L. J. and Scherer, S. S. (1998). Functional gap junctions in the Schwann cell myelin sheath. *J Cell Biol* **142**, 1095–104.

Ballesteros, J. A., Abrams, C. K., Oh, S., Verselis, V. K., Weinstein, H. and Bargiello, T. A. (1999). The role of a conserved proline residue in mediating conformational changes associated with voltage gating of Cx32 gap junctions. *Biophy J* **76**, 2887–98.

Bao, J., Wolpowitz, D., Role, L. W. and Talmage, D. A. (2003). Back signaling by the Nrg-1 intracellular domain. *J Cell Biol* **161**, 1133–41.

Barohn, R. J., Kissel, J. T., Warmolts, J. R. and Mendell, J. R. (1989). Chronic inflammatory demyelinating polyradiculoneuropathy. Clinical characteristics, course, and recommendations for diagnostic criteria. *Arch Neurol* **46**, 878–84.

Baron Van Evercooren, A., Kleiman, H. K., Seppa, J., Rentier, B. and Dubois-Dalcq, M. (1982). Fibronectin promotes rat Schwann cell growth and motility. *J Cell Biol* **93**, 211–16.

Baxter, R. V., Ben Othmane, K., Rochelle, J. M. et al. (2001). Ganglioside-induced differentiation-associated protein-1 is mutant in Charcot–Marie–Tooth disease type 4A/8q21. *Nat Genet* **30**, 21–2.

Be'eri, H., Reichert, F., Saada, A. and Rotshenker, S. (1998). The cytokine network of wallerian degeneration: IL-10 and GM-CSF. *Eur J Neurosci* **10**, 2707–13.

Beams, H. W. and Evans, T. C. (1953). Electron micrographs of motor end-plates. *Proc Soc Exp Biol Med* **82**, 344–6.

Bellone, E., Di Maria, E., Soriani, S. et al. (1999). A novel mutation (D305V) in the early growth response 2 gene is associated with severe Charcot–Marie–Tooth type 1 disease. *Hum Mutat* **14**, 353–4.

Belmadani, A., Tran, P. B., Ren, D., Assimacopoulos, S., Grove, E. A. and Miller, R. J. (2005). The chemokine stromal cell-derived factor-1 regulates the migration of sensory neuron progenitors. *J Neurosci* **25**, 3995–4003.

Bennett, V. and Baines, A. J. (2001). Spectrin and ankyrin-based pathways: metazoan inventions for integrating cells into tissues. *Physiol Rev* **81**, 1353–92.

Bennett, C. L., Shirk, A. J., Huynh, H. M. et al. (2004). SIMPLE mutation in demyelinating neuropathy and distribution in sciatic nerve. *Ann Neurol* **55**, 713–20.

Berciano, J., Garcia, A., Calleja, J. and Combarros, O. (2000). Clinico-electrophysiological correlation of extensor digitorum brevis muscle atrophy in children with Charcot–Marie–Tooth disease 1A duplication. *Neuromuscul Disord* **10**, 419–24.

Bergoffen, J., Scherer, S. S., Wang, S. et al. (1993). Connexin mutations in X-linked Charcot–Marie–Tooth disease. *Science* **262**, 2039–42.

Bergsteinsdottir, K., Kingston, A., Mirsky, R. and Jessen, K. R. (1991). Rat Schwann cells produce interleukin-1. *J Neuroimmunol* **34**, 15–23.

Bergsteinsdottir, K., Kingston, A. and Jessen, K. R. (1992). Rat Schwann cells can be induced to express major histocompatibility complex class II molecules in vivo. *J Neurocytol* **21**, 382–90.

Bermingham, J. R., Scherer, S. S., O'Connell, S. et al. (1996). Tst-1/Oct-6/SCIP regulates a unique step in peripheral myelination and is required for normal respiration. *Genes & Development* **10**, 1751–62.

Bermingham, J. R., Jr., Shearin, H., Pennington, J. et al. (2006). The claw paw mutation reveals a role for Lgi4 in peripheral nerve development. *Nat Neurosci* **9**, 76–84.

Berthold, C. H., Fraher, J. P., King, R. H. M. and Rydmark, M. (2005). Microscopic anatomy of the peripheral nervous system. Dyck, P. J., Thomas, P. K. (Eds.) *Peripheral Neuropathy*, 4th Edn, Elsevier Saunders, Philadelphia, PA, pp. 35–92.

Bigbee, J. W., Yoshino, J. E. and DeVries, G. H. (1987). Morphological and proliferative responses of cultured Schwann cells following rapid phagocytosis of a myelin-enriched fraction. *J Neurocytol* **16**, 487–96.

Bigotte, L., Arvidson, B. and Olsson, Y. (1982). Cytofluorescence localization of adriamycin in the nervous system. II. Distribution of the drug in the somatic and autonomic peripheral nervous systems of normal adult mice after intravenous injection. *Acta Neuropathol (Berl)* **57**, 130–6.

Birks, R., Huxley, H. E. and Katz, B. (1960a). The fine structure of the neuromuscular junction of the frog. *J Physiol* **150**, 134–44.

Birks, R., Katz, B. and Miledi, R. (1960b). Physiological and structural changes at the amphibian myoneural junction, in the course of nerve degeneration. *J Physiol* **150**, 145–68.

Bishop, D. L., Misgeld, T., Walsh, M. K., Gan, W. B. and Lichtman, J. W. (2004). Axon branch removal at developing synapses by axosome shedding. *Neuron* **44**, 651–61.

Bitgood, M. J. and McMahon, A. P. (1995). Hedgehog and Bmp genes are coexpressed at many diverse sites of cell–cell interaction in the mouse embryo. *Develop Biol* **172**, 126–38.

Bixby, S., Kruger, G. M., Mosher, J. T., Joseph, N. M. and Morrison, S. J. (2002). Cell-intrinsic differences between stem cells from different regions of the peripheral nervous system regulate the generation of neural diversity. *Neuron* **35**, 643–56.

Bjartmar, C., Wujek, J. R. and Trapp, B. D. (2003). Axonal loss in the pathology of MS: consequences for understanding the progressive phase of the disease. *J Neurol Sci* **206**, 165–71.

Blakemore, W. F. (2005). The case for a central nervous system (CNS) origin for the Schwann cells that remyelinate CNS axons following concurrent loss of oligodendrocytes and astrocytes. *Neuropathol Appl Neurobiol* **31**, 1–10.

Blanchard, A. D., Sinanan, A., Parmantier, E. et al. (1996). Oct-6 (SCIP/Tst-1) is expressed in Schwann cell precursors, embryonic Schwann cells, and postnatal myelinating Schwann cells: comparison with Oct-1, Krox-20, and Pax-3. *J Neurosci Res* **46**, 630–40.

Boeke, J. (1949). The sympathetic endformation, its synaptology, the interstitial cells, the periterminal network, and its bearing on the neurone theory. *Acta Anatomica* **8**, 18–61.

Boerkoel, C. F., Takashima, H., Stankiewicz, P. et al. (2001). Periaxin mutations cause recessive Dejerine–Sottas neuropathy. *Am J Hum Genet* **68**, 325–33.

Boiko, T., Rasband, M. N., Levinson, S. R. et al. (2001). Compact myelin dictates the differential targeting of two sodium channel isoforms in the same axon. *Neuron* **30**, 91–104.

Bolin, L. M., Verity, A. N., Silver, J. E., Shooter, E. M. and Abrams, J. S. (1995). Interleukin-6 production by Schwann cells and induction in sciatic nerve injury. *J Neurochem* **64**, 850–8.

Bolino, A., Muglia, M., Conforti, F. L. (2000). Charcot–Marie–Tooth type 4B is caused by mutations in the gene encoding myotubularin-related protein-2. *Nat Genet* **25**, 17–19.

Bonetti, B., Valdo, P., Stegagno, C. et al. (2000). Tumor necrosis factor alpha and human Schwann cells: signalling and phenotype modulation without cell death. *J Neuropathol Exp Neurol* **59**, 74–84.

Bonnon, C., Goutebroze, L., Denisenko Nehrbass, N., Girault, J. A. and Faivre Sarrailh, C. (2003). The paranodal complex of F3/Contactin and Caspr/Paranodin traffics to the cell surface via a non-conventional pathway. *J Biol Chem* **278**, 48339–47.

Bouchard, C., Lacroix, C., Plante, V. et al. (1999). Clinicopathologic findings and prognosis of chronic inflammatory demyelinating polyneuropathy. *Neurology* **52**, 498–503.

Bourde, O., Kiefer, R., Toyka, K. V. and Hartung, H. P. (1996). Quantification of interleukin-6 mRNA in wallerian degeneration by competitive reverse transcription polymerase chain reaction. *J Neuroimmunol* **69**, 135–40.

Bourque, M. J. and Robitaille, R. (1998). Endogenous peptidergic modulation of perisynaptic Schwann cells at the frog neuromuscular junction. *J Physiol* **512** (Pt 1), 197–209.

Boyles, J. K., Pitas, R. E., Wilson, E., Mahley, R. W. and Taylor, J. M. (1985). Apolipoprotein E associated with astrocytic glia of the central nervous system and with nonmyelinating glia of the peripheral nervous system. *J Clin Invest* **76**, 1501–13.

Bradley, W. G. and Jenkison, M. (1973). Abnormalities of peripheral nerves in murine muscular dystrophy. *J Neurol Sci* **18**, 227–47.

Bradley, W. G., Jaros, E. and Jenkison, M. (1977). The nodes of Ranvier in the nerves of mice with muscular dystrophy. *J Neuropathol Exp Neurol* **36**, 797–806.

Brady, S. T., Witt, A. S., Kirkpatrick, L. L. et al. (1999). Formation of compact myelin is required for maturation of the axonal cytoskeleton. *J Neurosci* **19**, 7278–88.

Braun, N., Sevigny, J., Robson, S. C., Hammer, K., Hanani, M. and Zimmermann, H. (2004). Association of the ecto-ATPase NTPDase2 with glial cells of the peripheral nervous system. *Glia* **45**, 124–32.

Braunewell, K. H., Martini, R., LeBaron, R. et al. (1995). Up-regulation of a chondroitin sulphate epitope during regeneration of mouse sciatic nerve: evidence that the immunoreactive molecules are related to the chondroitin sulphate proteoglycans decorin and versican. *Eur J Neurosci* **7**, 792–804.

Bray, G. M., Perkins, S., Peterson, A. C. and Aguayo, A. J. (1977). Schwann cell multiplication deficit in nerve roots of newborn dystrophic mice. A radioautographic and ultrastructural study. *J Neurol Sci* **32**, 203–12.

Brechenmacher, C., Vital, C., Deminiere, C. et al. (1987). Guillain–Barré syndrome: an ultrastructural study of peripheral nerve in 65 patients. *Clin Neuropathol* **6**, 19–24.

Brennan, A., Dean, C. H., Zhang, A. L., Cass, D. T., Mirsky, R. and Jessen, K. R. (2000). Endothelins control the timing of Schwann cell generation in vitro and in vivo. *Develop Biol* **227**, 545–57.

Brett, F. M., Costigan, D., Farrell, M. A. et al. (1998). Merosin-deficient congenital muscular dystrophy and cortical dysplasia. *Eur J Paediatr Neurol* **2**, 77–82.

Britsch, S., Li, L., Kirchhoff, S. et al. (1998). The ErbB2 and ErbB3 receptors and their ligand, neuregulin-1, are essential for development of the sympathetic nervous system. *Genes & Development* **12**, 1825–36.

Britsch, S., Goerich, D. E., Riethmacher, D. et al. (2001). The transcription factor Sox10 is a key regulator of peripheral glial development. *Genes & Development* **15**, 66–78.

Brown, M. C., Holland, R. L. and Hopkins, W. G. (1981). Motor nerve sprouting. *Ann Rev Neurosci* **4**, 17–42.

Bruck, W. (1997). The role of macrophages in Wallerian degeneration. *Brain Pathol* **7**, 741–52.

Bruzzone, R. and Ressot, C. (1997). Connexins, gap junctions and cell–cell signalling in the nervous system. *Eur J Neurosci* **9**, 1–6.

Bruzzone, R., White, T. W., Scherer, S. S., Fischbeck, K. H. and Paul, D. L. (1994). Null mutations of connexin32 in patients with X-linked Charcot–Marie–Tooth disease. *Neuron* **13**, 1253–60.

Bruzzone, R., White, T. W. and Paul, D. L. (1996). Connections with connexins: the molecular basis of direct intercellular signaling. *Eur J Biochem* **238**, 1–27.

Bult, C., Kibbe, W. A., Snoddy, J. et al. (2004). A genome end-game: understanding gene function in the nervous system. *Nat Neurosci* **7**, 484–5.

Bunge, M. B. (1993a). Schwann cell regulation of extracellular matrix biosynthesis and assembly. Dyck, P. J., Thomas, P. K., Griffin, J., Low, P. A., Poduslo, J. F. (Eds.) *Peripheral Neuropathy*, 3rd Edn, W. B. Saunders, Philadelphia, pp. 299–316.

Bunge, R. P. (1993b). Expanding roles for the Schwann cell: ensheathment, myelination, trophism and regeneration. *Current Opin Neurobiol* **3**, 805–9.

Bunge, R. P. (1994). The role of the Schwann cell in trophic support and regeneration. *J Neurol* **242**, S19–S21.

Bunge, M. W. P. and Wood, P. (2006). Transplantation of Schwann cells and olfactory ensheathing cells to promote regeneration in the CNS. Selzer, M. et al. (Eds.) *Textbook of Neural Repair and Rehabilitation*. Cambridge University Press, Cambridge. pp. 513–31.

Bunge, R. P. and Bunge, M. B. (1983). Interrelationship between Schwann cell function and extracellular matrix production. *Trends Neurosci* 499–505.

Byers, T. J., Lidov, H. G. and Kunkel, L. M. (1993). An alternative dystrophin transcript specific to peripheral nerve. *Nat Genet* **4**, 77–81.

Caccamo, D. V., Ho, K. L. and Garcia, J. H. (1992). Cauda equina tumor with ependymal and paraganglionic differentiation. *Hum Pathol* **23**, 835–38.

Cajal, S. R., DeFelipe, J. and Jones, E. G. (1991). *Cajal's Degeneration and Regeneration of the Nervous System*. Oxford University Press, Oxford.

Calabresi, P. A., Fields, N. S., Maloni, H. W. et al. (1998). Phase 1 trial of transforming growth factor beta 2 in chronic progressive MS. *Neurology* **51**, 289–92.

Cambier, J. C., Littman, D. R. and Weiss, A. (2001). Antigen presentation to T lymphocytes. Janeway, J. (Ed.) *Immunobiology*. Garland Publishing, New York.

Cameron-Curry, P. (1995). Glial lineage of the peripheral nervous system. *C R Seances Soc Biol Fil* **189**, 253–61.

Cameron-Curry, P., Dulac, C. and Le Douarin, N. M. (1993). Negative regulation of Schwann cell myelin protein gene expression by the dorsal root ganglionic microenvironment. *Eur J Neurosci* **5**, 594–604.

Cammer, W. and Tansey, F. A. (1987). Immunocytochemical localization of carbonic anhydrase in myelinated fibers in peripheral nerves of rat and mouse. *J Histochem Cytochem* **35**, 865–70.

Campana, W. M. and Myers, R. R. (2003). Exogenous erythropoietin protects against dorsal root ganglion apoptosis and pain following peripheral nerve injury. *Eur J Neurosci* **18**, 1497–506.

Campana, W. M., Myers, R. R. and Rearden, A. (2003). Identification of PINCH in Schwann cells and DRG neurons: shuttling and signaling after nerve injury. *Glia* **41**, 213–23.

Cao, G. and Ko, C. P. (2001). Schwann cell-conditioned medium modulates synaptic activities at *Xenopus* neuromuscular junctions *in vitro*. *Society for Neuroscience Abstracts*, **711**, 12.

Carey, D. J. (1997). Syndecans: multifunctional cell-surface co-receptors. *Biochem J* **327** (Part 1), 1–16.

Carey, D. J., Todd, M. S. and Rafferty, C. M. (1986). Schwann cell myelination: induction by exogenous basement membrane-like extracellular matrix. *J Cell Biol* **102**, 2254–63.

Carey, D. J., Crumbling, D. M., Stahl, R. C. and Evans, D. M. (1990). Association of cell surface heparan sulfate proteoglycans of Schwann cells with extracellular matrix proteins. *J Biol Chem* **265**, 20627–33.

Carey, D. J., Evans, D. M., Stahl, R. C. et al. (1992). Molecular cloning and characterization of N-syndecan, a novel transmembrane heparan sulfate proteoglycan. *J Cell Biol* **117**, 191–201.

Carey, D. J., Stahl, R. C., Asundi, V. K. and Tucker, B. (1993). Processing and subcellular distribution of the Schwann cell lipid-anchored heparan sulfate proteoglycan and identification as glypican. *Exp Cell Res* **208**, 10–18.

Carey, D. J., Stahl, R. C., Cizmeci-Smith, G. and Asundi, V. K. (1994). Syndecan-1 expressed in Schwann cells causes morphological transformation and cytoskeletal reorganization and associates with actin during cell spreading. *J Cell Biol* **124**, 161–70.

Castonguay, A. and Robitaille, R. (2001). Differential regulation of transmitter release by presynaptic and glial Ca2+ internal stores at the neuromuscular synapse. *J Neurosci* **21**, 1911–22.

Castro, C., Gomez-Hernandez, J. M., Silander, K. and Barrio, L. C. (1999). Altered formation of hemichannels and gap junction channels caused by C-terminal connexin-32 mutations. *J Neurosci* **19**, 3752–60.

Causey, G. (1960). *The Cell of Schwann*. E&S Livingstone Ltd, Edinburgh.

Cavaletti, G., Fabbrica, D., Minoia, C., Frattola, L. and Tredici, G. (1998). Carboplatin toxic effects on the peripheral nervous system of the rat. *Ann Oncol* **9**, 443–7.

Ceuterick-de Groote, C., De Jonghe, P., Timmerman, V. (2001). Infantile demyelinating neuropathy associated with a de novo point mutation on Ser72 in PMP22 and basal lamina onion bulbs in skin biopsy. *Pathol Res Pract* **197**, 193–8.

Chalasani, S. H., Baribaud, F., Coughlan, C. M. et al. (2003a). The chemokine stromal cell-derived factor-1 promotes the survival of embryonic retinal ganglion cells. *J Neurosci* **23**, 4601–12.

Chalasani, S. H., Sabelko, K. A., Sunshine, M. J., Littman, D. R. and Raper, J. A. (2003b). A chemokine, SDF-1, reduces the effectiveness of multiple axonal

repellents and is required for normal axon pathfinding. *J Neurosci* **23**, 1360–71.
Chan et al. (2006). The polarity protein Par-3 directly interacts with p75NTR to regulate myelination. *Science* **314**, 832–6.
Chan, J.R., Cosgaya, J.M., Wu, Y.J. and Shooter, E.M. (2001). Neurotrophins are key mediators of the myelination program in the peripheral nervous system. *Proc Nat Acad Sci USA* **98**, 14661–8.
Chan, J.R., Watkins, T.A., Cosgaya, J.M. et al. (2004). NGF controls axonal receptivity to myelination by Schwann cells or oligodendrocytes. *Neuron* **43**, 183–91.
Chance, P.F., Alderson, M.K., Leppig, K.A. et al. (1993). DNA deletion associated with hereditary neuropathy with liability to pressure palsies. *Cell* **72**, 143–51.
Chandross, K.J., Spray, D.C., Cohen, R.I. et al. (1996). TNF alpha inhibits Schwann cell proliferation, connexin46 expression, and gap junctional communication. *Mol Cell Neurosci* **7**, 479–500.
Chao, C.C. and Hu, S. (1994). Tumor necrosis factor-alpha potentiates glutamate neurotoxicity in human fetal brain cell cultures. *Dev Neurosci* **16**, 172–9.
Charcot, JM. and Marie, P. (1886). Sure une forme particulaire d'atrophie musculaire progressive souvent familial debutant par les pieds et les jambes et atteingnant plus tard les mains. *Rev Med(Paris)* **6**, 97–138.
Chen, C., Bharucha, V., Chen, Y. et al. (2002). Reduced sodium channel density, altered voltage dependence of inactivation, and increased susceptibility to seizures in mice lacking sodium channel β2-subunits. *Proc Nat Acad Sci USA* **99**, 17072–7.
Chen, C.L., Westenbroek, R.E., Xu, X.R. et al. (2004). Mice lacking sodium channel beta 1 subunits display defects in neuronal excitability, sodium channel expression, and nodal architecture. *J Neurosci* **24**, 4030–42.
Chen, L. and Ko, C.P. (1994). Extension of synaptic extracellular matrix during nerve terminal sprouting in living frog neuromuscular junctions. *J Neurosci* **14**, 796–808.
Chen, L.E., Seaber, A.V., Wong, G.H. and Urbaniak, J.R. (1996). Tumor necrosis factor promotes motor functional recovery in crushed peripheral nerve. *Neurochem Int* **29**, 197–203.
Chen, L.L., Folsom, D.B. and Ko, C.P. (1991). The remodeling of synaptic extracellular matrix and its dynamic relationship with nerve terminals at living frog neuromuscular junctions. *J Neurosci* **11**, 2920–30.
Chen, L.M., Bailey, D. and Fernandez-Valle, C. (2000). Association of beta 1 integrin with focal adhesion kinase and paxillin in differentiating Schwann cells. *J Neurosci* **20**, 3776–84.
Chen, S., Rio, C., Ji, R.R., Dikkes, P., Coggeshall, R.E., Woolf, C.J. and Corfas, G. (2003). Disruption of ErbB receptor signaling in adult non-myelinating Schwann cells causes progressive sensory loss. *Nat Neurosci* **6**, 1186–93.
Chen, Z.L. and Strickland, S. (2003). Laminin gamma1 is critical for Schwann cell differentiation, axon myelination, and regeneration in the peripheral nerve. *J Cell Biol* **163**, 889–99.
Cheng, H.L., Russell, J.W. and Feldman, E.L. (1999). IGF-I promotes peripheral nervous system myelination. *Ann NY Acad Sci* **883**, 124–30.
Cheng, H.L., Steinway, M., Delaney, C.L., Franke, T.F. and Feldman, E.L. (2000). IGF-I promotes Schwann cell motility and survival via activation of Akt. *Mol Cell Endocrinol* **170**, 211–15.
Chernousov, M.A., Rothblum, K., Tyler, W.A., Stahl, R.C. and Carey, D.J. (2000). Schwann cells synthesize type V collagen that contains a novel alpha 4

chain. Molecular cloning, biochemical characterization, and high affinity heparin binding of alpha 4(V) collagen. *J Biol Chem* **275**, 28208–15.

Chernousov, M.A. and Carey, D.J. (2003). AlphaVbeta8 integrin is a Schwann cell receptor for fibrin. *Exp Cell Res* **291**, 514–24.

Chernousov, M.A., Stahl, R.C. and Carey, D.J. (2001). Schwann cell type V collagen inhibits axonal outgrowth and promotes Schwann cell migration via distinct adhesive activities of the collagen and noncollagen domains. *J Neurosci* **21**, 6125–35.

Chiba, A., Kusunoki, S., Shimizu, T. and Kanazawa, I. (1992). Serum IgG antibody to ganglioside GQ1b is a possible marker of Miller–Fisher syndrome. *Ann Neurol* **31**, 677–9.

Chiba, A., Matsumura, K., Yamada, H. *et al.* (1997). Structures of sialylated O-linked oligosaccharides of bovine peripheral nerve alpha-dystroglycan. The role of a novel O-mannosyl-type oligosaccharide in the binding of alpha-dystroglycan with laminin. *J Biol Chem* **272**, 2156–62.

Chiu, S.Y., Schrager, P. and Ritchie, J.M. (1984). Neuronal-type Na+ and K+ channels in rabbit cultured Schwann cells. *Nature* **311**, 156–7.

Chow, I. and Poo, M.M. (1985). Release of acetylcholine from embryonic neurons upon contact with muscle cell. *J Neurosci* **5**, 1076–82.

Christopherson, K.S., Ullian, E.M., Stokes, C.C. *et al.* (2005). Thrombospondins are astrocyte-secreted proteins that promote CNS synaptogenesis. *Cell* **120**, 421–33.

Chun, S.J., Rasband, M.N., Sidman, R.L., Habib, A.A. and Vartanian, T. (2003). Integrin-linked kinase is required for laminin-2-induced oligodendrocyte cell spreading and CNS myelination. *J Cell Biol* **163**, 397–408.

Cifuentes-Diaz, C., Velasco, E., Meunier, F.A. *et al.* (1998). The peripheral nerve and the neuromuscular junction are affected in the tenascin-C-deficient mouse. *Cell Mol Biol (Noisy-le-Grand)* **44**, 357–79.

Ciment, G. (1990). The melanocyte Schwann cell progenitor: a bipotent intermediate in the neural crest lineage. *Comments Dev Neurobiol* **1**, 207–23.

Clegg, D.O., Wingerd, K.L., Hikita, S.T. and Tolhurst, E.C. (2003). Integrins in the development, function and dysfunction of the nervous system. *Front Biosci* **8**, 723–50.

Colby, J., Nicholson, R., Dickson, K.M. *et al.* (2000). PMP22 carrying the trembler or trembler-J mutation is intracellularly retained in myelinating Schwann cells. *Neurobiol Dis* **7**, 561–73.

Colognato, H., Winkelmann, D.A. and Yurchenco, P.D. (1999). Laminin polymerization induces a receptor-cytoskeleton network. *J Cell Biol* **145**, 619–31.

Colomar, A. and Robitaille, R. (2004). Glial modulation of synaptic transmission at the neuromuscular junction. *Glia* **47**, 284–9.

Constable, A.L., Armati, P.J., Toyka, K.V. and Hartung, H.P. (1994). Production of prostanoids by Lewis rat Schwann cells in vitro. *Brain Res* **635**, 75–80.

Constable, A.L., Armati, P.J. and Hartung, H.P. (1999). DMSO induction of the leukotriene LTC4 by Lewis rat Schwann cells. *J Neurol Sci* **162**, 120–6.

Constantin, G., Piccio, L., Bussini, S. *et al.* (1999). Induction of adhesion molecules on human schwann cells by proinflammatory cytokines, an immunofluorescence study. *J Neurol Sci* **170**, 124–30.

Constantinescu, C.S., Hilliard, B., Lavi, E., Ventura, E., Venkatesh, V. and Rostami, A. (1996). Suppression of experimental autoimmune neuritis by phosphodiesterase inhibitor pentoxifylline. *J Neurol Sci* **143**, 14–18.

Cook, D.N., Pisetsky, D.S. and Schwartz, D.A. (2004). Toll-like receptors in the pathogenesis of human disease. *Nat Immunol* **5**, 975–9.

Corbin, J. G., Kelly, D., Rath, E. M., Baerwald, K. D., Suzuki, K. and Popko, B. (1996). Targeted CNS expression of interferon-gamma in transgenic mice leads to hypomyelination, reactive gliosis, and abnormal cerebellar development. *Mol Cell Neurosci* **7**, 354–70.

Corfas, G., Velardez, M. O., Ko, C. P., Ratner, N. and Peles, E. (2004). Mechanisms and roles of axon–Schwann cell interactions. *J Neurosci* **24**, 9250–60.

Court, F. A., Sherman, D. L., Pratt, T. *et al.* (2004). Restricted growth of Schwann cells lacking Cajal bands slows conduction in myelinated nerves. *Nature* **431**, 191–5.

Couteaux, R. (1938). Sur l'origine de la sole des plaques motrices. *C R Soc Biol* **127**, 218–21.

Couteaux, R. (1973). Motor end-plate structure. Bourne, G. H. (Ed.) *The Structure and Function of Muscle*, 2nd Edn. Academic Press, New York, pp. 483–530.

Coyle, A. J. and Gutierrez-Ramos, J. C. (2001). The expanding B7 superfamily: increasing complexity in costimulatory signals regulating T cell function. *Nat Immunol* **2**, 203–9.

Crawford, K. and Armati, P. J. (1982). The development of human fetal dorsal root ganglia in vitro: the first 20 days. *Neuropathol Appl Neurobiol* **8**, 477–88.

Cuesta, A., Pedrola, L., Sevilla, T. *et al.* (2001). The gene encoding ganglioside-induced differentiation-associated protein 1 is mutated in axonal Charcot–Marie–Tooth type 4A disease. *Nat Genet* **30**, 22–5.

Culican, S. M., Nelson, C. C. and Lichtman, J. W. (1998). Axon withdrawal during synapse elimination at the neuromuscular junction is accompanied by disassembly of the postsynaptic specialization and withdrawal of Schwann cell processes. *J Neurosci* **18**, 4953–65.

Cummings, J. F., de Lahunta, A. and Mitchell, W. J., Jr. (1983). Ganglioradiculitis in the dog. A clinical, light- and electron-microscopic study. *Acta Neuropathol (Berl)* **60**, 29–39.

Custer, A. W., Kazarinov-Noyes, K., Sakurai, T. *et al.* (2003). The role of the ankyrin-binding protein NrCAM in node of Ranvier formation. *J Neurosci* **23**, 10032–9.

D'Antonio, M., Drogitti, A., Feltri, M. L. *et al.* (2006). TGF beta type II receptor signalling controls Schwann cell death and proliferation in developing nerves. *J Neurosci* **26**, 8417–27.

D'Urso, D., Brophy, P. J., Staugaitis, S. M. *et al.* (1990). Protein zero of peripheral nerve myelin: biosynthesis, membrane insertion, and evidence for homotypic interaction. *Neuron* **4**, 449–60.

Dahl, G., Werner, R., Levine, E. and Rabadan-Diehl, C. (1992). Mutational analysis of gap junction formation. *Biophys J* **62**, 172–80.

Dahle, C., Ekerfelt, C., Vrethem, M., Samuelsson, M. and Ernerudh, J. (1997). T helper type 2 like cytokine responses to peptides from P0 and P2 myelin proteins during the recovery phase of Guillain–Barré syndrome. *J Neurol Sci* **153**, 54–60.

Dalakas, M. C. and Engel, W. K. (1981). Chronic relapsing (dysimmune) polyneuropathy: pathogenesis and treatment. *Ann Neurol* **9** Suppl, 134–45.

Daniloff, J. K., Crossin, K. L., Pincon-Raymond, M., Murawsky, M., Rieger, F. and Edelman, G. M. (1989). Expression of cytotactin in the normal and regenerating neuromuscular system. *J Cell Biol* **108**, 625–35.

Darbas, A., Jaegle, M., Walbeehm, E. *et al.* (2004). Cell autonomy of the mouse claw paw mutation. *Develop Biol* **272**, 470–82.

Day, N. C., Wood, S. J., Ince, P. G. *et al.* (1997). Differential localization of voltage-dependent calcium channel alpha1 subunits at the human and rat neuromuscular junction. *J Neurosci* **17**, 6226–35.

de Waegh, S. and Brady, S.T. (1990). Altered slow axonal transport and regeneration in a myelin-deficient mutant mouse: the trembler as an in vivo model for Schwann cell-axon interactions. *J Neurosci* **10**, 1855–65.

de Waegh, S.M., Lee, V.M. and Brady, S.T. (1992). Local modulation of neurofilament phosphorylation, axonal caliber, and slow axonal transport by myelinating Schwann cells. *Cell* **68**, 451–63.

Dedek, K., Kunath, B., Kananura, C., Reuner, U., Jentsch, T. and Steinlein, O.K. (2001). Myokymia and neonatal epilepsy caused by a mutation in the voltage sensory of the KCNQ2 K^- channel. *Proc Nat Acad Sci USA* **98**, 12272–7.

Delpech, A., Girard, N. and Delpech, B. (1982). Localization of hyaluronectin in the nervous system. *Brain Res* **245**, 251–7.

Dennis, M.J. and Miledi, R. (1974). Electrically induced release of acetylcholine from denervated Schwann cells. *J Physiol* **237**, 431–52.

Deodato, F., Sabatelli, M., Ricci, E. et al. (2002). Hypermyelinating neuropathy, mental retardation and epilepsy in a case of merosin deficiency. *Neuromuscul Dis* **12**, 329–98.

Desaki, J. and Uehara, Y. (1981). The overall morphology of neuromuscular junctions as revealed by scanning electron microscopy. *J Neurocytol* **10**, 101–10.

Descarries, L.M., Cai, S., Robitaille, R., Josephson, E.M. and Morest, D.K. (1998). Localization and characterization of nitric oxide synthase at the frog neuromuscular junction. *J Neurocytol* **27**, 829–40.

Deschenes, S.M., Walcott, J.L., Wexler, T.L., Scherer, S.S. and Fischbeck, K.H. (1997). Altered trafficking of mutant connexin32. *J Neurosci* **17**, 9077–84.

Devaux, J.J. and Scherer, S.S. (2005). Altered ion channels in an animal model of Charcot–Marie–Tooth disease type IA. *J Neurosci* **25**, 1470–80.

Devaux, J., Alcaraz, G., Grinspan, J. et al. (2003). Kv3.1 is a novel component of CNS nodes. *J Neurosci* **23**, 4509–18.

Devaux, J.J., Kleopas, A.K., Cooper, E.C. and Scherer, S.S. (2004). KCNQ2 is a nodal K^+ channel. *J Neurosci* **24**, 1236–44.

Di Muzio, A., De Angelis, M.V., Di Fulvio, P. et al. (2003). Dysmyelinating sensory-motor neuropathy in merosin-deficient congenital muscular dystrophy. *Muscle Nerve* **27**, 500–6.

Dickens, P., Hill, P. and Bennett, M.R. (2003). Schwann cell dynamics with respect to newly formed motor-nerve terminal branches on mature (*Bufo marinus*) muscle fibers. *J Neurocytol* **32**, 381–92.

Doetsch, F. (2003). The glial identity of neural stem cells. *Nature Neurosci* **6**, 1127–34.

Donaghy, M., Gray, J.A., Squier, W. et al. (1989). Recurrent Guillain–Barré syndrome after multiple exposures to cytomegalovirus. *Am J Med* **87**, 339–41.

Dong, Z., Brennan, A., Liu, N. et al. (1995). Neu differentiation factor is a neuron-glia signal and regulates survival, proliferation, and maturation of rat Schwann cell precursors. *Neuron* **15**, 585–96.

Dong, Z., Sinanan, A., Parkinson, D., Parmantier, E., Mirsky, R. and Jessen, K.R. (1999). Schwann cell development in embryonic mouse nerves. *J Neurosci Res* **56**, 334–48.

Dore-Duffy, P., Balabanov, R., Washington, R. and Swanborg, R.H. (1994). Transforming growth factor beta 1 inhibits cytokine-induced CNS endothelial cell activation. *Mol Chem Neuropathol* **22**, 161–75.

Dowsing, B.J., Morrison, W.A., Nicola, N.A., Starkey, G.P., Bucci, T. and Kilpatrick, T.J. (1999). Leukemia inhibitory factor is an autocrine survival factor for Schwann cells. *J Neurochem* **73**, 96–104.

Duggins, A. J., McLeod, J. G., Pollard, J. D. et al. (1999). Spinal root and plexus hypertrophy in chronic inflammatory demyelinating polyneuropathy. *Brain* **122** (Part 7), 1383–90.

Dulac, C. and Le Douarin, N. M. (1991). Phenotypic plasticity of Schwann cells and enteric glial cells in response to the microenvironment. *Proc Nat Acad Sci USA* **88**, 6358–62.

Dunaevsky, A. and Connor, E. A. (1995). Long-term maintenance of presynaptic function in the absence of target muscle fibers. *J Neurosci* **15**, 6137–44.

Dunaevsky, A. and Connor, E. A. (1998). Stability of frog motor nerve terminals in the absence of target muscle fibers. *Dev Biol* **194**, 61–71.

Dunaevsky, A. and Connor, E. A. (2000). F-actin is concentrated in nonrelease domains at frog neuromuscular junctions. *J Neurosci* **20**, 6007–12.

Dupin, E., Baroffio, A., Dulac, C., Cameron-Curry, P. and Le Douarin, N. M. (1990). Schwann-cell differentiation in clonal cultures of the neural crest, as evidenced by the anti-Schwann cell myelin protein monoclonal antibody. *Proc Nat Acad Sci USA* **87**, 1119–23.

Dupin, E., Glavieux, C., Vaigot, P. and Le Douarin, N. M. (2000). Endothelin 3 induces the reversion of melanocytes to glia through a neural crest-derived glial-melanocytic progenitor. *Proc Nat Acad Sci USA* **97**, 7882–7.

Dupin, E., Real, C., Glavieux-Pardanaud, C., Vaigot, P. and Le Douarin, N. M. (2003). Reversal of developmental restrictions in neural crest lineages: transition from Schwann cells to glial-melanocytic precursors in vitro. *Proc Nat Acad Sci USA* **100**, 5229–33.

Dyck, P. J. (1975). Inherited neuronal degeneration and atrophy affecting peripheral motor, sensory and autonomic neurons. Dyck, P. J. (Ed.) *Peripheral Neuropathy*, 1st Edn. W.B. Saunders, Philadelphia, p. 825.

Dyck, P. J. and Lambert, E. H. (1968a). Lower motor and primary sensory neuron diseases with peroneal muscular atrophy. I. Neurologic, genetic, and electrophysiologic findings in hereditary polyneuropathies. *Arch Neurol* **18**, 603–18.

Dyck, P. J. and Lambert, E. H. (1968b). Lower motor and primary sensory neuron diseases with peroneal muscular atrophy. II. Neurologic, genetic, and electrophysiologic findings in various neuronal degenerations. *Arch Neurol* **18**, 619–25.

Dyck, P. J., Lais, A. C., Ohta, M., Bastron, J. A., Okazaki, H. and Groover, R. V. (1975). Chronic inflammatory polyradiculoneuropathy. *Mayo Clin Proc* **50**, 621–37.

Dytrych, L., Sherman, D. L., Gillespie, C. S. and Brophy, P. J. (1998). Two PDZ domain proteins encoded by the murine periaxin gene are the result of alternative intron retention and are differentially targeted in Schwann cells. *J Biol Chem* **273**, 5794–800.

Eccleston, P. A. (1992). Regulation of Schwann cell proliferation: mechanisms involved in peripheral nerve development. *Expe Cell Res* **199**, 1–9.

Eccleston, P. A., Jessen, K. R. and Mirsky, R. (1989). Transforming growth factor-beta and gamma-interferon have dual effects on growth of peripheral glia. *J Neurosci Res* **24**, 524–30.

EFN and PNS Task Force (2005). European Federation of Neurological Societies/Peripheral Nerve Society guideline on management of chronic inflammatory demyelinating polyradiculoneuropathy. Report of a joint task force of the European Federation of Neurological Societies and the Peripheral Nerve Society. *J Peripher Nerv Syst* **10**, 220–8.

Eichberg, J. and Iyer, S. (1996). Phosphorylation of myelin protein: recent advances. *Neurochem Res* **21**, 527–35.

Einheber, S., Milner, T. A., Giancotti, F. and Salzer, J. L. (1993). Axonal regulation of Schwann cell integrin expression suggests a role for alpha 6 beta 4 in myelination. *J Cell Biol* **123**, 1223–6.

Einheber, S., Hannocks, M. J., Metz, C. N., Rifkin, D. B. and Salzer, J. L. (1995). Transforming growth factor-beta 1 regulates axon/Schwann cell interactions. *J Cell Biol* **129**, 443–58.

Eldridge, C. F., Bunge, M. B., Bunge, R. P. and Wood, P. M. (1987). Differentiation of axon-related Schwann cells in vitro. I. Ascorbic acid regulates basal lamina assembly and myelin formation. *J Cell Biol* **105**, 1023–34.

Eldridge, C. F., Bunge, M. B. and Bunge, R. P. (1989). Differentiation of axon-related Schwann cell *in vitro*: II. Control of myelin formation by basal lamina. *J Neurosci* **9**, 625–38.

Elson, K., Ribeiro, R. M., Perelson, A. S., Simmons, A. and Speck, P. (2004). The life span of ganglionic glia in murine sensory ganglia estimated by uptake of bromodeoxyuridine. *Exp Neurol* **186**, 99–103.

Empl, M., Renaud, S., Erne, B. et al. (2001). TNF-alpha expression in painful and nonpainful neuropathies. *Neurology* **56**, 1371–7.

Engel, A. G. (1994). The neuromuscular junction. Engel, A. G., Franzini-Armstrong, C. (Eds.) *Myology*. McGraw-Hill Professional, New York.

English, A. W. (2003). Cytokines, growth factors and sprouting at the neuromuscular junction. *J Neurocytol* **32**, 943–60.

Erdman, R., Stahl, R. C., Rothblum, K., Chernousov, M. A. and Carey, D. J. (2002). Schwann cell adhesion to a novel heparan sulfate binding site in the N-terminal domain of alpha 4 type V collagen is mediated by syndecan-3. *J Biol Chem* **277**, 7619–25.

Ersdal, C., Ulvund, M. J., Espenes, A., Benestad, S. L., Sarradin, P. and Landsverk, T. (2005). Mapping PrPSc propagation in experimental and natural scrapie in sheep with different PrP genotypes. *Vet Pathol* **42**, 258–74.

Ervasti, J. M. and Campbell, K. P. (1993). A role for the dystrophin-glycoprotein complex as a transmembrane linker between laminin and actin. *J Cell Biol* **122**, 809–23.

Exley, A. R., Smith, N. and Winer, J. B. (1994). Tumour necrosis factor-alpha and other cytokines in Guillain–Barré syndrome. *J Neurol Neurosurg Psych* **57**, 1118–20.

Eylar, E. H., Uyemura, K., Brostoff, S. W., Kitamura, K., Ishaque, A. and Greenfield, S. (1979). Proposed nomenclature for PNS myelin proteins. *Neurochem Res* **4**, 289–93.

Fabry, Z., Topham, D. J., Fee, D. et al. (1995). TGF-beta 2 decreases migration of lymphocytes in vitro and homing of cells into the central nervous system in vivo. *J Immunol* **155**, 325–32.

Falls, D. L. (2003). Neuregulins: functions, forms, and signaling strategies. *Exp Cell Res* **284**, 14–30.

Fannon, A. M., Sherman, D. L., Ilyina-Gragerova, G., Brophy, P. J., Friedrich, V. L., Jr. and Colman, D. R. (1995). Novel E-cadherin-mediated adhesion in peripheral nerve: Schwann cell architecture is stabilized by autotypic adherens junctions. *J Cell Biol* **129**, 189–202.

Feasby, T. E., Hahn, A. F. and Gilbert, J. J. (1982). Passive transfer studies in Guillain–Barré polyneuropathy. *Neurology* **32**, 1159–67.

Feasby, T. E., Gilbert, J. J., Brown, W. F. et al. (1986). An acute axonal form of Guillain–Barré polyneuropathy. *Brain* **109** (Part 6), 1115–26.

Feasby, T. E., Hahn, A. F., Brown, W. F., Bolton, C. F. and Gilbert, J. J. (1988). Two types of axonal degeneration in acute Guillain–Barré syndrome. *J Neurol* **235**, S15.

Feder, N. (1971). Microperoxidase – an ultrastructural tracer of low molecular weight. *J Cell Biol* **51**, 339–43.

Feltri, M. L. and Wrabetz, L. (2005). Laminins and their receptors in Schwann cells and hereditary neuropathies. *J Peripher Nerv Syst* **10**, 128–43.

Feltri, M. L., Scherer, S. S., Nemni, R. et al. (1994). β4 integrin expression in myelinating Schwann cells is polarized, developmentally regulated and axonally dependent. *Development* **120**, 1287–1301.

Feltri, M. L., Arona, M., Scherer, S. S. and Wrabetz, L. (1997). Cloning and sequence of the cDNA encoding the beta 4 integrin subunit in rat peripheral nerve. *Gene* **186**, 299–304.

Feltri, M. L., Graus, P. D., Previtali, S. C. et al. (2002). Conditional disruption of beta 1 integrin in Schwann cells impedes interactions with axons. *J Cell Biol* **156**, 199–209.

Feng, Z. and Ko, C. P. (2004). Transforming growth factor (TGF)-beta 1 mediates Schwann cell-induced synaptogenesis at the neuromuscular junction *in vitro*. Program No 385 18, Abstract Viewer/Itinerary Planner Washington, DC: Society for Neuroscience.

Feng, G., Mellor, R. H., Bernstein, M. et al. (2000). Imaging neuronal subsets in transgenic mice expressing multiple spectral variants of GFP. *Neuron* **28**, 41–51.

Feng, Z., Koirala, S. and Ko, C. P. (2005). Synapse–glia interactions at the vertebrate neuromuscular junction. *Neuroscientist* **11**, 503–13.

Fenzi, F., Benedetti, M. D., Moretto, G. and Rizzuto, N. (2001). Glial cell and macrophage reactions in rat spinal ganglion after peripheral nerve lesions: an immunocytochemical and morphometric study. *Arch Ital Biol* **139**, 357–65.

Fernandez-Valle, C., Gwynn, L., Wood, P. M., Carbonetto, S. and Bunge, M. B. (1994). Anti-beta 1 integrin antibody inhibits Schwann cell myelination. *J Neurobiol* **25**, 1207–26.

Fernandez-Valle, C., Tang, Y., Ricard, J. et al. (2002). Paxillin binds schwannomin and regulates its density-dependent localization and effect on cell morphology. *Nat Genet* **31**, 354–62.

Fields, R. D. and Stevens, B. (2000). ATP: an extracellular signaling molecule between neurons and glia. *Trends Neurosci* **23**, 625–33.

Fields, R. D. and Stevens-Graham, B. (2002). New insights into neuron–glia communication. *Science* **298**, 556–62.

Filbin, M. T., Walsh, F. S., Trapp, B. D., Pizzey, J. A. and Tennekoon, G. I. (1990). Role of myelin P0 protein as a homophilic adhesion molecule. *Nature* **344**, 871–2.

Filbin, M. T., Zhang, K., Li, W. and Gao, Y. (1999). Characterization of the effect on adhesion of different mutations in myelin P0 protein. *Ann NY Acad Sci* **883**, 160–7.

Fisher, M. (1956). Syndrome of ophthalmoplegia, ataxia and areflexia. *New Engl J Med* **255**, 57–65.

Foote, A. K. and Blakemore, W. F. (2005). Inflammation stimulates remyelination in areas of chronic demyelination. *Brain* **128**, 528–39.

Forsberg, E., Ek, B., Engstrom, A. and Johansson, S. (1994). Purification and characterization of integrin alpha 9 beta 1. *Exp Cell Res* **213**, 183–90.

Fortun, J., Dunn, W. A., Jr., Joy, S., Li, J. and Notterpek, L. (2003). Emerging role for autophagy in the removal of aggresomes in Schwann cells. *J Neurosci* **23**, 10672–80.

Fortun, J., Li, J., Go, J., Fenstermaker, A., Fletcher, B. S. and Notterpek, L. (2005). Impaired proteasome activity and accumulation of ubiquitinated substrates in a hereditary neuropathy model. *J Neurochem* **92**, 1531–41.

Fragoso, G., Robertson, J., Athlan, E., Tam, E., Almazan, G. and Mushynski, W. E. (2003). Inhibition of p38 mitogen-activated protein kinase interferes with cell shape changes and gene expression associated with Schwann cell myelination. *Exp Neurol* **183**, 34–46.

Frampton, G., Winer, J.B., Cameron, J.S. and Hughes, R.A. (1988). Severe Guillain–Barré syndrome: an association with IgA anti-cardiolipin antibody in a series of 92 patients. *J Neuroimmunol* **19**, 133–9.

Franklin, R.J. (2002). Remyelination of the demyelinated CNS: the case for and against transplantation of central, peripheral and olfactory glia. *Brain Res Bull* **57**, 827–32.

Franzke, C.W., Tasanen, K., Schumann, H. and Bruckner-Tuderman, L. (2003). Collagenous transmembrane proteins: collagen XVII as a prototype. *Matrix Biol* **22**, 299–309.

Frei, R., Dowling, J., Carenini, S., Fuchs, E. and Martini, R. (1999). Myelin formation by Schwann cells in the absence of beta4 integrin. *Glia* **27**, 269–74.

French Cooperative Group on Plasma Exchange in Guillain–Barré syndrome (1987). Efficiency of plasma exchange in Guillain–Barré syndrome: role of replacement fluids. *Ann Neurol* **22**, 753–61.

Frostick, S.P., Yin, Q. and Kemp, G.J. (1998). Schwann cells, neurotrophic factors, and peripheral nerve regeneration. *Microsurgery* **18**, 397–405.

Fu, S.Y. and Gordon, T. (1997). The cellular and molecular basis of peripheral nerve regeneration. *Mol Neurobiol* **14**, 67–116.

Fujii, K., Tsuji, M. and Murota, K. (1986). Isolation of peripheral nerve collagen. *Neurochem Res* **11**, 1439–46.

Fujioka, T., Purev, E. and Rostami, A. (1999a). Chemokine mRNA expression in the cauda equina of Lewis rats with experimental allergic neuritis. *J Neuroimmunol* **97**, 51–9.

Fujioka, T., Kolson, D.L. and Rostami, A.M. (1999b). Chemokines and peripheral nerve demyelination. *J Neurovirol* **5**, 27–31.

Fuller, G.N., Spies, J.M., Pollard, J.D. and McLeod, J.G. (1994). Demyelinating neuropathies triggered by melanoma immunotherapy. *Neurology* **44**, 2404–5.

Gaboreanu, A., Hrstka, R., Xu, W. et al. (2004). *A New Protein that Interacts with the Cytoplasmic Domain of P0*. Am Society for Cell Biology Annual Meeting, Washington DC.

Gabriel, C.M., Hughes, R.A., Moore, S.E., Smith, K.J. and Walsh, F.S. (1998). Induction of experimental autoimmune neuritis with peripheral myelin protein-22. *Brain* **121** (Part 10), 1895–902.

Gabriel, C.M., Gregson, N.A. and Hughes, R.A. (2000). Anti-PMP22 antibodies in patients with inflammatory neuropathy. *J Neuroimmunol* **104**, 139–46.

Gadient, R.A. and Otten, U. (1996). Postnatal expression of interleukin-6 (IL-6) and IL-6 receptor (IL-6R) mRNAs in rat sympathetic and sensory ganglia. *Brain Res* **724**, 41–6.

Gambardella, A., Bono, F., Muglia, M., Valentino, P. and Quattrone, A. (1999). Autosomal recessive hereditary motor and sensory neuropathy with focally folded myelin sheaths (CMT4B). *Ann NY Acad Sci* **883**, 47–55.

Garbay, B., Heape, A.M., Sargueil, F. and Cassagne, C. (2000). Myelin synthesis in the peripheral nervous system. *Prog Neurobiol* **61**, 267–304.

Garbern, J.Y., Cambi, F., Tang, X.M. et al. (1997). Proteolipid protein is necessary in peripheral as well as central myelin. *Neuron* **19**, 205–18.

Garratt, A.N., Britsch, S. and Birchmeier, C. (2000a). Neuregulin, a factor with many functions in the life of a Schwann cell. *Bioessays* **22**, 987–96.

Garratt, A. N., Voiculescu, O., Topilko, P., Charnay, P. and Birchmeier, C. (2000b). A dual role of erbB2 in myelination and in expansion of the Schwann cell precursor pool. *J Cell Biol* **148**, 1035–46.

Gatto, C. L., Walker, B. J. and Lambert, S. (2003). Local ERM activation and dynamic growth cones at Schwann cell tips implicated in efficient formation of nodes of Ranvier. *J Cell Biol* **162**, 489–98.

George, A., Schmidt, C., Weishaupt, A., Toyka, K. V. and Sommer, C. (1999). Serial determination of tumor necrosis factor-alpha content in rat sciatic nerve after chronic constriction injury. *Exp Neurol* **160**, 124–32.

George, E. L., Georges-Labouesse, E. N., Patel-King, R. S., Rayburn, H. and Hynes, R. O. (1993). Defects in mesoderm, neural tube and vascular development in mouse embryos lacking fibronectin. *Development* **119**, 1079–91.

Georges-Labouesse, E. N., George, E. L., Rayburn, H. and Hynes, R. O. (1996). Mesodermal development in mouse embryos mutant for fibronectin. *Dev Dyn* **207**, 145–56.

Georgiou, J. and Charlton, M. P. (1999). Non-myelin-forming perisynaptic Schwann cells express protein zero and myelin-associated glycoprotein. *Glia* **27**, 101–9.

Georgiou, J., Robitaille, R., Trimble, W. S. and Charlton, M. P. (1994). Synaptic regulation of glial protein expression in vivo. *Neuron* **12**, 443–55.

Ghabriel, M. N. and Allt, G. (1981). Incisures of Schmidt–Lanterman. *Prog Neurobiol* **17**, 25–58.

Ghazvini, M., Mandemakers, W., Jaegle, M. et al. (2002). A cell type-specific allele of the POU gene Oct-6 reveals Schwann cell autonomous function in nerve development and regeneration. *EMBO J* **21**, 4612–20.

Ghislain, J., Desmarquet-Trin-Dinh, C., Jaegle, M., Meijer, D., Charnay, P. and Frain, M. (2002). Characterisation of cis-acting sequences reveals a biphasic, axon-dependent regulation of Krox20 during Schwann cell development. *Development* **129**, 155–66.

Giese, K. P., Martini, R., Lemke, G., Soriano, P. and Schachner, M. (1992). Mouse P0 gene disruption leads to hypomyelination, abnormal expression of recognition molecules, and degeneration of myelin and axons. *Cell* **71**, 565–76.

Gillespie, C. S., Sherman, D. L., Blair, G. E. and Brophy, P. J. (1994). Periaxin, a novel protein of myelinating Schwann cells with a possible role in axonal ensheathment. *Neuron* **12**, 497–508.

Girard, C., Bemelmans, A. P., Dufour, N. et al. (2005). Grafts of brain-derived neurotrophic factor and neurotrophin 3-transduced primate Schwann cells lead to functional recovery of the demyelinated mouse spinal cord. *J Neurosci* **25**, 7924–33.

Girault, J. A., Oguievetskaia, K., Carnaud, M., Denisenko-Nehrbass, N. and Goutebroze, L. (2003). Transmembrane scaffolding proteins in the formation and stability of nodes of Ranvier. *Biol Cell* **95**, 447–52.

Gleichmann, M., Gillen, C., Czardybon, M. et al. (2000). Cloning and characterization of SDF-1gamma, a novel SDF-1 chemokine transcript with developmentally regulated expression in the nervous system. *Eur J Neurosci* **12**, 1857–66.

Godschalk, P. C., Heikema, A. P., Gilbert, M. et al. (2004). The crucial role of *Campylobacter jejuni* genes in anti-ganglioside antibody induction in Guillain–Barré syndrome. *J Clin Invest* **114**, 1659–65.

Gold, R., Toyka, K. V. and Hartung, H. P. (1995). Synergistic effect of IFN-gamma and TNF-alpha on expression of immune molecules and antigen presentation by Schwann cells. *Cell Immunol* **165**, 65–70.

Gold, R., Archelos, J. J. and Hartung, H. P. (1999). Mechanisms of immune regulation in the peripheral nervous system. *Brain Pathol* **9**, 343–60.

Gold, R., Hartung, H. P. and Toyka, K. V. (2000). Animal models for autoimmune demyelinating disorders of the nervous system. *Mol Med Today* **6**, 88–91.

Gollan, L., Salomon, D., Salzer, J. L. and Peles, E. (2003). Caspr regulates the processing of contactin and inhibits its binding to neurofascin. *J Cell Biol* **163**, 1213–18.

Gonzalez-Martinez, T., Perez-Pinera, P., Diaz-Esnal, B. and Vega, J. A. (2003). S-100 proteins in the human peripheral nervous system. *Microsc Res Tech* **60**, 633–8.

Goodyear, C. S., O'Hanlon, G. M., Plomp, J. J. et al. (1999). Monoclonal antibodies raised against Guillain–Barré syndrome-associated *Campylobacter jejuni* lipopolysaccharides react with neuronal gangliosides and paralyze muscle–nerve preparations. *J Clin Invest* **104**, 697–708.

Gorson, K. C., Allam, G. and Ropper, A. H. (1997). Chronic inflammatory demyelinating polyneuropathy: clinical features and response to treatment in 67 consecutive patients with and without a monoclonal gammopathy. *Neurology* **48**, 321–8.

Gorson, K. C. and Chaudhry, V. (1999). Chronic inflammatory demyelinating polyneuropathy. *Curr Treat Options Neurol* **1**, 251–62.

Gotz, M. (2003). Glial cells generate neurons – master control within CNS regions: developmental perspectives on neural stem cells. *Neuroscientist* **9**, 379–97.

Gotz, M. and Barde, Y. A. (2005). Radial glial cells: defined and major intermediates between embryonic stem cells and CNS neurons. *Neuron* **46**, 369–72.

Goutebroze, L., Carnaud, M., Denisenko, N., Boutterin, M. C. and Girault, J. A. (2003). Syndecan-3 and syndecan-4 are enriched in Schwann cell perinodal processes. *BMC Neurosci* **4**, 29.

Greenfield, S., Brostoff, S., Eylar, E. H. and Morell, P. (1973). Protein composition of myelin of the peripheral nervous system. *J Neurochem* **20**, 1207–16.

Gregorian, S. K., Lee, W. P., Beck, L. S., Rostami, A. and Amento, E. P. (1994). Regulation of experimental autoimmune neuritis by transforming growth factor-beta 1. *Cell Immunol* **156**, 102–12.

Grewal, P. K., Holzfeind, P. J., Bittner, R. E. and Hewitt, J. E. (2001). Mutant glycosyltransferase and altered glycosylation of alpha-dystroglycan in the myodystrophy mouse. *Nat Genet* **28**, 151–4.

Grieco, T. M., Malhotra, J. D., Chen, C., Isom, L. L. and Raman, I. M. (2005). Open-channel block by the cytoplasmic tail of sodium channel β4 as a mechanism for resurgent sodium current. *Neuron* **45**, 233–44.

Griffin, J. W. and Sheikh, K. (2005). The Guillain–Barré syndromes. Dyck, P. (Ed.) *Peripheral Neuropathy*, 4th Edn. Elsevier Saunders, Philadelphia, pp. 2197–219.

Griffin, J. W., Stoll, G., Li, C. Y., Tyor, W. and Cornblath, D. R. (1990). Macrophage responses in inflammatory demyelinating neuropathies. *Ann Neurol* **27** Suppl, S64–S68.

Griffin, J. W., George, R., Lobato, C., Tyor, W. R., Yan, L. C. and Glass, J. D. (1992). Macrophage responses and myelin clearance during Wallerian degeneration: relevance to immune-mediated demyelination. *J Neuroimmunol* **40**, 153–65.

Griffin, J. W., George, R. and Ho, T. (1993a). Macrophage systems in peripheral nerves. A review. *J Neuropathol Exp Neurol* **52**, 553–60.

Griffin, J. W., Kidd, G. J. and Trapp, B. D. (1993b). Interactions between axons and Schwann cells. Dyck, P. J., Thomas, P. K. (Eds.) *Peripheral Neuropathy*, 3rd Edn. WB Saunders, Philadelphia, pp. 317–30.

Griffin, J.W., Li, C.Y., Ho, T.W. et al. (1995). Guillain–Barré syndrome in northern China. The spectrum of neuropathological changes in clinically defined cases. *Brain* **118** (Part 3), 577–95.
Griffin, J.W., Li, C.Y., Ho, T.W. et al. (1996a). Pathology of the motor-sensory axonal Guillain–Barré syndrome. *Ann Neurol* **39**, 17–28.
Griffin, J.W., Li, C.Y., Macko, C. et al. (1996b). Early nodal changes in the acute motor axonal neuropathy pattern of the Guillain–Barré syndrome. *J Neurocytol* **25**, 33–51.
Griffiths, I., Dickinson, P. and Montague, P. (1995). Expression of the proteolipid protein gene in glial cells of the post-natal peripheral nervous system of rodents. *Neuropathol Appl Neurobiol* **21**, 97–110.
Griffiths, I., Klugmann, M., Anderson, T., Thomson, C., Vouyiouklis, D. and Nave, K.A. (1998). Current concepts of PLP and its role in the nervous system. *Microsc Res Technique* **41**, 344–58.
Grim, M., Halata, Z. and Franz, T. (1992). Schwann cells are not required for guidance of motor nerves in the hindlimb in Splotch mutant mouse embryos. *Anat Embryol* **186**, 311–18.
Grinspan, J.B., Marchionni, M.A., Reeves, M., Coulaloglou, M. and Scherer, S.S. (1996). Axonal interactions regulate Schwann cell apoptosis in developing peripheral nerve: neuregulin receptors and the role of neuregulins. *J Neurosci* **16**, 6107–18.
Groneberg, D.A., Doring, F., Nickolaus, M., Daniel, H. and Fischer, A. (2001). Expression of PEPT2 peptide transporter mRNA and protein in glial cells of rat dorsal root ganglia. *Neurosci Lett* **304**, 181–4.
Groschup, M.H., Beekes, M., McBride, P.A., Hardt, M., Hainfellner, J.A. and Budka, H. (1999). Deposition of disease-associated prion protein involves the peripheral nervous system in experimental scrapie. *Acta Neuropathol (Berl)* **98**, 453–7.
Grothe, C., Meisinger, C., Hertenstein, A., Kurz, H. and Wewetzer, K. (1997). Expression of fibroblast growth factor-2 and fibroblast growth factor receptor 1 messenger RNAs in spinal ganglia and sciatic nerve: regulation after peripheral nerve lesion. *Neuroscience* **76**, 123–35.
Guenard, V., Dinarello, C.A., Weston, P.J. and Aebischer, P. (1991). Peripheral nerve regeneration is impeded by interleukin-1 receptor antagonist released from a polymeric guidance channel. *J Neurosci Res* **29**, 396–400.
Guenard, V., Rosenbaum, T., Gwynn, L.A., Doetschman, T., Ratner, N. and Wood, P.M. (1995a). Effect of transforming growth factor-beta 1 and -beta 2 on Schwann cell proliferation on neurites. *Glia* **13**, 309–18.
Guenard, V., Gwynn, L.A. and Wood, P.M. (1995b). Transforming growth factor-beta blocks myelination but not ensheathment of axons by Schwann cells in vitro. *J Neurosci* **15**, 419–28.
Guilbot, A., Williams, A., Ravise, N. et al. (2001). A mutation in periaxin is responsible for CMT4F, an autosomal recessive form of Charcot–Marie–Tooth disease. *Hum Mol Genet* **10**, 415–21.
Guillain, G., Barré, J.A. and Strohl, A. (1916). Sur un syndrome de radiculo-névrite avec hyperalbuminose du liquide céphalo-rachidien sans réaction cellulaire. Remarques sur les caractéres cliniques et graphiques des réflexes tendineux. *Bull Soc Méd Hôp Paris* **40**, 1462–70.
Guillain–Barré Syndrome Study Group (1985). Plasmapheresis and acute Guillain–Barré syndrome. *Neurology* **35**, 1096–104.
Guiloff, R.J. (1977). Peripheral nerve conduction in Miller–Fisher syndrome. *J Neurol Neurosurg Psychiatry* **40**, 801–7.

Haas, L. F. (1999). Neurological stamp. Theodore Schwann (1810–82). *J Neurol Neurosurg Psychiatry* **66**, 103.

Hadden, R. D., Karch, H., Hartung, H. P. et al. (2001). Preceding infections, immune factors, and outcome in Guillain–Barré syndrome. *Neurology* **56**, 758–65.

Hafer-Macko, C., Hsieh, S. T., Li, C. Y. et al. (1996a). Acute motor axonal neuropathy: an antibody-mediated attack on axolemma. *Ann Neurol* **40**, 635–44.

Hafer-Macko, C. E., Sheikh, K. A., Li, C. Y. et al. (1996b). Immune attack on the Schwann cell surface in acute inflammatory demyelinating polyneuropathy. *Ann Neurol* **39**, 625–35.

Hagedorn, L., Suter, U. and Sommer, L. (1999). P0 and PMP22 mark a multipotent neural crest-derived cell type that displays community effects in response to TGF-beta family factors. *Development* **126**, 3781–94.

Hagedorn, L., Paratore, C., Brugnoli, G. et al. (2000). The Ets domain transcription factor Erm distinguishes rat satellite glia from Schwann cells and is regulated in satellite cells by neuregulin signaling. *Dev Biol* **219**, 44–58.

Hahn, A. F. (1998). Guillain–Barré syndrome. *Lancet* **352**, 635–41.

Hahn, A. F., Bolton, C. F., White, C. M. et al. (1999). Genotype/phenotype correlations in X-linked dominant Charcot–Marie–Tooth disease. *Ann NY Acad Sci* **883**, 366–82.

Hahn, A. F., Ainsworth, P. J., Naus, C. C., Mao, J. and Bolton, C. F. (2000). Clinical and pathological observations in men lacking the gap junction protein connexin 32. *Muscle Nerve* **23**, S39–S48.

Hahn, A. F., Ainsworth, P. J., Bolton, C. F., Bilbao, J. M. and Vallat, J. M. (2001). Pathological findings in the x-linked form of Charcot–Marie–Tooth disease: a morphometric and ultrastructural analysis. *Acta Neuropathol (Berl)* **101**, 129–39.

Hahn, A., Ciskind, C., Krajewski, K., Lewis, R. and Me, S. (2005). Genotype–phenotype correlations in CMTX1 (Abstract). *J Peripher Nerv Syst* **10** (Suppl 1), 31–2.

Halfter, W., Dong, S., Schurer, B. and Cole, G. J. (1998). Collagen XVIII is a basement membrane heparan sulfate proteoglycan. *J Biol Chem* **273**, 25404–12.

Hall, S. M. and Williams, P. L. (1970). Studies on the "incisures" of Schmidt and Lanterman. *J Cell Sci* **6**, 767–91.

Hallows, J. L. and Tempel, B. L. (1998). Expression of Kv1.1, a Shaker-like potassium channel, is temporally regulated in embryonic neurons and glia. *J Neurosci* **18**, 5682–91.

Halstead, S. K., O'Hanlon, G. M., Humphreys, P. D. et al. (2004). Anti-disialoside antibodies kill perisynaptic Schwann cells and damage motor nerve terminals via membrane attack complex in a murine model of neuropathy. *Brain* **127**, 2109–23.

Halstead, S. K., Morrison, I., O'Hanlon, G. M. et al. (2005). Anti-disialosyl antibodies mediate selective neuronal or Schwann cell injury at mouse neuromuscular junctions. *Glia* **52**, 177–89.

Hammarberg, H., Lidman, O., Lundberg, C. et al. (2000). Neuroprotection by encephalomyelitis: rescue of mechanically injured neurons and neurotrophin production by CNS-infiltrating T and natural killer cells. *J Neurosci* **20**, 5283–91.

Hanani, M., Huang, T. Y., Cherkas, P. S., Ledda, M. and Pannese, E. (2002). Glial cell plasticity in sensory ganglia induced by nerve damage. *Neuroscience* **114**, 279–83.

Hanani, M. (2005). Satellite glial cells in sensory ganglia: from form to function. *Brain Res Brain Res Rev* **48**, 457–76.

Hanemann, C.O., Bergmann, C., Senderek, J., Zerres, K. and Sperfeld, A.D. (2003). Transient, recurrent, white matter lesions in X-linked Charcot–Marie–Tooth disease with novel connexin 32 mutation. *Arch Neurol* **60**, 605–9.

Harboe, M., Aseffa, A. and Leekassa, R. (2005). Challenges presented by nerve damage in leprosy. *Lepr Rev* **76**, 5–13.

Harding, A.E. and Thomas, P.K. (1980a). Genetic aspects of hereditary motor and sensory neuropathy (types I and II). *J Med Genet* **17**, 329–36.

Harding, A.E. and Thomas, P.K. (1980b). The clinical features of hereditary motor and sensory neuropathy types I and II. *Brain* **103**, 259–80.

Harrison, B.M., Hansen, L.A., Pollard, J.D. and McLeod, J.G. (1984). Demyelination induced by serum from patients with Guillain–Barré syndrome. *Ann Neurol* **15**, 163–70.

Harrison, R.G. (1908). Embryonic transplantation and development of nervous system. *Anat Rec* **2**, 385–410.

Harrison, R.G. (1924). Neuroblast versus sheath cell in the development of peripheral nerves. *J Comp Neurol* **37**, 123–205.

Hartung, H.P., Pollard, J.D., Harvey, G.K. and Toyka, K.V. (1995a). Immunopathogenesis and treatment of the Guillain–Barré syndrome– Part II. *Muscle Nerve* **18**, 154–64.

Hartung, H.P., Pollard, J.D., Harvey, G.K. and Toyka, K.V. (1995b). Immunopathogenesis and treatment of the Guillain–Barré syndrome– Part I. *Muscle Nerve* **18**, 137–153.

Hartung, H.P., Zielasek, J., Jung, S. and Toyka, K.V. (1996a). Effector mechanisms in demyelinating neuropathies. *Rev Neurol (Paris)* **152**, 320–7.

Hartung, H.P., Kiefer, R., Gold, R. and Toyka, K.V. (1996b). Autoimmunity in the peripheral nervous system. *Baillieres Clin Neurol* **5**, 1–45.

Hartung, H.P., Willison, H., Jung, S., Pette, M., Toyka, K.V. and Giegerich, G. (1996c). Autoimmune responses in peripheral nerve. *Springer Semin Immunopathol* **18**, 97–123.

Hartung, H.P., van der Meche, F.G. and Pollard, J.D. (1998). Guillain–Barré syndrome, CIDP and other chronic immune-mediated neuropathies. *Curr Opin Neurol* **11**, 497–513.

Harvey, G.K., Pollard, J.D., Schindhelm, K. and Antony, J. (1987). Chronic experimental allergic neuritis. An electrophysiological and histological study in the rabbit. *J Neurol Sci* **81**, 215–25.

Harvey, G.K., Toyka, K.V., Zielasek, J., Kiefer, R., Simonis, C. and Hartung, H.P. (1995). Failure of anti-GM1 IgG or IgM to induce conduction block following intraneural transfer. *Muscle Nerve* **18**, 388–94.

Hase, A., Saito, F., Yamada, H., Arai, K., Shimizu, T. and Matsumura, K. (2005). Characterization of glial cell line-derived neurotrophic factor family receptor alpha-1 in peripheral nerve Schwann cells. *J Neurochem* **95**, 537–43.

Hassan, S.M., Jennekens, F.G., Veldman, H. and Oestreicher, B.A. (1994). GAP-43 and p75NGFR immunoreactivity in presynaptic cells following neuromuscular blockade by botulinum toxin in rat. *J Neurocytol* **23**, 354–63.

Hatton, G.I. and Parpura, V. (2004). *Glial Neuronal Signaling*. Kluwer Academic Publishers, Boston.

Haydon, P.G. (2001). Glia: listening and talking to the synapse. *Nat Rev Neurosci* **2**, 185–93.

Hays, A.P., Lee, S.S. and Latov, N. (1988). Immune reactive C3d on the surface of myelin sheaths in neuropathy. *J Neuroimmunol* **18**, 231–44.

Hayworth, C. R., Moody, S. E., Chodosh, L. A., Krieg, P. A., Rimer, M. and Thompson, W. (2006). Induction of neuregulin signaling in mouse Schwann cells in vivo mimics responses to denervation *J Neurosci* **26**, 6873–84.

Herrera, A. A., Banner, L. R. and Nagaya, N. (1990). Repeated, in vivo observation of frog neuromuscular junctions: remodelling involves concurrent growth and retraction. *J Neurocytol* **19**, 85–99.

Herrera, A. A., Qiang, H. and Ko, C. P. (2000). The role of perisynaptic Schwann cells in development of neuromuscular junctions in the frog (*Xenopus laevis*). *J Neurobiol* **45**, 237–54.

Heumann, R., Lindholm, D., Bandtlow, C. et al. (1987a). Differential regulation of mRNA encoding nerve growth factor and its receptor in rat sciatic nerve during development, degeneration, and regeneration: role of macrophages. *Proc Nat Acad Sci USA* **84**, 8735–9.

Heumann, R., Korsching, S., Bandtlow, C. and Thoenen, H. (1987b). Changes of nerve growth factor synthesis in nonneuronal cells in response to sciatic nerve transection. *J Cell Biol* **104**, 1623–31.

Heuser, J. E., Reese, T. S. and Landis, D. M. (1976). Preservation of synaptic structure by rapid freezing. *Cold Spring Harb Symp Quant Biol* **40**, 17–24.

Heuss, D., Engelhardt, A., Gobel, H. and Neundorfer, B. (1995). Light-microscopic study of phosphoprotein B-50 in myopathies. *Virchows Arch* **426**, 69–76.

Hill, C. E., Moon, L. D., Wood, P. M. and Bunge, M. B. (2006). Labeled Schwann cell transplantation: cell loss, host Schwann cell replacement, and strategies to enhance survival. *Glia* **53**, 338–43.

Hirata, K., Zhou, C., Nakamura, K. and Kawabuchi, M. (1997). Postnatal development of Schwann cells at neuromuscular junctions, with special reference to synapse elimination. *J Neurocytol* **26**, 799–809.

Hirata, K., Mitoma, H., Ueno, N., He, J. W. and Kawabuchi, M. (1999). Differential response of macrophage subpopulations to myelin degradation in the injured rat sciatic nerve. *J Neurocytol* **28**, 685–95.

Hirota, H., Kiyama, H., Kishimoto, T. and Taga, T. (1996). Accelerated nerve regeneration in mice by upregulated expression of interleukin (IL) 6 and IL-6 receptor after trauma. *J Exp Med* **183**, 2627–34.

Hisahara, S., Shoji, S., Okano, H. and Miura, M. (1997). ICE/CED-3 family executes oligodendrocyte apoptosis by tumor necrosis factor. *J Neurochem* **69**, 10–20.

Ho, T. W., Mishu, B., Li, C. Y. et al. (1995). Guillain–Barré syndrome in northern China. Relationship to *Campylobacter jejuni* infection and anti-glycolipid antibodies. *Brain* **118** (Part 3), 597–605.

Ho, T. W., Hsieh, S. T., Nachamkin, I. et al. (1997a). Motor nerve terminal degeneration provides a potential mechanism for rapid recovery in acute motor axonal neuropathy after *Campylobacter* infection. *Neurology* **48**, 717–24.

Ho, T. W., Li, C. Y., Cornblath, D. R. et al. (1997b). Patterns of recovery in the Guillain–Barré syndromes. *Neurology* **48**, 695–700.

Ho, T. W., McKhann, G. M. and Griffin, J. W. (1998). Human autoimmune neuropathies. *Ann Rev Neurosci* **21**, 187–226.

Ho, T. W., Willison, H. J., Nachamkin, I. et al. (1999). Anti-GD1a antibody is associated with axonal but not demyelinating forms of Guillain–Barré syndrome. *Ann Neurol* **45**, 168–73.

Hodgkinson, S. J., Westland, K. W. and Pollard, J. D. (1994). Transfer of experimental allergic neuritis by intra neural injection of sensitized lymphocytes. *J Neurol Sci* **123**, 162–72.

Hoebe, K., Janssen, E. and Beutler, B. (2004). The interface between innate and adaptive immunity. *Nat Immunol* **5**, 971−4.

Hofler, H., Walter, G.F. and Denk, H. (1984). Immunohistochemistry of folliculostellate cells in normal human adenohypophysis and in pituitary adenomas. *Acta Neuropathol (Berl)* **65**, 35−40.

Hoke, A. and Keswani, S.C. (2005). Neuroprotection in the PNS: erythropoietin and immunophilin ligands. *Ann NY Acad Sci* **1053**, 491−501.

Hoke, A., Ho, T., Crawford, T.O., Lebel, C., Hilt, D. and Griffin, J.W. (2003). Glial cell line derived neurotrophic factor alters axon Schwann cell units and promotes myelination in unmyelinated nerve fibers. *J Neurosci* **23**, 561−7.

Holen, T. and Mobbs, C.V. (2004). Lobotomy of genes: use of RNA interference in neuroscience. *Neurosci* **126**, 1−7.

Holland, N.R., Crawford, T.O., Hauer, P., Cornblath, D.R., Griffin, J.W. and McArthur, J.C. (1998). Small-fiber sensory neuropathies: clinical course and neuropathology of idiopathic cases. *Ann Neurol* **44**, 47−59.

Homma, S., Yaginuma, H. and Oppenheim, R.W. (1994). Programmed cell death during the earliest stages of spinal cord development in the chick embryo: a possible means of early phenotypic selection. *J Comp Neurol* **345**, 377−95.

Hong, C.S. and Saint-Jeannet, J.P. (2005). Sox proteins and neural crest development. *Semin Cell Dev Biol* **16**, 694−703.

Hoogendijk, J.E., Janssen, E.A., Gabreels-Festen, A.A. *et al.* (1993). Allelic heterogeneity in hereditary motor and sensory neuropathy type Ia (Charcot−Marie−Tooth disease type 1a). *Neurology* **43**, 1010−15.

Houlden, H., King, R.H., Wood, N.W., Thomas, P.K. and Reilly, M.M. (2001). Mutations in the 5′ region of the myotubularin-related protein 2 (MTMR2) gene in autosomal recessive hereditary neuropathy with focally folded myelin. *Brain* **124**, 907−15.

Hsiao, L.L., Peltonen, J., Jaakkola, S., Gralnick, H., Uitto, J. (1991). Plasticity of integrin expression by nerve-derived connective tissue cells. Human Schwann cells, perineurial cells, and fibroblasts express markedly different patterns of beta 1 integrins during nerve development, neoplasia, and in vitro. *J Clin Invest* **87**, 811−20.

Hsieh, S.T., Kidd, G.J., Crawford, T.O. *et al.* (1994). Regional modulation of neurofilament organization by myelination in normal axons. *J Neurosci* **14**, 6392−401.

Hu, H.M., O'Rourke, K., Boguski, M.S. and Dixit, V.M. (1994). A novel RING finger protein interacts with the cytoplasmic domain of CD40. *J Biol Chem* **269**, 30069−72.

Hu, W., Ramacher, M., Hartung, H.-P. and Kieseier, B.C. (2004). Schwann cells express Toll-like receptors. *J Neuroimmunol* **154**, 48.

Huang, E.J. and Reichardt, L.F. (2001). Neurotrophins: roles in neuronal development and function. *Ann Rev Neurosci* **24**, 677−736.

Hughes, B.W., Kusner, L.L. and Kaminski, H.J. (2005a). Molecular architecture of the neuromuscular junction. *Muscle Nerve* **33**, 445−61.

Hughes, P.M., Wells, G.M., Clements, J.M. *et al.* (1998). Matrix metalloproteinase expression during experimental autoimmune neuritis. *Brain* **121** (Part 3), 481−94.

Hughes R. (1990). *Guillain−Barré Syndrome*. Springer-Verlag, Heidelberg.

Hughes, R. (2004). Treatment of Guillain−Barré syndrome with corticosteroids: lack of benefit? *Lancet* **363**, 181−2.

Hughes, R. and van der Meche, F.G. (2000). Corticosteroids for treating Guillain−Barré syndrome. *Cochrane Database Syst Rev* CD001446.

Hughes, R., Sanders, E., Hall, S., Atkinson, P., Colchester, A. and Payan, P. (1992). Subacute idiopathic demyelinating polyradiculoneuropathy. *Arch Neurol* **49**, 612–16.

Hughes, R., Atkinson, P.F., Gray, I.A. and Taylor, W.A. (1987). Major histocompatibility antigens and lymphocyte subsets during experimental allergic neuritis in the Lewis rat. *J Neurol* **234**, 390–5.

Hughes, R., Hadden, R.D., Gregson, N.A. and Smith, K.J. (1999). Pathogenesis of Guillain–Barré syndrome. *J Neuroimmunol* **100**, 74–97.

Hughes, R., Raphael, J.C., Swan, A.V. and van Doorn, P.A. (2001). Intravenous immunoglobulin for Guillain–Barré syndrome. *Cochrane Database Syst Rev* CD002063.

Hughes, R., Swan, A.V. and van Doorn, P.A. (2003). Cytotoxic drugs and interferons for chronic inflammatory demyelinating polyradiculoneuropathy. *Cochrane Database Syst Rev* CD003280.

Hughes, R.A., Wijdicks, E.F., Benson, E. *et al.* (2005a). Supportive care for patients with Guillain–Barré syndrome. *Arch Neurol* **62**, 1194–8.

Hughes, R.A.C., Bouche, P., Cornblath, D.R. *et al.* (2005b). European Federation of Neurological Societies/Peripheral Nerve Society guideline on management of chronic inflammatory demyelinating polyradiculoneuropathy. Report of a joint task force of the European Federation of Neurological Societies and the Peripheral Nerve Society. *J Peripher Nerv Syst* **10**, 220–8.

Huxley, C., Passage, E., Manson, A. *et al.* (1996). Construction of a mouse model of Charcot–Marie–Tooth disease type 1A by pronuclear injection of human YAC DNA. *Hum Mol Genet* **5**, 563–9.

Hyland, H.H. and Russell, W.R. (1930). Chronic progressive polyneuritis with report of a fatal case. *Brain* **53**, 278–9.

Hynes, R.O. (1992). Integrins: versatility, modulation, and signaling in cell adhesion. *Cell* **69**, 11–25.

Ichimura, T. and Ellisman, M.H. (1991). Three-dimensional fine structure of cytoskeletal–membrane interactions at nodes of Ranvier. *J Neurocytol* **20**, 667–81.

Ilyas, A.A., Mithen, F.A., Dalakas, M.C. *et al.* (1991). Antibodies to sulfated glycolipids in Guillain–Barré syndrome. *J Neurol Sci* **105**, 108–17.

Ilyas, A.A., Mithen, F.A., Dalakas, M.C., Chen, Z.W. and Cook, S.D. (1992). Antibodies to acidic glycolipids in Guillain–Barré syndrome and chronic inflammatory demyelinating polyneuropathy. *J Neurol Sci* **107**, 111–21.

Imamura, M., Araishi, K., Noguchi, S. and Ozawa, E. (2000). A sarcoglycan–dystroglycan complex anchors Dp116 and utrophin in the peripheral nervous system. *Hum Mol Genet* **9**, 3091–100.

Inoue, K., Tanabe, Y. and Lupski, J.R. (1999). Myelin deficiencies in both the central and the peripheral nervous systems associated with a SOX10 mutation. *Ann Neurol* **46**, 313–18.

Inoue, K., Shilo, K., Boerkoel, C.F. *et al.* (2002). Congenital hypomyelinating neuropathy, central dysmyelination, and Waardenburg–Hirschsprung disease: phenotypes linked by SOX10 mutation. *Ann Neurol* **52**, 836–42.

Inoue, K., Khajavi, M., Ohyama, T. *et al.* (2004). Molecular mechanism for distinct neurological phenotypes conveyed by allelic truncating mutations. *Nat Genet* **36**, 361–9.

Ionasescu, V.V., Ionasescu, R., Searby, C. and Neahring, R. (1995). Dejerine–Sottas disease with de novo dominant point mutation of the PMP22 gene. *Neurology* **45**, 1766–7.

Ishikawa, S., Ohshima, Y., Suzuki, T. and Oboshi, S. (1979). Primitive neuroectodermal tumor (neuroepithelioma) of spinal nerve root – report of an adult case and establishment of a cell line. *Acta Pathol Jpn* **29**, 289–301.
Ivanova, A. and Nachev, S. (1990). Morphological changes in the nervous system in lead poisoning. I. Experimentally induced lead neuropathy. *Eksp Med Morfol* **29**, 18–23.
Iwasaki, A. and Medzhitov, R. (2004). Toll-like receptor control of the adaptive immune responses. *Nat Immunol* **5**, 987–95.
Iwase, T., Jung, C.G., Bae, H., Zhang, M. and Soliven, B. (2005). Glial cell line-derived neurotrophic factor-induced signaling in Schwann cells. *J Neurochem* **94**, 1488–99.
Jaakkola, S., Peltonen, J. and Uitto, J.J. (1989). Perineurial cells coexpress genes encoding interstitial collagens and basement membrane zone components. *J Cell Biol* **108**, 1157–63.
Jacobs, B.C., Endtz, H., van der Meche, F.G., Hazenberg, M.P., Achtereekt, H.A. and van Doorn, P.A. (1995). Serum anti-GQ1b IgG antibodies recognize surface epitopes on *Campylobacter jejuni* from patients with Miller–Fisher syndrome. *Ann Neurol* **37**, 260–4.
Jacobs, J.M. and Love, S. (1985). Qualitative and quantitative morphology of human sural nerve at different ages. *Brain* **108** (Part 4), 897–924.
Jaegle, M., Mandemakers, W., Broos, L. et al. (1996). The POU factor Oct-6 and Schwann cell differentiation. *Science* **273**, 507–10.
Jaegle, M., Ghazvini, M., Mandemakers, W. et al. (2003). The POU proteins Brn-2 and Oct-6 share important functions in Schwann cell development. *Genes & Development* **17**, 1380–91.
Jahromi, B.S., Robitaille, R. and Charlton, M.P. (1992). Transmitter release increases intracellular calcium in perisynaptic Schwann cells in situ. *Neuron* **8**, 1069–77.
James, S., Patel, N.J., Thomas, P.K. and Burnstock, G. (1993). Immunocytochemical localisation of insulin receptors on rat superior cervical ganglion neurons in dissociated cell culture. *J Anat* **182** (Part 1), 95–100.
Jander, S., Bussini, S., Neuen-Jacob, E. et al. (2002). Osteopontin: a novel axon-regulated Schwann cell gene. *J Neurosci Res* **67**, 156–66.
Janeway, C.A., Jr. (1992). The immune system evolved to discriminate infectious nonself from noninfectious self. *Immunol Today* **13**, 11–16.
Jaros, E. and Bradley, W.G. (1978). Development of the amyelinated lesion in the ventral root of the dystrophic mouse. Ultrastructural, quantitative and autoradiographic study. *J Neurol Sci* **36**, 317–39.
Jaros, E. and Bradley, W.G. (1979). Atypical axon–Schwann cell relationships in the common peroneal nerve of the dystrophic mouse: an ultrastructural study. *Neuropathol Appl Neurobiol* **5**, 133–47.
Jessen, K.R. (2004). Glial cells. *Int J Biochem Cell Biol* **36**, 1861–7.
Jessen, K.R. and Mirsky, R. (1980). Glial cells in the enteric nervous system contain glial fibrillary acidic protein. *Nature* **286**, 736–7.
Jessen, K.R. and Mirsky, R. (1985). Glial fibrillary acidic polypeptides in peripheral glia. Molecular weight, heterogeneity and distribution. *J Neuroimmunol* **8**, 377–93.
Jessen, K.R. and Mirsky, R. (1999). Schwann cells and their precursors emerge as major regulators of nerve development. *Trends Neurosci* **22**, 402–10.
Jessen, K.R. and Mirsky, R. (2002). Signals that determine Schwann cell identity. *J Anat* **200**, 367–75.

Jessen, K. R. and Mirsky, R. (2004). Schwann cell development. Lazzarini, R. A. (Ed.) *Myelin Biology and Disorders*. Elsevier, Philadelphia, pp. 329–59.

Jessen, K. R. and Mirsky, R. (2005a). The origin and development of glial cells in peripheral nerves. *Nature Rev Neurosci* **6**, 671–82.

Jessen, K. R. and Mirsky, R. (2005b). The Schwann cell lineage. Kettenmann, H., Ransom, B. (Eds.) *Neuroglia*, 2 Edn. Oxford University Press, Oxford, pp. 85–100.

Jessen, K. R., Morgan, L., Stewart, H. J. S. and Mirsky, R. (1990). Three markers of adult non-myelin-forming Schwann cells, 217c (Ran-1), A5E3 and GFAP: development and regulation by neuron–Schwann cell interactions. *Development* **109**, 91–103.

Jessen, K. R., Brennan, A., Morgan, L. et al. (1994). The Schwann cell precursor and its fate: a study of cell death and differentiation during gliogenesis in rat embryonic nerves. *Neuron* **12**, 509–27.

John, G. R., Shankar, S. L., Shafit-Zagardo, B. et al. (2002). Multiple sclerosis: re-expression of a developmental pathway that restricts oligodendrocyte maturation. *Nat Med* **8**, 1115–21.

Johnson, A. N. and Newfeld, S. J. (2002). The TGF-beta family: signaling pathways, developmental roles, and tumor suppressor activities. *Scientific World J* **2**, 892–925.

Jones, L. L., Sajed, D. and Tuszynski, M. H. (2003). Axonal regeneration through regions of chondroitin sulfate proteoglycan deposition after spinal cord injury: a balance of permissiveness and inhibition. *J Neurosci* **23**, 9276–88.

Jones, P. L. and Jones, F. S. (2000). Tenascin-C in development and disease: gene regulation and cell function. *Matrix Biol* **19**, 581–96.

Jordan, C. L. and Williams, T. J. (2001). Testosterone regulates terminal Schwann cell number and junctional size during developmental synapse elimination. *Dev Neurosci* **23**, 441–51.

Joseph, N. M., Mukouyama, Y. S., Mosher, J. T. et al. (2004). Neural crest stem cells undergo multilineage differentiation in developing peripheral nerves to generate endoneurial fibroblasts in addition to Schwann cells. *Development* **131**, 5599–612.

Kadlubowski, M. and Hughes, R. A. (1979). Identification of the neuritogen for experimental allergic neuritis. *Nature* **277**, 140–1.

Kadlubowski, M. and Hughes, R. A. (1980). The neuritogenicity and encephalitogenicity of P2 in the rat, guinea-pig and rabbit. *J Neurol Sci* **48**, 171–8.

Kaelin-Lang, A., Lauterburg, T. and Burgunder, J. M. (1998). Expression of adenosine A2a receptor gene in rat dorsal root and autonomic ganglia. *Neurosci Lett* **246**, 21–4.

Kameda, Y. (1996). Immunoelectron microscopic localization of vimentin in sustentacular cells of the carotid body and the adrenal medulla of guinea pigs. *J Histochem Cytochem* **44**, 1439–49.

Kang, H., Tian, L. and Thompson, W. (2003). Terminal Schwann cells guide the reinnervation of muscle after nerve injury. *J Neurocytol* **32**, 975–85.

Katz, R. L. (1966). *Nerve, Muscle and Synapse*. McGraw-Hill, New York.

Kawabuchi, M., Zhou, C. J., Wang, S., Nakamura, K., Liu, W. T. and Hirata, K. (2001). The spatiotemporal relationship among Schwann cells, axons and postsynaptic acetylcholine receptor regions during muscle reinnervation in aged rats. *Anat Rec* **264**, 183–202.

Kelly, A. M. and Zacks, S. I. (1969). The fine structure of motor endplate morphogenesis. *J Cell Biol* **42**, 154–69.

Kennedy, W. R. (2004). Opportunities afforded by the study of unmyelinated nerves in skin and other organs. *Muscle Nerve* **29**, 756–67.

Kennedy, W. R., Wendelschafer-Crabb, G. and Johnson, T. (1996). Quantitation of epidermal nerves in diabetic neuropathy. *Neurology* **47**, 1042–8.

Kerschensteiner, M., Stadelmann, C., Dechant, G., Wekerle, H. and Hohlfeld, R. (2003). Neurotrophic cross-talk between the nervous and immune systems: implications for neurological diseases. *Ann Neurol* **53**, 292–304.

Kettenmann, H. and Ransom, B. R. (2005). *Neuroglia*, 2nd Edn. Oxford University Press, New York.

Keynes, R. J. (1987). Schwann cells during neural development and regeneration: leaders or followers? *Trends Neurosci* **10**, 137–9.

Khalili-Shirazi, A., Hughes, R. A., Brostoff, S. W., Linington, C. and Gregson, N. (1992). T cell responses to myelin proteins in Guillain–Barré syndrome. *J Neurol Sci* **111**, 200–3.

Khalili-Shirazi, A., Atkinson, P., Gregson, N. and Hughes, R. A. (1993). Antibody responses to P0 and P2 myelin proteins in Guillain–Barré syndrome and chronic idiopathic demyelinating polyradiculoneuropathy. *J Neuroimmunol* **46**, 245–51.

Kiefer, R., Streit, W. J., Toyka, K. V., Kreutzberg, G. W. and Hartung, H. P. (1995). Transforming growth factor-beta 1: a lesion-associated cytokine of the nervous system. *Int J Dev Neurosci* **13**, 331–9.

Kiefer, R., Funa, K., Schweitzer, T. *et al.* (1996). Transforming growth factor-beta 1 in experimental autoimmune neuritis. Cellular localization and time course. *Am J Pathol* **148**, 211–23.

Kiefer, R., Kieseier, B. C., Stoll, G. and Hartung, H. P. (2001). The role of macrophages in immune-mediated damage to the peripheral nervous system. *Prog Neurobiol* **64**, 109–27.

Kieseier, B. C., Krivacic, K., Jung, S. *et al.* (2000). Sequential expression of chemokines in experimental autoimmune neuritis. *J Neuroimmunol* **110**, 121–9.

Kieseier, B. C., Dalakas, M. C. and Hartung, H. P. (2002a). Immune mechanisms in chronic inflammatory demyelinating neuropathy. *Neurology* **59**, S7–S12.

Kieseier, B. C., Tani, M., Mahad, D. *et al.* (2002b). Chemokines and chemokine receptors in inflammatory demyelinating neuropathies: a central role for IP-10. *Brain* **125**, 823–34.

Kieseier, B. C., Kiefer, R., Gold, R., Hemmer, B., Willison, H. J. and Hartung, H. P. (2004). Advances in understanding and treatment of immune-mediated disorders of the peripheral nervous system. *Muscle Nerve* **30**, 131–56.

Kim, S. A., Vacratsis, P. O., Firestein, R., Cleary, M. L. and Dixon, J. E. (2003). Regulation of myotubularin-related (MTMR)2 phosphatidylinositol phosphatase by MTMR5, a catalytically inactive phosphatase. *Proc Nat Acad Sci USA* **100**, 4492–7.

Kim, S. I., Voshol, H., van, O. J., Hastings, T. G., Cascio, M. and Glucksman, M. J. (2004). Neuroproteomics: expression profiling of the brain's proteomes in health and disease. *Neurochem Res* **29**, 1317–31.

Kingsley, D. M. (1994). The TGF-beta superfamily: new members, new receptors, and new genetic tests of function in different organisms. *Genes Dev* **8**, 133–46.

Kingston, A. E., Bergsteinsdottir, K., Jessen, K. R., van Der Meide, P. H., Colston, M. J. and Mirsky, R. (1989). Schwann cells co-cultured with stimulated T cells and antigen express major histocompatibility complex (MHC) class II determinants without interferon-gamma pretreatment: synergistic effects of interferon-gamma and tumor necrosis factor on MHC class II induction. *Eur J Immunol* **19**, 177–83.

Kinugasa, Y., Ishiguro, H., Tokita, Y., Oohira, A., Ohmoto, H. and Higashiyama, S. (2004). Neuroglycan C, a novel member of the neuregulin family. *Biochem Biophys Res Commun* **321**, 1045–9.

Kioussi, C., Gross, M.K. and Gruss, P. (1995). Pax3: A paired domain gene as a regulator in PNS myelination. *Neuron* **15**, 553–62.

Kirschner, D.A., Wrabetz, L. and Feltri, M.L. (2004). The P0 gene. Lazzarini, R.L. (Ed.) *Myelin Biology and Disorders*. Elsevier, Philadelphia, pp. 523–45.

Ko, C.P. (1981). Electrophysiological and freeze-fracture studies of changes following denervation at frog neuromuscular junctions. *J Physiol* **321**, 627–39.

Ko, C.P. (1987). A lectin, peanut agglutinin, as a probe for the extracellular matrix in living neuromuscular junctions. *J Neurocytol* **16**, 567–76.

Ko, C.P. and Chen, L. (1996). Synaptic remodeling revealed by repeated in vivo observations and electron microscopy of identified frog neuromuscular junctions. *J Neurosci* **16**, 1780–90.

Ko, C.P. and Thompson, W. (2003). Special issue – the neuromuscular junction. *J Neurocytol* **32**, 423–1037.

Kocsis, J.D., Akiyama, Y. and Radtke, C. (2004). Neural precursors as a cell source to repair the demyelinated spinal cord. *J Neurotrauma* **21**, 441–9.

Koenig, H.L., Schumacher, M., Ferzaz, B. *et al*. (1995). Progesterone synthesis and myelin formation by Schwann cells. *Science* **268**, 1500–3.

Koirala, S. and Ko, C.P. (2004). Pruning an axon piece by piece: a new mode of synapse elimination. *Neuron* **44**, 578–80.

Koirala, S., Qiang, H. and Ko, C.P. (2000). Reciprocal interactions between perisynaptic Schwann cells and regenerating nerve terminals at the frog neuromuscular junction. *J Neurobiol* **44**, 343–60.

Koirala, S., Reddy, L.V. and Ko, C.P. (2003). Roles of glial cells in the formation, function, and maintenance of the neuromuscular junction. *J Neurocytol* **32**, 987–1002.

Koller, H., von Giesen, H.J. and Siebler, M. (1995). Impairment of electrophysiological function of astrocytes by cerebrospinal fluid from a patient with Waldenstrom's macroglobulinemia. *J Neuroimmunol* **61**, 35–9.

Koller, H., Kieseier, B.C., Jander, S. and Hartung, H.P. (2005). Chronic inflammatory demyelinating polyneuropathy. *N Engl J Med* **352**, 1343–56.

Koski, C.L. (1997). Mechanisms of Schwann cell damage in inflammatory neuropathy. *J Infect Dis* **176** (Suppl 2), S169–S172.

Koski, C.L., Humphrey, R. and Shin, M.L. (1985). Anti-peripheral myelin antibody in patients with demyelinating neuropathy: quantitative and kinetic determination of serum antibody by complement component 1 fixation. *Proc Nat Acad Sci USA* **82**, 905–9.

Koski, C.L., Sanders, M.E., Swoveland, P.T. *et al*. (1987). Activation of terminal components of complement in patients with Guillain–Barré syndrome and other demyelinating neuropathies. *J Clin Invest* **80**, 1492–7.

Krajewski, K., Turansky, C., Lewis, R. *et al*. (1999). Correlation between weakness and axonal loss in patients with CMT1A. *Ann NY Acad Sci* **883**, 490–2.

Krajewski, K.M., Lewis, R.A., Fuerst, D.R. *et al*. (2000). Neurological dysfunction and axonal degeneration in Charcot–Marie–Tooth disease type 1A. *Brain* **123**, 1516–27.

Kramer, R.H., Cheng, Y.F. and Clyman, R. (1990). Human microvascular endothelial cells use beta 1 and beta 3 integrin receptor complexes to attach to laminin. *J Cell Biol* **111**, 1233–43.

Krammer, P.H. (2000). CD95's deadly mission in the immune system. *Nature* **407**, 789–95.

Kreusch, A., Pfaffinger, P. J., Stevens, C. F. and Choe, S. (1998). Crystal structure of the tetramerization domain of the *Shaker* potassium channel. *Nature* **392**, 945–8.

Kritas, S. K., Pensaert, M. B., Nauwynck, H. J. and Kyriakis, S. C. (1999). Neural invasion of two virulent suid herpesvirus 1 strains in neonatal pigs with or without maternal immunity. *Vet Microbiol* **69**, 143–56.

Krucke, W. (1955). Evkrankungen des peripheren nerven systems. Evkrankungen der peripheren nerve. Lubarsch, O., Henke, F., Rossle, R. (Eds.) Handbuck der Speziellen Pathologischen Anatomie und Histologie, 13 Springer-Verlag, Berlin, pp. 164–83.

Kubu, C. J., Orimoto, K., Morrison, S. J., Weinmaster, G., Anderson, D. J. and Verdi, J. M. (2002). Developmental changes in Notch1 and Numb expression mediated by local cell–cell interactions underlie progressively increasing delta sensitivity in neural crest stem cells. *Develop Biol* **244**, 199–214.

Kuffler, D. P. (1986). Accurate reinnervation of motor end plates after disruption of sheath cells and muscle fibers. *J Comp Neurol* **250**, 228–35.

Kullberg, R. W., Lentz, T. L. and Cohen, M. W. (1977). Development of the myotomal neuromuscular junction in *Xenopus laevis*: an electrophysiological and fine-structural study. *Dev Biol* **60**, 101–29.

Kumpulainen, T. and Korhonen, L. K. (1982). Immunohistochemical localization of carbonic anhydrase isoenzyme C in the central and peripheral nervous system of the mouse. *J Histochem Cytochem* **30**, 283–92.

Kurtz, A., Zimmer, A., Schnutgen, F., Bruning, G., Spener, F. and Muller, T. (1994). The expression pattern of a novel gene encoding brain-fatty acid binding protein correlates with neuronal and glial cell development. *Development* **120**, 2637–49.

Kury, P., Koller, H., Hamacher, M., Cornely, C., Hasse, B. and Muller, H. W. (2003). Cyclic AMP and tumor necrosis factor-alpha regulate CXCR4 gene expression in Schwann cells. *Mol Cell Neurosci* **24**, 1–9.

Kusunoki, S., Chiba, A., Hitoshi, S., Takizawa, H. and Kanazawa, I. (1995). Anti-Gal-C antibody in autoimmune neuropathies subsequent to mycoplasma infection. *Muscle Nerve* **18**, 409–13.

Kuwabara, S., Ogawara, K., Misawa, S. *et al.* (2004). Does *Campylobacter jejuni* infection elicit 'demyelinating' Guillain–Barré syndrome? *Neurology* **63**, 529–33.

Kwa, M. S., van, S. I., Brand, A., Baas, F. and Vermeulen, M. (2001). Investigation of serum response to PMP22, connexin 32 and P(0) in inflammatory neuropathies. *J Neuroimmunol* **116**, 220–25.

Kwa, M. S., Van, S. I., de Jonge, R. R. *et al.* (2003). Autoimmunoreactivity to Schwann cells in patients with inflammatory neuropathies. *Brain* **126**, 361–75.

La, F. M., Underwood, J. L., Rappolee, D. A. and Werb, Z. (1996). Basement membrane and repair of injury to peripheral nerve: defining a potential role for macrophages, matrix metalloproteinases, and tissue inhibitor of metalloproteinases-1. *J Exp Med* **184**, 2311–26.

Lacas-Gervais, S., Guo, J., Strenzke, N. *et al.* (2004). βIVΣ1 spectrin stabilizes the nodes of Ranvier and axon initial segments. *J Cell Biol* **166**, 983–90.

Lambert, S., Davis, J. Q. and Bennett, V. (1997). Morphogenesis of the node of Ranvier: co-clusters of ankyrin and ankyrin-binding integral proteins define early developmental intermediates. *J Neurosci* **17**, 7025–36.

Lampert, P. W. (1969). Mechanism of demyelination in experimental allergic neuritis. Electron microscopic studies. *Lab Invest* **20**, 127–38.

Laporte, J., Blondeau, F., Buj-Bello, A. and Mandel, J. L. (2001). The myotubularin family: from genetic disease to phosphoinositide metabolism. *Trends Genet* **17**, 221–8.

Latov, N. (2002). Diagnosis of CIDP. *Neurology* **59**, S2–S6.

Lazarini, F., Tham, T. N., Casanova, P., Renzana-Seisdedos, F. and Dubois-Dalcq, M. (2003). Role of the alpha-chemokine stromal cell-derived factor (SDF-1) in the developing and mature central nervous system. *Glia* **42**, 139–48.

Le Douarin, N. M. and Dupin, E. (1993). Cell lineage analysis in neural crest ontogeny. *J Neurobiol* **24**, 146–61.

Le Douarin, N. M. and Kalcheim, C. (1999). *The Neural Crest*. Cambridge University Press, Cambridge.

Le, N., Nagarajan, R., Wang, J. Y. T. et al. (2005a). Nab proteins are essential for peripheral nervous system myelination. *Nat Neurosci* **8**, 932–40.

Le, N., Nagarajan, R., Wang, J. Y. T., Araki, T., Schmidt, R. E. and Milbrandt, J. (2005b). Analysis of congenital hypomyelinating Egr2Lo/Lo nerves identifies Sox2 as an inhibitor of Schwann cell differentiation and myelination. *Proc Nat Acad Sci USA* **102**, 2596–601.

Lee, M. J., Brennan, A., Blanchard, A. et al. (1997). P0 is constitutively expressed in the rat neural crest and embryonic nerves and is negatively and positively regulated by axons to generate non-myelin-forming and myelin-forming Schwann cells, respectively. *Mol Cell Neurosci* **8**, 336–50.

Lefcort, F., Venstrom, K., McDonald, J. A. and Reichardt, L. F. (1992). Regulation of expression of fibronectin and its receptor, alpha 5 beta 1, during development and regeneration of peripheral nerve. *Development* **116**, 767–82.

Leimeroth, R., Lobsiger, C., Lussi, A., Taylor, V., Suter, U. and Sommer, L. (2002). Membrane-bound neuregulin1 type III actively promotes Schwann cell differentiation of multipotent progenitor cells. *Develop Biol* **246**, 245–58.

Lemke, G. (2001). Glial control of neuronal development. *Ann Rev Neurosci* **24**, 87–105.

Lemke, G. and Axel, R. (1985). Isolation and sequence of a cDNA encoding the major structural protein of peripheral myelin. *Cell* **40**, 501–8.

Lentz, S. I., Miner, J. H., Sanes, J. R. and Snider, W. D. (1997). Distribution of the ten known laminin chains in the pathways and targets of developing sensory axons. *J Comp Neurol* **378**, 547–61.

Leppert, D., Hughes, P., Huber, S. et al. (1999). Matrix metalloproteinase upregulation in chronic inflammatory demyelinating polyneuropathy and nonsystemic vasculitic neuropathy. *Neurology* **53**, 62–70.

Letinsky, M. S., Fischbeck, K. H. and McMahan, U. J. (1976). Precision of reinnervation of original postsynaptic sites in frog muscle after a nerve crush. *J Neurocytol* **5**, 691–718.

Levedakou, E. N., Chen, X. J., Soliven, B. and Popko, B. (2005). Disruption of the mouse Large gene in the enr and myd mutants results in nerve, muscle, and neuromuscular junction defects. *Mol Cell Neurosci* **28**, 757–69.

Lewis, J., Al-Ghaith, L., Swanson, G. and Khan, A. (1983). The control of axon outgrowth in the developing chick wing. *Prog Clin Biol Res* **110** (Part A), 195–205.

Lewis, R. A. and Shy, M. E. (1999). Electrodiagnostic findings in CMTX: a disorder of the Schwann cell and peripheral nerve myelin. *Ann NY Acad Sci* **883**, 504–7.

Li, J., Krajewski, K., Shy, M. E. and Lewis, R. A. (2002a). Hereditary neuropathy with liability to pressure palsy: the electrophysiology fits the name. *Neurology* **58**, 1769–73.

Li, J., Krajewski, K., Lewis, R. A. and Shy, M. E. (2004). Loss-of-function phenotype of hereditary neuropathy with liability to pressure palsies. *Muscle Nerve* **29**, 205–10.

Li, J., Bai, Y., Ghandour, K., Qin, P. et al. (2005a). Skin biopsies in myelin-related neuropathies: bringing molecular pathology to the bedside. *Brain* **128**, 1168–77.

Li, J., Grandis, M., Trostinskaia, A. et al. 32765. Intralaminar protein accumulation of myelin and axonal degeneration in a human MPZ mutation: an autopsy study.

Li, L., Xian, C. J., Zhong, J. H. and Zhou, X. F. (2003). Lumbar 5 ventral root transection-induced upregulation of nerve growth factor in sensory neurons and their target tissues: a mechanism in neuropathic pain. *Mol Cell Neurosci* **23**, 232–50.

Li, Q., Shirabe, K., Thisse, C. et al. (2005b). Chemokine signaling guides axons within the retina in zebrafish. *J Neurosci* **25**, 1711–17.

Li, S., Liquari, P., McKee, K. K. et al. (2005c). Laminin-sulfatide binding initiates basement membrane assembly and enables receptor signaling in Schwann cells and fibroblasts. *J Cell Biol* **169**, 179–89.

Li, X., Lynn, B. D., Olson, C. et al. (2002b). Connexin29 expression, immunocytochemistry and freeze-fracture replica immunogold labelling (FRIL) in sciatic nerve. *Eur J Neurosci* **16**, 795–806.

Li, X., Gonias, S. L. and Campana, W. M. (2005d). Schwann cells express erythropoietin receptor and represent a major target for Epo in peripheral nerve injury. *Glia* **51**, 254–65.

Li, J., Bai, Y. and Ianakova, E. et al. (2006). Major myelin protein gene (P0) mutation causes a novel form of axonal degeneration. *J Comp Neurol* **498**, 252–65.

Lichtman, J. W. and Sanes, J. R. (2003). Watching the neuromuscular junction. *J Neurocytol* **32**, 767–75.

Lichtman, J. W., Magrassi, L. and Purves, D. (1987). Visualization of neuromuscular junctions over periods of several months in living mice. *J Neurosci* **7**, 1215–22.

Lilje, O. (2002). The processing and presentation of endogenous and exogenous antigen by Schwann cells in vitro. *Cell Mol Life Sci* **59**, 2191–8.

Lilje, O. and Armati, P. J. (1997). The distribution and abundance of MHC and ICAM-1 on Schwann cells in vitro. *J Neuroimmunol* **77**, 75–84.

Lilje, O. and Armati, P. J. (1999). Restimulation of resting autoreactive T cells by Schwann cells in vitro. *Exp Mol Pathol* **67**, 164–74.

Lin, C., Numakura, C., Ikegami, T. et al. (1999). Deletion and nonsense mutations of the connexin 32 gene associated with Charcot–Marie–Tooth disease. *Tohoku J Exp Med* **188**, 239–44.

Lin, W., Sanchez, H. B., Deerinck, T., Morris, J. K., Ellisman, M. and Lee, K. F. (2000). Aberrant development of motor axons and neuromuscular synapses in erbB2-deficient mice. *Proc Nat Acad Sci USA* **97**, 1299–304.

Linden, D. C., Jerian, S. M. and Letinsky, M. S. (1988). Neuromuscular junction development in the cutaneous pectoris muscle of *Rana catesbeiana*. *Exp Neurol* **99**, 735–60.

Lindholm, D., Heumann, R., Meyer, M. and Thoenen, H. (1987). Interleukin-1 regulates synthesis of nerve growth factor in non-neuronal cells of rat sciatic nerve. *Nature* **330**, 658–9.

Linington, C., Izumo, S., Suzuki, M., Uyemura, K., Meyermann, R. and Wekerle, H. (1984). A permanent rat T cell line that mediates experimental allergic neuritis in the Lewis rat in vivo. *J Immunol* **133**, 1946–50.

Linington, C., Bradl, M., Lassmann, H., Brunner, C. and Vass, K. (1988). Augmentation of demyelination in rat acute allergic encephalomyelitis by circulating mouse monoclonal antibodies directed against a myelin/oligodendrocyte glycoprotein. *Am J Pathol* **130**, 443−54.

Linington, C., Lassmann, H., Ozawa, K., Kosin, S. and Mongan, L. (1992). Cell adhesion molecules of the immunoglobulin supergene family as tissue-specific autoantigens: induction of experimental allergic neuritis (EAN) by P0 protein-specific T cell lines. *Eur J Immunol* **22**, 1813−17.

Linington, C., Berger, T., Perry, L. et al. (1993). T cells specific for the myelin oligodendrocyte glycoprotein mediate an unusual autoimmune inflammatory response in the central nervous system. *Eur J Immunol* **23**, 1364−72.

Lisak, R. and Brown, M.J. (1987). Acquired demyelinating polyneuropathies. *Semin Neurol* **7**, 40−8.

Lisak, R. and Bealmear, B. (1991). Antibodies to interleukin-1 inhibit cytokine-induced proliferation of neonatal rat Schwann cells in vitro. *J Neuroimmunol* **31**, 123−32.

Lisak, R. and Bealmear, B. (1992). Differences in the capacity of gamma-interferons from different species to induce class I and II major histocompatibility complex antigens on neonatal rat Schwann cells in vitro. *Pathobiology* **60**, 322−9.

Lisak, R. and Bealmear, B. (1994). Antibodies to interleukin-6 inhibit Schwann cell proliferation induced by unfractionated cytokines. *J Neuroimmunol* **50**, 127−32.

Lisak, R. and Bealmear, B. (1995). Transforming growth factor-β (TGF-β) is co-mitogenic for Schwann cells (SC) with interleukin-α (IL-α). *Neurology* **45**, A164−A165.

Lisak, R. and Bealmear, B. (1997). Upregulation of intercellular adhesion molecule-1 (ICAM-1) on rat Schwann cells in vitro: comparison of interferon-gamma, tumor necrosis factor-alpha and interleukin-1. *J Peripher Nerv Syst* **2**, 233−43.

Lisak, R., Hirayama, M., Kuchmy, D. et al. (1983). Cultured human and rat oligodendrocytes and rat Schwann cells do not have immune response gene associated antigen (Ia) on their surface. *Brain Res* **289**, 285−92.

Lisak, R., Kuchmy, D., Rmati-Gulson, P.J., Brown, M.J. and Sumner, A.J. (1984). Serum-mediated Schwann cell cytotoxicity in the Guillain−Barré syndrome. *Neurology* **34**, 1240−3.

Lisak, R., Sobue, G., Kuchmy, D., Burns, J.B. and Pleasure, D.E. (1985). Products of activated lymphocytes stimulate Schwann cell mitosis in vitro. *Neurosci Lett* **57**, 105−11.

Lisak, R., Bealmear, B. and Ragheb, S. (1994). Interleukin-1 alpha, but not interleukin-1 beta, is a co-mitogen for neonatal rat Schwann cells in vitro and acts via interleukin-1 receptors. *J Neuroimmunol* **55**, 171−7.

Lisak, R., Bealmear, B., Benjamins, J., Yu, C. and Skoff, R. (1996). Transforming growth factor-β (TGF-β) has different effects on proliferation of neonatal central (CNS) and peripheral (PNS) macroglia in vitro. *Neurology* **46**, A190.

Lisak, R., Skundric, D., Bealmear, B. and Ragheb, S. (1997). The role of cytokines in Schwann cell damage, protection, and repair. *J Infect Dis* **176** (Suppl 2), S173−S179.

Lisak, R., Bealmear, B., Benjamins, J. and Skoff, A. (1998). Inflammatory cytokines inhibit upregulation of glycolipid expression by Schwann cells in vitro. *Neurology* **51**, 1661−5.

Lisak, R., Bealmear, B., Benjamins, J. and Skoff, A. (1999). Tumor necrosis factor-α (TNF-α) upregulation of intracellular adhesion molecule-1 by Schwann cells is predominantly mediated by TNF receptor type I. *Soc Neurosci Abstr* **25**, 294.

Lisak, R., Bealmear, B., Benjamins, J. A. and Skoff, A. M. (2001). Interferon-gamma, tumor necrosis factor-alpha, and transforming growth factor-beta inhibit cyclic AMP-induced Schwann cell differentiation. *Glia* **36**, 354–63.

Lisak, R., Bealmear, B., Nedelkoska, L. and Benjamins, J. A. (2006). Secretory products of central nervous system glial cells induce Schwann cell proliferation and protect from cytokine-mediated death. *J Neurosci Res* **83**, 1425–31.

Liu, H., Nakagawa, T., Kanematsu, T., Uchida, T. and Tsuji, S. (1999). Isolation of 10 differentially expressed cDNAs in differentiated Neuro2a ceFlls induced through controlled expression of the GD3 synthase gene. *J Neurochem* **72**, 1781–90.

Lobsiger, C. S., Taylor, V. and Suter, U. (2002). The early life of a Schwann cell. *Biol Chem* **383**, 245–53.

Loeb, J. A., Khurana, T. S., Robbins, J. T., Yee, A. G. and Fischbach, G. D. (1999). Expression patterns of transmembrane and released forms of neuregulin during spinal cord and neuromuscular synapse development. *Development* **126**, 781–91.

Love, F. M. and Thompson, W. J. (1998). Schwann cells proliferate at rat neuromuscular junctions during development and regeneration. *J Neurosci* **18**, 9376–85.

Love, F. M. and Thompson, W. J. (1999). Glial cells promote muscle reinnervation by responding to activity-dependent postsynaptic signals. *J Neurosci* **19**, 10390–6.

Love, F. M., Son, Y. J. and Thompson, W. J. (2003). Activity alters muscle reinnervation and terminal sprouting by reducing the number of Schwann cell pathways that grow to link synaptic sites. *J Neurobiol* **54**, 566–76.

Lubischer, J. L. and Bebinger, D. M. (1999). Regulation of terminal Schwann cell number at the adult neuromuscular junction. *J Neurosci* **19**, RC46.

Lubischer, J. L. and Thompson, W. J. (1999). Neonatal partial denervation results in nodal but not terminal sprouting and a decrease in efficacy of remaining neuromuscular junctions in rat soleus muscle. *J Neurosci* **19**, 8931–44.

Luitjen, J. A. F. M. and Faille-Kuyper, E. H. B. (1972). The occurrence of IgM and complement factors along myelin sheaths of peripheral nerves. An immunohistochemical study of the Guillain–Barré syndrome. *J Neurol Sci* **15**, 219–24.

Lunn, M. P., Manji, H., Choudhary, P. P., Hughes, R. A. and Thomas, P. K. (1999). Chronic inflammatory demyelinating polyradiculoneuropathy: a prevalence study in south east England. *J Neurol Neurosurg Psychiatry* **66**, 677–80.

Lupski, J. R., de Oca-Luna, R. M., Slaugenhaupt, S. *et al.* (1991). DNA duplication associated with Charcot–Marie–Tooth disease type 1A. *Cell* **66**, 219–32.

Lustig, M., Zanazzi, G., Sakurai, T. *et al.* (2001). Nr–CAM and neurofascin interactions regulate ankyrin G and sodium channel clustering at the node of Ranvier. *Curr Biol* **11**, 1864–9.

Mackie, E. J. and Tucker, R. P. (1999). The tenascin-C knockout revisited. *J Cell Sci* **112** (Part 22), 3847–53.

Macleod, G. T., Dickens, P. A. and Bennett, M. R. (2001). Formation and function of synapses with respect to Schwann cells at the end of motor nerve terminal branches on mature amphibian (*Bufo marinus*) muscle. *J Neurosci* **21**, 2380–92.

Madrid, R.E., Jaros, E., Cullen, M.J. and Bradley, W.G. (1975). Genetically determined defect of Schwann cell basement membrane in dystrophic mouse. *Nature* **257**, 319–21.

Magda, P., Latov, N., Brannagan, T.H., III, Weimer, L.H., Chin, R.L. and Sander, H.W. (2003). Comparison of electrodiagnostic abnormalities and criteria in a cohort of patients with chronic inflammatory demyelinating polyneuropathy. *Arch Neurol* **60**, 1755–9.

Magyar, J.P., Martini, R., Ruelicke, T. *et al.* (1996). Impaired differentiation of Schwann cells in transgenic mice with increased PMP22 gene dosage. *J Neurosci* **16**, 5351–60.

Maier, M., Berger, P. and Suter, U. (2002). Understanding Schwann cell–neurone interactions: the key to Charcot–Marie–Tooth disease? *J Anat* **200**, 357–66.

Mancardi, G.L., Cadoni, A., Zicca, A. *et al.* (1988). HLA-DR Schwann cell reactivity in peripheral neuropathies of different origins. *Neurology* **38**, 848–51.

Mandich, P., Mancardi, G.L., Varese, A. *et al.* (1999). Congenital hypomyelination due to myelin protein zero Q215X mutation. *Ann Neurol* **45**, 676–8.

Marchionni, M.A., Goodearl, A.D.J., Chen, M.S. *et al.* (1993). Glial growth factors are alternatively spliced erbB2 ligands expressed in the nervous system. *Nature* **362**, 312–18.

Marcus, J. and Popko, B. (2002). Galactolipids are molecular determinants of myelin development and axo-glial organization. *BBA Gen Subjects* **1573**, 406–13.

Maro, G.S., Vermeren, M., Voiculescu, O. *et al.* (2004). Neural crest boundary cap cells constitute a source of neuronal and glial cells of the PNS. *Nat Neurosci* **7**, 930–8.

Marques, W., Jr., Thomas, P.K., Sweeney, M.G., Carr, L. and Wood, N.W. (1998). Dejerine–Sottas neuropathy and PMP22 point mutations: a new base pair substitution and a possible 'hot spot' on Ser72. *Ann Neurol* **43**, 680–3.

Martin, J.R. and Suzuki, S. (1987). Inflammatory sensory polyradiculopathy and reactivated peripheral nervous system infection in a genital herpes model. *J Neurol Sci* **79**, 155–71.

Martin, S., Levine, A.K., Chen, Z.J., Ughrin, Y. and Levine, J.M. (2001). Deposition of the NG2 proteoglycan at nodes of Ranvier in the peripheral nervous system. *J Neurosci* **21**, 8119–28.

Martini, R., Schachner, M. and Faissner, A. (1990). Enhanced expression of the extracellular matrix molecule J1/tenascin in the regenerating adult mouse sciatic nerve. *J Neurocytol* **19**, 601–16.

Masaki, T., Matsumura, K., Saito, F. *et al.* (2000). Expression of dystroglycan and laminin-2 in peripheral nerve under axonal degeneration and regeneration. *Acta Neuropathol (Berl)* **99**, 289–95.

Masaki, T., Matsumura, K., Hirata, A. *et al.* (2002). Expression of dystroglycan and the laminin-alpha 2 chain in the rat peripheral nerve during development. *Exp Neurol* **174**, 109–17.

Massague, J., Andres, J., Attisano, L. *et al.* (1992). TGF-beta receptors. *Mol Reprod Dev* **32**, 99–104.

Massicotte, C. and Scherer, S. (2004). *Neuropathies – Translating Causes into Treatments*. Waxman, S.G. (Ed.), Elsevier, Philadelphia, pp. 401–14.

Mata, M., Siegel, G.J., Hieber, V., Beaty, M.W. and Fink, D.J. (1991). Differential distribution of (Na,K)-ATPase alpha isoform mRNAs in the peripheral nervous system. *Brain Res* **546**, 47–54.

Mathey, E.K., Pollard, J.D. and Armati, P.J. (1999). TNF alpha, IFN gamma and IL-2 mRNA expression in CIDP sural nerve biopsies. *J Neurol Sci* **163**, 47–52.

Matsumoto, K., Sawa, H., Sato, M., Orba, Y., Nagashima, K. and Ariga, H. (2002). Distribution of extracellular matrix tenascin-X in sciatic nerves. *Acta Neuropathol (Berl)* **104**, 448–54.

Maurel, P. and Salzer, J. L. (2000). Axonal regulation of Schwann cell proliferation and survival and the initial events of myelination requires PI 3-kinase activity. *J Neurosci* **20**, 4635–45.

Maurer, M., Toyka, K. V. and Gold, R. (2002). Cellular immunity in inflammatory autoimmune neuropathies. *Rev Neurol (Paris)* **158**, S7–S15.

McCarty, J. H., Lacy Hulbert, A., Charest, A. *et al.* (2004). Selective ablation of {alpha}v integrins in the central nervous system leads to cerebral hemorrhage, seizures, axonal degeneration and premature death. *Development*.

McCombe, P. A., Clark, P., Frith, J. A. *et al.* (1985). Alpha-1 antitrypsin phenotypes in demyelinating disease: an association between demyelinating disease and the allele PiM3. *Ann Neurol* **18**, 514–16.

McCombe, P. A., McManis, P. G., Frith, J. A., Pollard, J. D. and McLeod, J. G. (1987a). Chronic inflammatory demyelinating polyradiculoneuropathy associated with pregnancy. *Ann Neurol* **21**, 102–4.

McCombe, P. A., Pollard, J. D. and McLeod, J. G. (1987b). Chronic inflammatory demyelinating polyradiculoneuropathy. A clinical and electrophysiological study of 92 cases. *Brain* **110** (Part 6), 1617–30.

McGrath, K. E., Koniski, A. D., Maltby, K. M., McGann, J. K. and Palis, J. (1999). Embryonic expression and function of the chemokine SDF-1 and its receptor, CXCR4. *Dev Biol* **213**, 442–56.

McKhann, G. M., Cornblath, D. R., Ho, T. *et al.* (1991). Clinical and electrophysiological aspects of acute paralytic disease of children and young adults in northern China. *Lancet* **338**, 593–7.

McKhann, G. M., Cornblath, D., Griffin, J. W. *et al.* (1999). Acute motor axonal neuropathy – a frequent cause of acute flaccid paralysis in China. *Ann Neurol* **33**, 333–42.

McLennan, I. S. and Koishi, K. (2002). The transforming growth factor-betas: multifaceted regulators of the development and maintenance of skeletal muscles, motoneurons and Schwann cells. *Int J Dev Biol* **46**, 559–67.

McLeod, J. G., Pollard, J. D., Macaskill, P., Mohamed, A., Spring, P. and Khurana, V. (1999). Prevalence of chronic inflammatory demyelinating polyneuropathy in New South Wales, Australia. *Ann Neurol* **46**, 910–13.

McMahan, U. J. (1990). The agrin hypothesis. *Cold Spring Harb Symp Quant Biol* **55**, 407–18.

Medzhitov, R. and Janeway, C., Jr. (2000). Innate immunity. *N Engl J Med* **343**, 338–44.

Meier, C., Parmantier, E., Brennan, A., Mirsky, R. and Jessen, K. R. (1999). Developing Schwann cells acquire the ability to survive without axons by establishing an autocrine circuit involving insulin-like growth factor, neurotrophin-3, and platelet-derived growth factor-BB. *J Neurosci* **19**, 3847–59.

Meier, C., Dermietzel, R., Davidson, K. G. V., Yasumura, T. and Rash, J. E. (2004). Connexin32-containing gap junctions in Schwann cells at the internodal zone of partial myelin compaction and in Schmidt–Lanterman incisures. *J Neurosci* **24**, 3186–98.

Meintanis, S., Thomaidou, D., Jessen, K. R., Mirsky, R. and Matsas, R. (2001). The neuron-glia signal beta-neuregulin promotes Schwann cell motility via the MAPK pathway. *Glia* **34**, 39–51.

Melcangi, R.C., Cavarretta, I.T.R., Ballabio, M. et al. (2005). Peripheral nerves: a target for the action of neuroactive steroids. *Brain Res Rev* **48**, 328−38.

Melendez-Vasquez, C., Redford, J., Choudhary, P.P. et al. (1997). Immunological investigation of chronic inflammatory demyelinating polyradiculoneuropathy. *J Neuroimmunol* **73**, 124−34.

Melendez-Vasquez, C.V., Rios, J.C., Zanazzi, G., Lambert, S., Bretscher, A. and Salzer, J.L. (2001). Nodes of Ranvier form in association with ezrin-radixin-moesin (ERM)-positive Schwann cell processes. *Proc Nat Acad Sci USA* **98**, 1235−40.

Melendez-Vasquez, C.V., Einheber, S. and Salzer, J.L. (2004). Rho kinase regulates Schwann cell myelination and formation of associated axonal domains. *J Neurosci* **24**, 3953−63.

Menichella, D.M., Arroyo, E.J., Awatramani, R. et al. (2001). Protein zero is necessary for e-cadherin-mediated adherens junction formation in schwann cells. *Mol Cell Neurosci* **18**, 606−18.

Menichella, D.M., Goodenough, D.A., Sirkowski, E., Scherer, S.S. and Paul, D.L. (2003). Connexins are critical for normal myelination in the CNS. *J Neurosci* **23**, 5963−73.

Metral, S., Raphael, J.C., Hort-Legrand, C.I. and Elkharrat, D. (1989). Serum demyelinating activity and Guillain−Barré syndrome: favorable effect of plasma exchange. *Rev Neurol (Paris)* **145**, 312−19.

Mews, M. and Meyer, M. (1993). Modulation of Schwann cell phenotype by TGF-beta 1: inhibition of P0 mRNA expression and downregulation of the low affinity NGF receptor. *Glia* **8**, 208−17.

Meyer, D., Yamaai, T., Garratt, A. et al. (1997). Isoform-specific expression and function of neuregulin. *Development* **124**, 3575−86.

Michailov, G.V., Sereda, M.W., Brinkmann, B.G. et al. (2004). Axonal neuregulin-1 regulates myelin sheath thickness. *Science* **304**, 700−3.

Miledi, R. and Slater, C.R. (1968). Electrophysiology and electron-microscopy of rat neuromuscular junctions after nerve degeneration. *Proc R Soc Lond B Biol Sci* **169**, 289−306.

Miledi, R. and Slater, C.R. (1970). On the degeneration of rat neuromuscular junctions after nerve section. *J Physiol* **207**, 507−28.

Miller, D.J., Njenga, M.K., Parisi, J.E. and Rodriguez, M. (1996). Multi-organ reactivity of a monoclonal natural autoantibody that promotes remyelination in a mouse model of multiple sclerosis. *J Histochem Cytochem* **44**, 1005−11.

Miller, K.E., Richards, B.A. and Kriebel, R.M. (2002). Glutamine-, glutamine synthetase-, glutamate dehydrogenase- and pyruvate carboxylase-immunoreactivities in the rat dorsal root ganglion and peripheral nerve. *Brain Res* **945**, 202−11.

Milner, P., Lovelidge, C.A., Taylor, W.A. and Hughes, R.A. (1987). P0 myelin protein produces experimental allergic neuritis in Lewis rats. *J Neurol Sci* **79**, 275−85.

Milner, R., Wilby, M., Nishimura, S. et al. (1997). Division of labor of Schwann cell integrins during migration on peripheral nerve extracellular matrix ligands. *Dev Biol* **185**, 215−28.

Miner, J.H. and Yurchenco, P.D. (2004). Laminin functions in tissue morphogenesis. *Ann Rev Cell Dev Biol* **20**, 255−84.

Mirsky, R. and Jessen, K.R. (1983). A cell surface protein of astrocytes, Ran-2, distinguishes non-myelin-forming Schwann cells from myelin-forming Schwann cells. *Dev Neurosci* **6**, 304−16.

Mirsky, R. and Jessen, K.R. (1999). The neurobiology of Schwann cells. *Brain Pathol* **9**, 293−311.

Mirsky, R. and Jessen, K. R. (2005). Molecular signaling in Schwann cell development. Dyck, P. J., Thomas, P. K. (Eds.) *Peripheral Neuropathy*, 4th Edn. Elsevier, Philadelphia, pp. 341–76.

Mirsky, R. and Jessen, K. R. (1996). Schwann cell development, differentiation and myelination. *Curr Opin Neurobiol* **6**, 89–96.

Mirsky, R., Stewart, H. J., Tabernero, A. *et al.* (1996). Development and differentiation of Schwann cells. *Rev Neurol (Paris)* **152**, 308–13.

Mirsky, R., Parkinson, D. B., Dong, Z. *et al.* (2001). Regulation of genes involved in Schwann cell development and differentiation. *Prog Brain Res* **132**, 3–11.

Mithen, F. A., Colburn, S. and Birchem, R. (1990). Human alpha tumor necrosis factor does not damage cultures containing rat Schwann cells and sensory neurons. *Neurosci Res* **9**, 59–63.

Mithen, F. A., Ilyas, A. A., Birchem, R. and Cook, S. D. (1992). Effects of Guillain–Barré sera containing antibodies against glycolipids in cultures of rat Schwann cells and sensory neurons. *J Neurol Sci* **112**, 223–32.

Modlin, R. L. (2002). Learning from leprosy: insights into contemporary immunology from an ancient disease. *Skin Pharmacol Appl Skin Physiol* **15**, 1–6.

Mokuno, K., Sobue, G., Reddy, U. R. *et al.* (1988). Regulation of Schwann cell nerve growth factor receptor by cyclic adenosine 3′,5′-monophosphate. *J Neurosci Res* **21**, 465–72.

Mollaaghababa, R. and Pavan, W. J. (2003). The importance of having your SOX on: role of SOX10 in the development of neural crest-derived melanocytes and glia. *Oncogene* **22**, 3024–34.

Morgan, L., Jessen, K. R. and Mirsky, R. (1991). The effects of cAMP on differentiation of cultured Schwann cells: progression from an early phenotype (04+) to a myelin phenotype (P0+, GFAP-, N-CAM-, NGF-receptor-) depends on growth inhibition. *J Cell Biol* **112**, 457–67.

Mori, K., Chano, T., Yamamoto, K., Matsusue, Y. and Okabe, H. (2004). Expression of macrophage inflammatory protein-1alpha in Schwann cell tumors. *Neuropathology* **24**, 131–5.

Morris, J. K., Lin, W., Hauser, C., Marchuk, Y., Getman, D. and Lee, K. F. (1999). Rescue of the cardiac defect in ErbB2 mutant mice reveals essential roles of ErbB2 in peripheral nervous system development. *Neuron* **23**, 273–83.

Morrison, S. J., White, P. M., Zock, C. and Anderson, D. J. (1999). Prospective identification, isolation by flow cytometry, and in vivo self-renewal of multipotent mammalian neural crest stem cells. *Cell* **96**, 737–49.

Morrison, S. J., Perez, S. E., Qiao, Z. *et al.* (2000). Transient Notch activation initiates an irreversible switch from neurogenesis to gliogenesis by neural crest stem cells. *Cell* **101**, 499–510.

Morrissey, T. K., Levi, A. D. O., Nuijens, A., Sliwkowski, M. X. and Bunge, R. P. (1995). Axon-induced mitogenesis of human Schwann cells involves heregulin and p185erbB2. *Proc Nat Acad Sci USA* **92**, 1431–5.

Muntoni, F., Brockington, M., Torelli, S. and Brown, S. C. (2004). Defective glycosylation in congenital muscular dystrophies. *Curr Opin Neurol* **17**, 205–9.

Murata, K. and Dalakas, M. C. (2000). Expression of the co-stimulatory molecule BB-1, the ligands CTLA-4 and CD28 and their mRNAs in chronic inflammatory demyelinating polyneuropathy. *Brain* **123** (Part 8), 1660–6.

Murnane, R. D., Ahern-Rindell, A. J. and Prieur, D. J. (1991). Ultrastructural lesions of ovine GM1 gangliosidosis. *Mod Pathol* **4**, 755–62.

Murwani, R., Hodgkinson, S. and Armati, P. (1996). Tumor necrosis factor alpha and interleukin-6 mRNA expression in neonatal Lewis rat Schwann cells and a neonatal rat Schwann cell line following interferon gamma stimulation. *J Neuroimmunol* **71**, 65–71.

Musarella, M., Alcaraz, G., Caillol, G., Boudier, J. L., Couraud, F. and Autillo-Touati, A. (2006). Expression of Nav1.6 sodium channels by Schwann cells at neuromuscular junctions: role in the motor endplate disease phenotype. *Glia* **53**, 13–23.

Nacimiento, W., Schoen, S. W., Nacimiento, A. C. and Kreutzberg, G. W. (1991). Cytochemistry of 5′-nucleotidase in the superior cervical ganglion of cat and guinea pig. *Brain Res* **567**, 283–9.

Nagarajan, R., Svaren, J., Le, N., Araki, T., Watson, M. and Milbrandt, J. (2001). EGR2 mutations in inherited neuropathies dominant-negatively inhibit myelin gene expression. *Neuron* **30**, 355–68.

Nakagawa, M., Miyagoe-Suzuki, Y., Ikezoe, K. *et al.* (2001a). Schwann cell myelination occurred without basal lamina formation in laminin alpha2 chain-null mutant (dy3K/dy3K) mice. *Glia* **35**, 101–10.

Nakagawa, M., Takashima, H., Umehara, F. *et al.* (2001b). Clinical phentoype in X-linked Charcot–Marie–Tooth disease with an entire deletion of the connexin 32 coding sequence. *J Neurol Sci* **185**, 31–6.

Nardelli, E., Bassi, A., Mazzi, G., Anzini, P. and Rizzuto, N. (1995). Systemic passive transfer studies using IgM monoclonal antibodies to sulfatide. *J Neuroimmunol* **63**, 29–37.

Navon, R., Seifried, B., Gal-On, N. S. and Sadeh, M. (1996). A new point mutation affecting the fourth transmembrane domain of PMP22 results in severe de novo Charcot–Marie–Tooth disease. *Hum Genet* **97**, 685–7.

Nedergaard, M., Ransom, B. and Goldman, S. A. (2003). New roles for astrocytes: redefining the functional architecture of the brain. *Trends Neurosci* **26**, 523–30.

Nelis, E., Timmerman, V., De Jonghe, P. and Van Broeckhoven, C. (1994). Identification of a 5′ splice site mutation in the PMP-22 gene in autosomal dominant Charcot–Marie–Tooth disease type 1. *Hum Mol Genet* **3**, 515–16.

Nelis, E., Van Broeckhoven, C., De Jonghe, P. *et al.* (1996). Estimation of the mutation frequencies in Charcot–Marie–Tooth disease type 1 and hereditary neuropathy with liability to pressure palsies: a European collaborative study. *Eur J Hum Genet* **4**, 25–33.

Newman, E. A. and Volterra, A. (2004). Glial control of synaptic function. *Glia* **47**, 207–8.

Nguyen, Q. T., Sanes, J. R. and Lichtman, J. W. (2002). Pre-existing pathways promote precise projection patterns. *Nat Neurosci* **5**, 861–7.

Nicholson, G. and Corbett, A. (1996). Slowing of central conduction in X-linked Charcot–Marie–Tooth neuropathy shown by brain stem auditory evoked responses. *J Neurol Neurosurg Psychiatry* **61**, 43–6.

Nicholson, G. and Nash, J. (1993). Intermediate nerve conduction velocities define X-linked Charcot–Marie–Tooth neuropathy families. *Neurology* **43**, 2558–64.

Nicholson, G. A., Valentijn, L. J., Cherryson, A. K. *et al.* (1994). A frame shift mutation in the PMP22 gene in hereditary neuropathy with liability to pressure palsies. *Nat Genet* **6**, 263–6.

Nickols, J. C., Valentine, W., Kanwal, S. and Carter, B. D. (2003). Activation of the transcription factor NF-kappaB in Schwann cells is required for peripheral myelin formation. *Nature Neuroscience* **6**, 161–7.

Niessen, C. M., Cremona, O., Daams, H., Ferraresi, S., Sonnenberg, A. and Marchisio, P. C. (1994). Expression of the integrin α6β4 in peripheral nerves: localization in Schwann and perineurial cells and different variants of the β4 subunit. *J Cell Sci* **107**, 543–52.

Noakes, P. G. and Bennett, M. R. (1987). Growth of axons into developing muscles of the chick forelimb is preceded by cells that stain with Schwann cell antibodies. *J Comp Neurol* **259**, 330–47.

Noakes, P. G., Bennett, M. R. and Stratford, J. (1988). Migration of Schwann cells and axons into developing chick forelimb muscles following removal of either the neural tube or the neural crest. *J Comp Neurol* **277**, 214–33.

Nolano, M., Provitera, V., Crisci, C. et al. (2003). Quantification of myelinated endings and mechanoreceptors in human digital skin. *Ann Neurol* **54**, 197–205.

Nyland, H., Matre, R. and Mork, S. (1981). Immunological characterization of sural nerve biopsies from patients with Guillain–Barré syndrome. *Ann Neurol* **9** (Suppl), 80–6.

O'Hanlon, G. M., Plomp, J. J., Chakrabarti, M. et al. (2001). Anti-GQ1b ganglioside antibodies mediate complement-dependent destruction of the motor nerve terminal. *Brain* **124**, 893–906.

O'Malley, J. P., Waran, M. T. and Balice-Gordon, R. J. (1999). In vivo observations of terminal Schwann cells at normal, denervated, and reinnervated mouse neuromuscular junctions. *J Neurobiol* **38**, 270–86.

O'Reilly, M. S., Boehm, T., Shing, Y. et al. (1997). Endostatin: an endogenous inhibitor of angiogenesis and tumor growth. *Cell* **88**, 277–85.

Occhi, S., Zambroni, d., Del Carro, U. et al. (2005). Both laminin and Schwann cell dystroglycan are necessary for proper clustering of sodium channels at nodes of Ranvier. *Journal of Neuroscience.* **25**, 9418–27.

Odenthal, U., Haehn, S., Tunggal, P. et al. (2004). Molecular analysis of laminin N-terminal domains mediating self-interactions. *J Biol Chem* **279**, 44504–12.

Oh, S., Ri, Y., Bennett, M. V., Trexler, E. B., Verselis, V. K. and Bargiello, T. A. (1997). Changes in permeability caused by connexin 32 mutations underlie X-linked Charcot–Marie–Tooth disease. *Neuron* **19**, 927–38.

Oh, S. J., Kurokawa, K., de Almeida, D. F., Ryan, H. F., Jr. and Claussen, G. C. (2003). Subacute inflammatory demyelinating polyneuropathy. *Neurology* **61**, 1507–12.

Oliveira, R. B., Ochoa, M. T., Sieling, P. A. et al. (2003). Expression of Toll-like receptor 2 on human Schwann cells: a mechanism of nerve damage in leprosy. *Infect Immun* **71**, 1427–33.

Oliveira, R. B., Sampaio, E. P., Aarestrup, F. et al. (2005). Cytokines and *Mycobacterium leprae* induce apoptosis in human Schwann cells. *J Neuropathol Exp Neurol* **64**, 882–90.

Olsson, T., Holmdahl, R., Klareskog, L. and Forsum, U. (1983). Ia-expressing cells and T lymphocytes of different subsets in peripheral nerve tissue during experimental allergic neuritis in Lewis rats. *Scand J Immunol* **18**, 339–43.

Omori, Y., Mesnil, M. and Yamasaki, H. (1996). Connexin 32 mutations from X-linked Charcot–Marie–Tooth disease patients: functional defects and dominant negative effects. *Mol Biol Cell* **7**, 907–16.

Oomes, P. G., van der Meche, F. G., Markus-Silvis, L., Meulstee, J. and Kleyweg, R. P. (1991). In vivo effects of sera from Guillain–Barré subgroups: an electrophysiological and histological study on rat nerves. *Muscle Nerve* **14**, 1013–20.

Orlikowski, D., Chazaud, B., Plonquet, A. et al. (2003). Monocyte chemoattractant protein 1 and chemokine receptor CCR2 productions in Guillain–Barré syndrome and experimental autoimmune neuritis. *J Neuroimmunol* **134**, 118–27.

Ota, K., Irie, H. and Takahashi, K. (1987). T cell subsets and Ia-positive cells in the sciatic nerve during the course of experimental allergic neuritis. *J Neuroimmunol* **13**, 283–92.

Ottani, V., Martini, D., Franchi, M., Ruggeri, A. and Raspanti, M. (2002). Hierarchical structures in fibrillar collagens. *Micron* **33**, 587–96.

Otten, U., Ehrhard, P. and Peck, R. (1989). Nerve growth factor induces growth and differentiation of human B lymphocytes. *Proc Nat Acad Sci USA* **86**, 10059–63.

Palade, G.E. and Palay, S.L. (1954). Electron microscope observations of interneuronal and neuromuscular synapses. *Anat Rec* **118**, 335–6.

Palm, S.L. and Furcht, L.T. (1983). Production of laminin and fibronectin by Schwannoma cells: cell–protein interactions in vitro and protein localization in peripheral nerve in vivo. *J Cell Biol* **96**, 1218–26.

Palumbo, C., Massa, R., Panico, M.B. et al. (2002). Peripheral nerve extracellular matrix remodeling in Charcot–Marie–Tooth type I disease. *Acta Neuropathol (Berl)* **104**, 287–96.

Panas, M., Kalfakis, N., Karadimas, C. and Vassilopoulos, D. (2001). Episodes of generalized weakness in two sibs with the C164T mutation of the connexin 32 gene. *Neurology* **57**, 1906–8.

Paratore, C., Goerich, D.E., Suter, U., Wegner, M. and Sommer, L. (2001). Survival and glial fate acquisition of neural crest cells are regulated by an interplay between the transcription factor Sox10 and extrinsic combinatorial signaling. *Development* **128**, 3949–61.

Pareek, S., Notterpek, L., Snipes, G.J. et al. (1997). Neurons promote the translocation of peripheral myelin protein 22 into myelin. *J Neurosci* **17**, 7754–62.

Pareyson, D., Taroni, F., Botti, S. et al. (2000). Cranial nerve involvement in CMT disease type 1 due to early growth response 2 gene mutation. *Neurology* **54**, 1696–8.

Parkinson, D.B., Dong, Z., Bunting, H. et al. (2001). Transforming growth factor beta (TGFbeta) mediates Schwann cell death in vitro and in vivo: examination of c-Jun activation, interactions with survival signals, and the relationship of TGFbeta-mediated death to Schwann cell differentiation. *J Neurosci* **21**, 8572–85.

Parkinson, D.B., Langner, K., Namini, S.S., Jessen, K.R. and Mirsky, R. (2002). beta-neuregulin and autocrine mediated survival of Schwann cells requires activity of Ets family transcription factors. *Mol Cell Neurosci* **20**, 154–67.

Parkinson, D.B., Bhaskaran, A., Droggiti, A. et al. (2004). Krox-20 inhibits Jun-NH2-terminal kinase/c-Jun to control Schwann cell proliferation and death. *J Cell Biol* **164**, 385–94.

Parkinson, D.B., Bhaskaran, A., Mirsky, R. and Jessen, K.R. (2005) Regulation of the myelinating phenotype of Schwann cells by Krox-20. Medimond International Proc. VII Eur. Meeting on Glial Cell Function in Health and Disease. Amsterdam, pp. 139–43.

Parmantier, E., Lynn, B., Lawson, D. et al. (1999). Schwann cell-derived desert hedgehog controls the development of peripheral nerve sheaths. *Neuron* **23**, 713–24.

Patton, B.L. (2003). Basal lamina and the organization of neuromuscular synapses. *J Neurocytol* **32**, 883–903.

Patton, B.L., Miner, J.H., Chiu, A.Y. and Sanes, J.R. (1997). Distribution and function of laminins in the neuromuscular system of developing, adult, and mutant mice. *J Cell Biol* **139**, 1507–21.

Patton, B. L., Chiu, A. Y. and Sanes, J. R. (1998). Synaptic laminin prevents glial entry into the synaptic cleft. *Nature* **393**, 698–701.

Paulson, H. L., Garbern, J. Y., Hoban, T. F. *et al.* (2002). Transient central nervous system white matter abnormality in X-linked Charcot–Marie–Tooth disease. *Ann Neurol* **52**, 429–34.

Pedrola, L., Espert, A., Wu, X., Claramunt, R., Shy, M. E. and Palau, F. (2005). GDAP1, the protein causing Charcot–Marie–Tooth disease type 4A, is expressed in neurons and is associated with mitochondria. *Hum Mol Genet* **14**, 1087 94.

Pelidou, S. H., Deretzi, G., Zou, L. P., Quiding, C. and Zhu, J. (1999). Inflammation and severe demyelination in the peripheral nervous system induced by the intraneural injection of recombinant mouse interleukin-12. *Scand J Immunol* **50**, 39–44.

Peltonen, J., Jaakkola, S., Hsiao, L. L., Timpl, R., Chu, M. L. and Uitto, J. (1990). Type VI collagen. In situ hybridizations and immunohistochemistry reveal abundant mRNA and protein levels in human neurofibroma, schwannoma and normal peripheral nerve tissues. *Lab Invest* **62**, 487–92.

Peng, H. B., Yang, J. F., Dai, Z. *et al.* (2003). Differential effects of neurotrophins and Schwann cell-derived signals on neuronal survival/growth and synaptogenesis. *J Neurosci* **23**, 5050–60.

Pereira, R. A., Tscharke, D. C. and Simmons, A. (1994). Upregulation of class I major histocompatibility complex gene expression in primary sensory neurons, satellite cells, and Schwann cells of mice in response to acute but not latent herpes simplex virus infection in vivo. *J Exp Med* **180**, 841–50.

Pereira, R. M., Calegari-Silva, T. C., Hernandez, M. O. *et al.* (2005). Mycobacterium leprae induces NF-kappaB-dependent transcription repression in human Schwann cells. *Biochem Biophys Res Commun* **335**, 20–6.

Perrin, F. E., Lacroix, S., Viles-Trigueros, M. and David, S. (2005). Involvement of monocyte chemoattractant protein-1, macrophage inflammatory protein-1alpha and interleukin-1beta in Wallerian degeneration. *Brain* **128**, 854–66.

Personius, K. E. and Sawyer, R. P. (2005). Terminal Schwann cell structure is altered in diaphragm of mdx mice. *Muscle Nerve* **32**, 656–63.

Pfrieger, F. W. and Barres, B. A. (1997). Synaptic efficacy enhanced by glial cells in vitro. *Science* **277**, 1684–7.

Pietri, T., Eder, O., Breau, M. A. *et al.* (2004). Conditional beta 1-integrin gene deletion in neural crest cells causes severe developmental alterations of the peripheral nervous system. *Development* **131**, 3871–83.

Pingault, V., Guiochon-Mantel, A., Bondurand, N. *et al.* (2000). Peripheral neuropathy with hypomyelination, chronic intestinal pseudo-obstruction and deafness: a developmental 'neural crest syndrome' related to a SOX10 mutation. *Ann Neurol* **48**, 671–6.

Plante-Bordeneuve, V., Guiochon-Mantel, A., Lacroix, C., Lapresle, J. and Said, G. (1999). The Roussy–Levy family: from the original description to the gene. *Ann Neurol* **46**, 770–3.

Plomp, J. J., Molenaar, P. C., O'Hanlon, G. M. *et al.* (1999). Miller–Fisher anti-GQ1b antibodies: alpha-latrotoxin-like effects on motor end plates. *Ann Neurol* **45**, 189–99.

Podratz, J. L., Rodriguez, E. H., Di Nonno, E. S. and Windebank, A. J. (1998). Myelination by Schwann cells in the absence of extracellular matrix assembly. *Glia* **23**, 383–8.

Podratz, J. L., Rodriguez, E. and Windebank, A. J. (2001). Role of the extracellular matrix in myelination of peripheral nerve. *Glia* **35**, 35–40.

Podratz, J.L., Rodriguez, E.H. and Windebank, A.J. (2004). Antioxidants are necessary for myelination of dorsal root ganglion neurons, in vitro. *Glia* **45**, 54–8.

Poliak, S. and Peles, E. (2003). The local differentiation of myelinated axons at nodes of Ranvier. *Nat Rev Neurosci* **4**, 968–80.

Poliak, S., Salomon, S., Elhanany, H. *et al.* (2003). Juxtaparanodal clustering of *Shaker*-like K^+ channels in myelinated axons depends on Caspr2 and TAG-1. *J Cell Biol* **162**, 1149–60.

Pollard, J.D. (1994). Chronic inflammatory demyelinating polyradiculoneuropathy. *Baillieres Clin Neurol* **3**, 107–27.

Pollard, J.D. (2002). Chronic inflammatory demyelinating polyradiculoneuropathy. *Curr Opin Neurol* **15**, 279–83.

Pollard, J.D., McLeod, J.G., Gatenby, P. and Kronenberg, H. (1983). Prediction of response to plasma exchange in chronic relapsing polyneuropathy. A clinico-pathological correlation. *J Neurol Sci* **58**, 269–87.

Pollard, J.D., McCombe, P.A., Baverstock, J., Gatenby, P.A. and McLeod, J.G. (1986). Class II antigen expression and T lymphocyte subsets in chronic inflammatory demyelinating polyneuropathy. *J Neuroimmunol* **13**, 123–34.

Pollard, J.D., Baverstock, J. and McLeod, J.G. (1987). Class II antigen expression and inflammatory cells in the Guillain–Barré syndrome. *Ann Neurol* **21**, 337–41.

Pollard, J.D., Westland, K.W., Harvey, G.K. *et al.* (1995). Activated T cells of nonneural specificity open the blood–nerve barrier to circulating antibody. *Ann Neurol* **37**, 467–75.

Polydefkis, M., Yiannoutsos, C.T., Cohen, B.A. *et al.* (2002). Reduced intraepidermal nerve fiber density in HIV-associated sensory neuropathy. *Neurology* **58**, 115–19.

Popko, B., Corbin, J.G., Baerwald, K.D., Dupree, J. and Garcia, A.M. (1997). The effects of interferon-gamma on the central nervous system. *Mol Neurobiol* **14**, 19–35.

Powell, H.C., Braheny, S.L., Hughes, R.A. and Lampert, P.W. (1984). Antigen-specific demyelination and significance of the bystander effect in peripheral nerves. *Am J Pathol* **114**, 443–53.

Previtali, S.C., Feltri, M.L., Archelos, J.J., Quattrini, A., Wrabetz, L. and Hartung, H. (2001). Role of integrins in the peripheral nervous system. *Prog Neurobiol* **64**, 35–49.

Previtali, S.C., Nodari, A., Taveggia, C. *et al.* (2003a). Expression of laminin receptors in schwann cell differentiation: evidence for distinct roles. *J Neurosci* **23**, 5520–30.

Previtali, S.C., Dina, G., Nodari, A. *et al.* (2003b). Schwann cells synthesize alpha7beta1 integrin which is dispensable for peripheral nerve development and myelination. *Mol Cell Neurosci* **23**, 210–18.

Prineas, J.W. (1971a). Demyelination and remyelination in recurrent idiopathic polyneuropathy. An electron microscope study. *Acta Neuropathol (Berl)* **18**, 34–57.

Prineas, J.W. (1971b). Ultrastructural changes in the peripheral nerves in experimental dying-back polyneuropathies. *Proc Aust Assoc Neurol* **8**, 121–3.

Prineas, J.W. (1972). Acute idiopathic polyneuritis. An electron microscope study. *Lab Invest* **26**, 133–47.

Prineas, J.W. (1981). Pathology of the Guillain–Barré syndrome. *Ann Neurol* **9** (Suppl), 6–19.

Prineas, J.W. and McLeod, J.G. (1976). Chronic relapsing polyneuritis. *J Neurol Sci* **27**, 427–58.

Pritchard, J. and Hughes, R.A. (2004). Guillain–Barré syndrome. *Lancet* **363**, 2186–8.

Pritchard, J., Hayday, A.C., Gregson, N.A. and Hughes, R.A.C. (2004). *Alterations in Circulating T Cell Populations in Guillain–Barré Syndrome*. **31**. 2004.

Probert, L., Akassoglou, K., Pasparakis, M., Kontogeorgos, G. and Kollias, G. (1995). Spontaneous inflammatory demyelinating disease in transgenic mice showing central nervous system-specific expression of tumor necrosis factor alpha. *Proc Nat Acad Sci USA* **92**, 11294–8.

Probstmeier, R., Nellen, J., Gloor, S., Wernig, A. and Pesheva, P. (2001). Tenascin-R is expressed by Schwann cells in the peripheral nervous system. *J Neurosci Res* **64**, 70–8.

Purves, D. and Lichtman, J.W. (1985). *Principles of Neural Development*. Sinauer Associates, Sunderland, Mass.

Quijano-Roy, S., Renault, F., Romero, N., Guicheney, P., Fardeau, M. and Estournet, B. (2004). EMG and nerve conduction studies in children with congenital muscular dystrophy. *Muscle Nerve* **29**, 292–9.

Rabadan-Diehl, C., Dahl, G. and Werner, R. (1994). A connexin-32 mutation associated with Charcot–Marie–Tooth disease does not affect channel formation in oocytes. *FEBS Lett* **351**, 90–4.

Raeymaekers, P., Timmerman, V., Nelis, E. et al. (1991). Duplication in chromosome 17p11.2 in Charcot–Marie–Tooth neuropathy type 1a (CMT 1a). The HMSN Collaborative Research Group. *Neuromuscul Disord* **1**, 93–7.

Ragozzino, D., Renzi, M., Giovannelli, A. and Eusebi, F. (2002). Stimulation of chemokine CXC receptor 4 induces synaptic depression of evoked parallel fibers inputs onto Purkinje neurons in mouse cerebellum. *J Neuroimmunol* **127**, 30–6.

Raine, C.S. and Cross, A.H. (1989). Axonal dystrophy as a consequence of long-term demyelination. *Lab Invest* **60**, 714–25.

Ranvier, L. (1878). *Lecons sur l'Histologies du Systeme Nerveux*. Savy 2, Paris.

Raphael, J.C., Chevret, S., Hughes, R.A. and Annane, D. (2002). *Plasma exchange for Guillain–Barré syndrome*. Cochrane Database Syst Rev CD001798.

Rasband, M.N. (2004). It's 'juxta' potassium channel! *J Neurosci Res* **76**, 749–57.

Rasband, M.N., Park, E.W., Zhen, D. et al. (2002). Clustering of neuronal potassium channels is independent of their interaction with PSD-95. *J Cell Biol* **159**, 663–72.

Reddy, L.V., Koirala, S., Sugiura, Y., Herrera, A.A. and Ko, C.P. (2003). Glial cells maintain synaptic structure and function and promote development of the neuromuscular junction in vivo. *Neuron* **40**, 563–80.

Redford, E.J., Hall, S.M. and Smith, K.J. (1995). Vascular changes and demyelination induced by the intraneural injection of tumour necrosis factor. *Brain* **118** (Part 4), 869–78.

Redford, E.J., Smith, K.J., Gregson, N.A. et al. (1997). A combined inhibitor of matrix metalloproteinase activity and tumour necrosis factor-alpha processing attenuates experimental autoimmune neuritis. *Brain* **120** (Part 10), 1895–1905.

Reger, J.F. (1955). Electron microscopy of the motor end-plate in rat intercostal muscle. *Anat Rec* **122**, 1–15.

Reichert, F., Levitzky, R. and Rotshenker, S. (1996). Interleukin 6 in intact and injured mouse peripheral nerves. *Eur J Neurosci* **8**, 530–5.

Reist, N.E. and Smith, S.J. (1992). Neurally evoked calcium transients in terminal Schwann cells at the neuromuscular junction. *Proc Nat Acad Sci USA* **89**, 7625–9.

Rentier, B., Piette, J., Baudoux, L. et al. (1996). Lessons to be learned from varicella-zoster virus. *Vet Microbiol* **53**, 55–66.

Reynolds, M. L. and Woolf, C. J. (1992). Terminal Schwann cells elaborate extensive processes following denervation of the motor endplate. *J Neurocytol* **21**, 50–66.

Rezajooi, K., Pavlides, M., Winterbottom, J. et al. (2004). NG2 proteoglycan expression in the peripheral nervous system: upregulation following injury and comparison with CNS lesions. *Mol Cell Neurosci* **25**, 572–84.

Rich, M. M. and Pinter, M. J. (2001). Sodium channel inactivation in an animal model of acute quadriplegic myopathy. *Ann Neurol* **50**, 26–33.

Rich, M. M., Pinter, M. J., Kraner, S. D. and Barchi, R. L. (1998). Loss of electrical excitability in an animal model of acute quadriplegic myopathy. *Ann Neurol* **43**, 171–9.

Ridley, A. J., Davis, J. B., Stroobant, P. and Land, H. (1989). Transforming growth factors-beta 1 and beta 2 are mitogens for rat Schwann cells. *J Cell Biol* **109**, 3419–24.

Rieger, F., Daniloff, J. K., Pincon-Raymond, M., Crossin, K. L., Grumet, M. and Edelman, G. M. (1986). Neuronal cell adhesion molecules and cytotactin are colocalized at the node of Ranvier. *J Cell Biol* **103**, 379–91.

Riethmacher, D., Sonnenberg-Riethmacher, E., Brinkmann, V., Yamaai, T., Lewin, G. R. and Birchmeier, C. (1997). Severe neuropathies in mice with targeted mutations in the ErbB3 receptor. *Nature* **389**, 725–30.

Rimer, M., Prieto, A. L., Weber, J. L. et al. (2004). Neuregulin-2 is synthesized by motor neurons and terminal Schwann cells and activates acetylcholine receptor transcription in muscle cells expressing ErbB4. *Mol Cell Neurosci* **26**, 271–81.

Rios, J. C., Rubin, M., Martin, M. S. et al. (2003). Paranodal interactions regulate expression of sodium channel subtypes and provide a diffusion barrier for the node of Ranvier. *J Neurosci* **23**, 7001–11.

Ritz, M. F., Lechner-Scott, J., Scott, R. J. et al. (2000). Characterisation of autoantibodies to peripheral myelin protein 22 in patients with hereditary and acquired neuropathies. *J Neuroimmunol* **104**, 155–63.

Rizvi, T. A., Huang, Y., Sidani, A. et al. (2002). A novel cytokine pathway suppresses glial cell melanogenesis after injury to adult nerve. *J Neurosci* **22**, 9831–40.

Roa, B. B., Garcia, C. A., Suter, U. et al. (1993a). Charcot–Marie–Tooth disease type 1A. Association with a spontaneous point mutation in the PMP22 gene. *N Engl J Med* **329**, 96–101.

Roa, B. B., Garcia, C. A., Pentao, L. et al. (1993b). Evidence for a recessive PMP22 point mutation in Charcot–Marie–Tooth disease type 1A. *Nat Genet* **5**, 189–94.

Roa, B. B., Dyck, P. J., Marks, H. G., Chance, P. F. and Lupski, J. R. (1993c). Dejerine–Sottas syndrome associated with point mutation in the peripheral myelin protein 22 (PMP22) gene. *Nat Genet* **5**, 269–73.

Robbins, N. and Polak, J. (1988). Filopodia, lamellipodia and retractions at mouse neuromuscular junctions. *J Neurocytol* **17**, 545–61.

Roberts, A. B. and Sporn, M. B. (1993). Physiological actions and clinical applications of transforming growth factor-beta (TGF-beta). *Growth Factors* **8**, 1–9.

Robertson, J. D. (1956). The ultrastructure of a reptilian myoneural junction. *J Biophys Biochem Cytol* **2**, 381–94.

Robitaille, R. (1995). Purinergic receptors and their activation by endogenous purines at perisynaptic glial cells of the frog neuromuscular junction. *J Neurosci* **15**, 7121–31.

Robitaille, R. (1998). Modulation of synaptic efficacy and synaptic depression by glial cells at the frog neuromuscular junction. *Neuron* **21**, 847–55.

Robitaille, R., Bourque, M.J. and Vandaele, S. (1996). Localization of L-type Ca2+ channels at perisynaptic glial cells of the frog neuromuscular junction. *J Neurosci* **16**, 148–58.

Robitaille, R., Jahromi, B.S. and Charlton, M.P. (1997). Muscarinic Ca2+ responses resistant to muscarinic antagonists at perisynaptic Schwann cells of the frog neuromuscular junction. *J Physiol* **504** (Part 2), 337–47.

Rochon, D., Rousse, I. and Robitaille, R. (2001). Synapse–glia interactions at the mammalian neuromuscular junction. *J Neurosci* **21**, 3819–29.

Rogers, T., Chandler, D., Angelicheva, D. et al. (2000). A novel locus for autosomal recessive peripheral neuropathy in the EGR2 region on 10q23. *Am J Hum Genet* **67**, 664–71.

Rosen, J.L., Brown, M.J. and Rostami, A. (1992). Evolution of the cellular response in P2-induced experimental allergic neuritis. *Pathobiology* **60**, 108–12.

Rosenbluth, J., Dupree, J.L. and Popko, B. (2003). Nodal sodium channel domain integrity depends on the conformation of the paranodal junction, not on the presence of transverse bands. *Glia* **41**, 318–25.

Rossi, D. and Zlotnik, A. (2000). The biology of chemokines and their receptors. *Ann Rev Immunol* **18**, 217–42.

Rostami, A., Burns, J.B., Brown, M.J. et al. (1985). Transfer of experimental allergic neuritis with P2-reactive T-cell lines. *Cell Immunol* **91**, 354–61.

Rostami, A., Gregorian, S.K., Brown, M.J. and Pleasure, D.E. (1990). Induction of severe experimental autoimmune neuritis with a synthetic peptide corresponding to the 53–78 amino acid sequence of the myelin P2 protein. *J Neuroimmunol* **30**, 145–51.

Rothblum, K., Stahl, R.C. and Carey, D.J. (2004). Constitutive release of alpha4 type V collagen N-terminal domain by Schwann cells and binding to cell surface and extracellular matrix heparan sulfate proteoglycans. *J Biol Chem* **279**, 51282–8.

Rothe, F., Langnaese, K. and Wolf, G. (2005). New aspects of the location of neuronal nitric oxide synthase in the skeletal muscle: a light and electron microscopic study. *Nitric Oxide* **13**, 21–35.

Rudel, C. and Rohrer, H. (1992). Analysis of glia cell differentiation in the developing chick peripheral nervous system: sensory and sympathetic satellite cells express different cell surface antigens. *Development* **115**, 519–26.

Rufer, M., Flanders, K. and Unsicker, K. (1994). Presence and regulation of transforming growth factor beta mRNA and protein in the normal and lesioned rat sciatic nerve. *J Neurosci Res* **39**, 412–23.

Rungby, J. (1986). Exogenous silver in dorsal root ganglia, peripheral nerve, enteric ganglia, and adrenal medulla. *Acta Neuropathol (Berl)* **69**, 45–53.

Russell, J.W., Gill, J.S., Sorenson, E.J., Schultz, D.A. and Windebank, A.J. (2001). Suramin-induced neuropathy in an animal model. *J Neurol Sci* **192**, 71–80.

Rutkowski, J.L., Kirk, C.J., Lerner, M.A. and Tennekoon, G.I. (1995). Purification and expansion of human Schwann cells in vitro. *Nat Med* **1**, 80–3.

Rutkowski, J.L., Tuite, G.F., Lincoln, P.M., Boyer, P.J., Tennekoon, G.I. and Kunkel, S.L. (1999). Signals for proinflammatory cytokine secretion by human Schwann cells. *J Neuroimmunol* **101**, 47–60.

Sahenk, Z., Chen, L. and Mendell, J.R. (1999). Effects of PMP22 duplication and deletions on the axonal cytoskeleton. *Ann Neurol* **45**, 16–24.

Sahenk, Z. (1999). Abnormal Schwann cell–axon interactions in CMT neuropathies. The effects of mutant Schwann cells on the axonal

cytoskeleton and regeneration-associated myelination. *Ann NY Acad Sci* **883**, 415–26.
Said, G., Goulon-Goeau, C., Lacroix, C. and Moulonguet, A. (1994). Nerve biopsy findings in different patterns of proximal diabetic neuropathy. *Ann Neurol* **35**, 559–69.
Saida, K., Saida, T., Brown, M.J., Silberberg, D.H. and Asbury, A.K. (1978a). Antiserum-mediated demyelination in vivo: a sequential study using intraneural injection of experimental allergic neuritis serum. *Lab Invest* **39**, 449–62.
Saida, T., Saida, K., Silberberg, D.H. and Brown, M.J. (1978b). Transfer of demyelination by intraneural injection of experimental allergic neuritis serum. *Nature* **272**, 639–41.
Saida, T., Saida, K., Dorfman, S.H. *et al.* (1979a). Experimental allergic neuritis induced by sensitization with galactocerebroside. *Science* **204**, 1103–6.
Saida, T., Saida, K., Brown, M.J. and Silberberg, D.H. (1979b). Peripheral nerve demyelination induced by intraneural injection of experimental allergic encephalomyelitis serum. *J Neuropathol Exp Neurol* **38**, 498–518.
Saida, T., Saida, K., Silberberg, D.H. and Brown, M.J. (1981). Experimental allergic neuritis induced by galactocerebroside. *Ann Neurol* **9** (Suppl), 87–101.
Saida, T., Saida, K., Lisak, R.P., Brown, M.J., Silberberg, D.H. and Asbury, A.K. (1982). In vivo demyelinating activity of sera from patients with Guillain–Barré syndrome. *Ann Neurol* **11**, 69–75.
Saito, A. and Zacks, S.I. (1969). Ultrastructure of Schwann and perineurial sheaths at the mouse neuromuscular junction. *Anat Rec* **164**, 379–90.
Saito, F., Masaki, T., Kamakura, K. *et al.* (1999). Characterization of the transmembrane molecular architecture of the dystroglycan complex in schwann cells. *J Biol Chem* **274**, 8240–6.
Saito, F., Moore, S.A., Barresi, R. *et al.* (2003). Unique role of dystroglycan in peripheral nerve myelination, nodal structure, and sodium channel stabilization. *Neuron* **38**, 747–58.
Salloway, S., Mermel, L.A., Seamans, M. *et al.* (1996). Miller–Fisher syndrome associated with *Campylobacter jejuni* bearing lipopolysaccharide molecules that mimic human ganglioside GD3. *Infect Immun* **64**, 2945–9.
Salpeter, M.M. (1987). *The Vertebrate Neuromuscular Junction*. A. R. Liss, New York.
Salzer, J.L. (2003). Polarized domains of myelinated axons. *Neuron* **40**, 297–318.
Salzer, J.L., Williams, A.K., Glaser, L., Bunge, R.P. (1980). Studies of Schwann cell proliferation. II. Characterization of the stimulation and specificity of the response to a neurite membrane fraction. *J Cell Biol* **84**, 753–66.
Salzer, J.L., Lovejoy, L., Linder, M.C. and Rosen, C. (1998). Ran-2, a glial lineage marker, is a GPI-anchored form of ceruloplasmin. *J Neurosci Res* **54**, 147–57.
Samii, A., Unger, J. and Lange, W. (1999). Vascular endothelial growth factor expression in peripheral nerves and dorsal root ganglia in diabetic neuropathy in rats. *Neurosci Lett* **262**, 159–62.
Samuel, N.M., Jessen, K.R., Grange, J.M. and Mirsky, R. (1987a). Gamma interferon, but not *Mycobacterium leprae*, induces major histocompatibility class II antigens on cultured rat Schwann cells. *J Neurocytol* **16**, 281–7.
Samuel, N.M., Mirsky, R., Grange, J.M. and Jessen, K.R. (1987b). Expression of major histocompatibility complex class I and class II antigens in human Schwann cell cultures and effects of infection with *Mycobacterium leprae*. *Clin Exp Immunol* **68**, 500–9.
Sancho, S., Magyar, J.P., Aguzzi, A. and Suter, U (1999). Distal axonopathy in peripheral nerves of PMP22-mutant mice. *Brain* **122** (Part 8), 1563–77.

Sanes, J.R. and Lichtman, J.W. (1999). Development of the vertebrate neuromuscular junction. *Ann Rev Neurosci* **22**, 389–442.

Sanes, J.R., Schachner, M. and Covault, J. (1986). Expression of several adhesive macromolecules (N-CAM, L1, J1, NILE, uvomorulin, laminin, fibronectin, and a heparan sulfate proteoglycan) in embryonic, adult, and denervated adult skeletal muscle. *J Cell Biol* **102**, 420–31.

Sanes, J.R., Engvall, E., Butkowski, R. and Hunter, D.D. (1990). Molecular heterogeneity of basal laminae: isoforms of laminin and collagen IV at the neuromuscular junction and elsewhere. *J Cell Biol* **111**, 1685–99.

Sawant-Mane, S., Clark, M.B. and Koski, C.L. (1991). In vitro demyelination by serum antibody from patients with Guillain–Barré syndrome requires terminal complement complexes. *Ann Neurol* **29**, 397–404.

Sawant-Mane, S., Estep, A., III and Koski, C.L. (1994). Antibody of patients with Guillain–Barré syndrome mediates complement-dependent cytolysis of rat Schwann cells: susceptibility to cytolysis reflects Schwann cell phenotype. *J Neuroimmunol* **49**, 145–52.

Scarpini, E., Lisak, R., Beretta, S. et al. (1989). Type II major histocompatibility antigens on normal and pathological human nerves. Scarpini, E (Ed.) *Peripheral Nerve Development and Regneration: Recent Advances and Clinical Applications.* Livinia Press, Padova, pp. 189–92.

Scarpini, E., Lisak, R.P., Beretta, S. et al. (1990). Quantitative assessment of class II molecules in normal and pathological nerves. Immunocytochemical studies in vivo and in tissue culture. *Brain* **113** (Part 3), 659–75.

Schafer, D.P., Bansal, R., Hedstrom, K.L., Pfeiffer, S.E. and Rasband, M.N. (2004). Does paranode formation and maintenance require partitioning of neurofascin 155 into lipid rafts? *J Neurosci* **24**, 3176–85.

Scherer, S. (1999). Axonal pathology in demyelinating diseases. *Ann Neurol* **45**, 6–7.

Scherer, S.S. (1997). The biology and pathobiology of Schwann cells. *Curr Opin Neurol* **10**, 386–97.

Scherer, S.S. and Salzer, J.L. (1996). Axon–Schwann cell interactions during peripheral nerve degeneration and regeneration. Jessen, K.R., Richardson, W.D. (Eds.) *Glial Cell Development: Basic Principles and Clinical Relevance.* Bios Scientific Publisher Ltd, Oxford, UK, pp. 165–96.

Scherer, S.S. and Arroyo, E.J. (2002). Recent progress on the molecular organization of myelinated axons. *J Peripher Nerv Syst* **7**, 1–12.

Scherer, S.S., Arroyo, E.J. and Peles, E. (2004). Functional organization of the nodes of Ranvier. Lazzarini, R.L. (Ed.) *Myelin Biology and Disorders.* Elsevier, San Diego, pp. 89–116.

Scherer, S.S. and Kleopa, K. (2005). X-linked Charcot–Marie–Tooth disease. Dyck, P.J. (Ed.) *Peripheral Neuropathy*, 4th Edn. Elsevier, Saunders, Philadelphia, pp. 1791–805.

Scherer, S.S. and Salzer, J.L. (2001). Axon–Schwann cell interactions during peripheral nerve degeneration and regeneration. Jessen, K.R., Richardson, W.D. (Eds.) *Glial Cell Development*, 2nd Edn. Oxford University Press, Oxford.

Scherer, S.S., Kamholz, J. and Jakowlew, S.B. (1993). Axons modulate the expression of transforming growth factor-betas in Schwann cells. *Glia* **8**, 265–76.

Scherer, S.S., Wang, D.Y., Kuhn, R., Lemke, G., Wrabetz, L. and Kamholz, J. (1994). Axons regulate Schwann cell expression of the POU transcription factor SCIP. *J Neurosci* **14**, 1930–42.

Scherer, S.S., Xu, Y.T., Bannerman, P.G., Sherman, D.L. and Brophy, P.J. (1995). Periaxin expression in myelinating Schwann cells: modulation by axon—glial interactions and polarized localization during development. *Development* **121**, 4265—73.

Scherer, S. S., Xu, Y. T., Nelles, E., Fischbeck, K., Willecke, K. and Bone, L. J. (1998). Connexin32-null mice develop demyelinating peripheral neuropathy. *Glia* **24**, 8—20.

Schmid, R. S., McGrath, B., Berechid, B. E. *et al.* (2003). Neuregulin 1-erbB2 signaling is required for the establishment of radial glia and their transformation into astrocytes in cerebral cortex. *Proc Nat Acad Sci USA* **100**, 4251—6.

Schmidbauer, M., Budka, H., Pilz, P., Kurata, T. and Hondo, R. (1992). Presence, distribution and spread of productive varicella zoster virus infection in nervous tissues. *Brain* **115** (Part 2), 383—98.

Schmidt, B., Stoll, G., Hartung, H. P., Heininger, K., Schafer, B. and Toyka, K. V. (1990). Macrophages but not Schwann cells express Ia antigen in experimental autoimmune neuritis. *Ann Neurol* **28**, 70—7.

Schmidt, B., Toyka, K. V., Kiefer, R., Full, J., Hartung, H. P. and Pollard, J. (1996). Inflammatory infiltrates in sural nerve biopsies in Guillain—Barré syndrome and chronic inflammatory demyelinating neuropathy. *Muscle Nerve* **19**, 474—87.

Schnaar, R. L. (2004). Glycolipid-mediated cell—cell recognition in inflammation and nerve regeneration. *Arch Biochem Biophys* **426**, 163—72.

Schneider-Schaulies, J., Kirchhoff, F., Archelos, J. and Schachner, M. (1991). Down-regulation of myelin-associated glycoprotein on Schwann cells by interferon-gamma and tumor necrosis factor-alpha affects neurite outgrowth. *Neuron* **7**, 995—1005.

Schneider, C., Wicht, H., Enderich, J., Wegner, M. and Rohrer, H. (1999). Bone morphogenetic proteins are required in vivo for the generation of sympathetic neurons. *Neuron* **24**, 861—70.

Schneider, S., Bosse, F., D'Urso, D. *et al.* (2001). The AN2 protein is a novel marker for the Schwann cell lineage expressed by immature and nonmyelinating Schwann cells. *J Neurosci* **21**, 920—33.

Schroder, J. M., Ceuterick, C. M. J. J., DeJonghe, P. *et al.* (2001). Separation of terminal myelin loops from axons in periaxin neuropathy (CNT4F). European Charcot—Marie—Tooth Consortium Annual Symposium, Antwerp, Belgium.

Schubert, D. (1992). Synergistic interactions between transforming growth factor beta and fibroblast growth factor regulate Schwann cell mitosis. *J Neurobiol* **23**, 143—8.

Schuster, N., Bender, H., Rossler, O. G. *et al.* (2003). Transforming growth factor-beta and tumor necrosis factor-alpha cooperate to induce apoptosis in the oligodendroglial cell line OLI-neu. *J Neurosci Res* **73**, 324—33.

Selmaj, K. and Raine, C. S. (1988). Tumor necrosis factor mediates myelin damage in organotypic cultures of nervous tissue. *Ann NY Acad Sci* **540**, 568—70.

Selmaj, K., Raine, C. S., Farooq, M., Norton, W. T. and Brosnan, C. F. (1991a). Cytokine cytotoxicity against oligodendrocytes. Apoptosis induced by lymphotoxin. *J Immunol* **147**, 1522—9.

Selmaj, K., Cross, A. H., Farooq, M., Brosnan, C. F. and Raine, C. S. (1991b). Nonspecific oligodendrocyte cytotoxicity mediated by soluble products of activated T cell lines. *J Neuroimmunol* **35**, 261—71.

Semenenko, F. M., Sidebottom, E. and Cuello, A. C. (1987). A monoclonal antibody against a novel intracellular neural antigen expressed differentially in neural cell types. *J Neuroimmunol* **13**, 243—58.

Senderek, J., Bergmann, C., Weber, S. et al. (2003). Mutation of the SBF2 gene, encoding a novel member of the myotubularin family, in Charcot–Marie–Tooth neuropathy type 4B2/11p15. *Hum Mol Genet* **12**, 349–56.

Sereda, M., Griffiths, I., Puhlhofer, A. et al. (1996). A transgenic rat model of Charcot–Marie–Tooth disease. *Neuron* **16**, 1049–60.

Sereda, M.W., Horste, G.M.Z., Suter, U., Uzma, N. and Nave, K.A. (2003). Therapeutic administration of progesterone antagonist in a model of Charcot–Marie–Tooth disease (CMT-1A). *Nat Med* **9**, 1533–7.

Setoguchi, R., Hori, S., Takahashi, T. and Sakaguchi, S. (2005). Homeostatic maintenance of natural Foxp3(+) CD25(+) CD4(+) regulatory T cells by interleukin (IL)-2 and induction of autoimmune disease by IL-2 neutralization. *J Exp Med* **201**, 723–35.

Seyer, J.M., Kang, A.H. and Whitaker, J.N. (1977). The characterization of type I and type III collagens from human peripheral nerve. *Biochim Biophys Acta* **492**, 415–25.

Shah, N.M., Marchionni, M.A., Isaacs, I., Stroobant, P. and Anderson, D.J. (1994). Glial growth factor restricts mammalian neural crest stem cells to a glial fate. *Cell* **77**, 349–60.

Shah, N.M., Groves, A.K. and Anderson, D.J. (1996). Alternative neural crest cell fates are instructively promoted by TGF beta superfamily members. *Cell* **85**, 331–43.

Shamash, S., Reichert, F. and Rotshenker, S. (2002). The cytokine network of Wallerian degeneration: tumor necrosis factor-alpha, interleukin-1alpha, and interleukin-1beta. *J Neurosci* **22**, 3052–60.

Shames, I., Fraser, A., Colby, J., Orfali, W. and Snipes, G.J. (2003). Phenotypic differences between peripheral myelin protein-22 (PMP22) and myelin protein zero (P0) mutations associated with Charcot–Marie–Tooth-related diseases. *J Neuropathol Exp Neurol* **62**, 751–64.

Shanthaveerappa, T.R. and Bourne, G.H. (1966). Perineural epithelium: a new concept of its role in the integrity of the peripheral nervous system. *Science* **154**, 1464–7.

Shanthaveerappa, T.R. and Bourne, G.H. (1967). Nature and origin of perisynaptic cells of the motor end plate. *Int Rev Cytol* **21**, 353–64.

Shapiro, L., Doyle, J.P., Hensley, P., Colman, D.R. and Hendrickson, W.A. (1996). Crystal structure of the extracellular domain from P0, the major structural protein of peripheral nerve myelin. *Neuron* **17**, 435–49.

Sharief, M.K., McLean, B. and Thompson, E.J. (1993). Elevated serum levels of tumor necrosis factor-alpha in Guillain–Barré syndrome. *Ann Neurol* **33**, 591–6.

Sharpe, A.H. and Freeman, G.J. (2002). The B7-CD28 superfamily. *Nat Rev Immunol* **2**, 116–26.

Sheikh, K.A., Ho, T.W., Nachamkin, I. et al. (1998). Molecular mimicry in Guillain–Barré syndrome. *Ann NY Acad Sci* **845**, 307–1.

Shellswell, G.B., Restall, D.J., Duance, V.C. and Bailey, A.J. (1979). Identification and differential distribution of collagen types in the central and peripheral nervous systems. *FEBS Lett* **106**, 305–8.

Sherman, D.L. and Brophy, P.J. (2000). A tripartite nuclear localization signal in the PDZ-domain protein L-periaxin. *J Biol Chem* **275**, 4537–40.

Sherman, D.L. and Brophy, P.J. (2005). Mechanisms of axon ensheathment and myelin growth. *Nat Rev Neurosci* **6**, 683–90.

Sherman, D.L., Fabrizi, C., Gillespie, C.S. and Brophy, P.J. (2001). Specific disruption of a Schwann cell dystrophin-related protein complex in a demyelinating neuropathy. *Neuron* **30**, 677–87.

Sherman, L., Stocker, K.M., Morrison, R. and Ciment, G. (1993). Basic Fibroblast Growth Factor (bFGF) acts intracellularly to cause the transdifferentiation of avian neural crest-derived Schwann cell precursors into melanocytes. *Development* **118**, 1313–26.

Shorer, Z., Philpot, J., Muntoni, F., Sewry, C. and Dubowitz, V. (1995). Demyelinating peripheral neuropathy in merosin-deficient congenital muscular dystrophy. *J Child Neurol* **10**, 472–5.

Shuman, S., Hardy, M., Sobue, G. and Pleasure, D. (1988). A cyclic AMP analogue induces synthesis of a myelin-specific glycoprotein by cultured Schwann cells. *J Neurochem* **50**, 190–4.

Shy, M. (2005). Hereditary motor and sensory neuropathies related to MPZ P0 mutations. Dyck, P.J. (Ed). *Peripheral Neuropathy*, 4th Edn. Saunders, Philadelphia, pp. 1681–716.

Shy, M.E., Jani, A., Krajewski, K.M. et al. (2004). Phenotypic clustering in MPZ mutations. *Brain* **127**, 371–84.

Shy, M.E., Shi, Y., Wrabetz, L., Kamholz, J. and Scherer, S.S. (1996). Axon–Schwann cell interactions regulate the expression of c-jun in Schwann cells. *J Neurosci Res* **43**, 511–25.

Shy, M.E., Garbern, J.Y. and Kamholz, J. (2002). Hereditary motor and sensory neuropathies: a biological perspective. *Lancet Neurol* **1**, 110–18.

Shy, M.E., Hobson, G., Jain, M. et al. (2003). Schwann cell expression of PLP1 but not DM20 is necessary to prevent neuropathy. *Ann Neurol* **53**, 354–65.

Shy, M., Lupski, J.R., Chance, P.F., Klein, C.J. and Dyck, P. (2005). The hereditary motor and sensory neuropathies: an overview of the clinical, genetic, electrophysiologic and pathlogic features. Dyck P.J. (Ed.) *Peripheral Neuropathy*, 4th Edn. Saunders, Philadelphia, pp. 1623–58.

Skoff, A.M., Lisak, R.P., Bealmear, B. and Benjamins, J.A. (1998). TNF-alpha and TGF-beta act synergistically to kill Schwann cells. *J Neurosci Res* **53**, 747–56.

Skre, H. (1974). Genetic and clinical aspects of Charcot–Marie–Tooth's disease. *Clin Genet* **6**, 98–118.

Skundric, D., Bealmear, B. and Lisak, R. (1996a). Inducible coexpression of IL-1, IL-6 and TNF-α in cultured Schwann cells. *J Neurochem* **66**, Suppl. 1, S48.

Skundric, D., Bealmear, B. and Lisak, R. (1996b). Inducible IL-1α, IL-1, IL-1R and IL-1 receptor antagonist (RA) expression in cultured rat Schwann cells (SC). *J Neurochem* **64**, Suppl., S69.

Skundric, D., Bealmear, B. and Lisak, R. (1997a). IL-1β, IL-6 and TNF-α upregulate expression of each other in cultured Schwann cells (SC). *J Neurochem* **69**, Suppl., S152.

Skundric, D.S., Bealmear, B. and Lisak, R.P. (1997b). Induced upregulation of IL-1, IL-1RA and IL-1R type I gene expression by Schwann cells. *J Neuroimmunol* **74**, 9–18.

Skundric, D.S., Lisak, R.P., Rouhi, M., Kieseier, B.C., Jung, S. and Hartung, H.P. (2001). Schwann cell-specific regulation of IL-1 and IL-1Ra during EAN: possible relevance for immune regulation at paranodal regions. *J Neuroimmunol* **116**, 74–82.

Skundric, D.S., Dai, R., James, J. and Lisak, R.P. (2002). Activation of IL-1 signaling pathway in Schwann cells during diabetic neuropathy. *Ann NY Acad Sci* **958**, 393–8.

Slezak, M. and Pfrieger, F.W. (2003). New roles for astrocytes: regulation of CNS synaptogenesis. *Trends Neurosci* **26**, 531–5.

Smart, S.L., Lopantsev, V., Zhang, C.L. et al. (1998). Deletion of the Kv1.1 potassium channel causes epilepsy in mice. *Neuron* **20**, 809–19.

Smith, K.J. and Hall, S.M. (1988). Peripheral demyelination and remyelination initiated by the calcium-selective ionophore ionomycin: in vivo observations. *J Neurol Sci* **83**, 37–53.

Sobue, G., Shuman, S. and Pleasure, D. (1986). Schwann cell responses to cyclic AMP: proliferation, change in shape, and appearance of surface galactocerebroside. *Brain Res* **362**, 23–32.

Sobue, G., Nakao, N., Murakami, K. et al. (1990). Type I familial amyloid polyneuropathy. A pathological study of the peripheral nervous system. *Brain* **113** (Part 4), 903–19.

Soilu-Hanninen, M., Ekert, P., Bucci, T., Syroid, D., Bartlett, P.F. and Kilpatrick, T.J. (1999). Nerve growth factor signaling through p75 induces apoptosis in Schwann cells via a Bcl-2-independent pathway. *J Neurosci* **19**, 4828–38.

Soliven, B., Szuchet, S. and Nelson, D.J. (1991). Tumor necrosis factor inhibits K+ current expression in cultured oligodendrocytes. *J Membr Biol* **124**, 127–37.

Son, Y.J. and Thompson, W.J. (1995a). Schwann cell processes guide regeneration of peripheral axons. *Neuron* **14**, 125–32.

Son, Y.J. and Thompson, W.J. (1995b). Nerve sprouting in muscle is induced and guided by processes extended by Schwann cells. *Neuron* **14**, 133–41.

Sondell, M., Lundborg, G. and Kanje, M. (1999). Vascular endothelial growth factor has neurotrophic activity and stimulates axonal outgrowth, enhancing cell survival and Schwann cell proliferation in the peripheral nervous system. *J Neurosci* **19**, 5731–40.

Southard-Smith, E.M., Kos, L. and Pavan, W.J. (1998). SOX10 mutation disrupts neural crest development in Dom Hirschsprung mouse model. *Nat Genet* **18**, 60–4.

Southwood, C.M., Garbern, J., Jiang, W. and Gow, A. (2002). The unfolded protein response modulates disease severity in Pelizaeus–Merzbacher disease. *Neuron* **36**, 585–96.

Spierings, E., deB.T., Zulianello, L. and Ottenhoff, T.H. (2000). Novel mechanisms in the immunopathogenesis of leprosy nerve damage: the role of Schwann cells, T cells and *Mycobacterium leprae*. *Immunol Cell Biol* **78**, 349–55.

Spierings, E., deB.T., Wieles, B., Adams, L.B., Marani, E. and Ottenhoff, T.H. (2001). *Mycobacterium leprae*-specific, HLA class II-restricted killing of human Schwann cells by CD4+ Th1 cells: a novel immunopathogenic mechanism of nerve damage in leprosy. *J Immunol* **166**, 5883–8.

Spies, J.M., Westland, K.W., Bonner, J.G. and Pollard, J.D. (1995a). Intraneural activated T cells cause focal breakdown of the blood–nerve barrier. *Brain* **118** (Part 4), 857–68.

Spies, J.M., Pollard, J.D., Bonner, J.G., Westland, K.W. and McLeod, J.G. (1995b). Synergy between antibody and P2-reactive T cells in experimental allergic neuritis. *J Neuroimmunol* **57**, 77–84.

Steck, A.J., Schaeren-Wiemers, N. and Hartung, H.P. (1998). Demyelinating inflammatory neuropathies, including Guillain–Barré syndrome. *Curr Opin Neurol* **11**, 311–18.

Steinhoff, U. and Kaufmann, S.H. (1988). Specific lysis by CD8+ T cells of Schwann cells expressing *Mycobacterium leprae* antigens. *Eur J Immunol* **18**, 969–72.

Steinhoff, U., Schoel, B. and Kaufmann, S.H. (1990). Lysis of interferon-gamma activated Schwann cell by cross-reactive CD8+ alpha/beta T cells with specificity for the mycobacterial 65 kd heat shock protein. *Int Immunol* **2**, 279–84.

Stevens, B., Ishibashi, T., Chen, J. F. and Fields, R. D. (2004). Adenosine: an activity-dependent axonal signal regulating MAP kinase and proliferation in developing Schwann cells. *Neuron Glia Biology* **1**, 23–34.

Stewart, G. J., Pollard, J. D., McLeod, J. G. and Wolnizer, C. M. (1978). HLA antigens in the Landry–Guillain–Barré syndrome and chronic relapsing polyneuritis. *Ann Neurol* **4**, 285–9.

Stewart, H. J., Rougon, G., Dong, Z., Dean, C., Jessen, K. R. and Mirsky, R. (1995a). TGF-betas upregulate NCAM and L1 expression in cultured Schwann cells, suppress cyclic AMP-induced expression of O4 and galactocerebroside, and are widely expressed in cells of the Schwann cell lineage in vivo. *Glia* **15**, 419–36.

Stewart, H. J., Curtis, R., Jessen, K. R. and Mirsky, R. (1995b). TGF-beta s and cAMP regulate GAP-43 expression in Schwann cells and reveal the association of this protein with the trans-Golgi network. *Eur J Neurosci* **7**, 1761–72.

Stewart, H. J., Turner, D., Jessen, K. R. and Mirsky, R. (1997). Expression and regulation of alpha1beta1 integrin in Schwann cells. *J Neurobiol* **33**, 914–28.

Stewart, H. J. S., Morgan, L., Jessen, K. R. and Mirsky, R. (1993). Changes in DNA synthesis rate in the Schwann cell lineage in vivo are correlated with the precursor-Schwann cell transition and myelination. *Eur J Neurosci* **5**, 1136–44.

Stewart, H. J. S., Bradke, F., Tabernero, A., Morrell, D., Jessen, K. R. and Mirsky, R. (1996). Regulation of rat Schwann cell P0 expression and DNA synthesis by insulin-like growth factors in vitro. *Eur J Neurosci* **8**, 553–64.

Stewart, H. J. S., Brennan, A., Rahman, M. *et al.* (2001). Developmental regulation and overexpression of the transcription factor AP-2, a potential regulator of the timing of Schwann cell generation. *Eur J Neurosci* **14**, 363–72.

Stirling, C. A. (1975). Abnormalities in Schwann cell sheaths in spinal nerve roots of dystrophic mice. *J Anat* **119**, 169–80.

Stolinski, C., Breathnach, A. S., Thomas, P. K., Gabriel, G. and King, R. M. H. (1985). Distribution of particle aggregates in the internodal axolemma and adaxonal Schwann cell membrane of rodent peripheral nerve. *J Neurol Sci* **67**, 213–22.

Stoll, G. and Muller, H. W. (1999). Nerve injury, axonal degeneration and neural regeneration: basic insights. *Brain Pathol* **9**, 313–25.

Stoll, G., Schwendemann, G., Heininger, K. *et al.* (1986). Relation of clinical, serological, morphological, and electrophysiological findings in galactocerebroside-induced experimental allergic neuritis. *J Neurol Neurosurg Psychiatry* **49**, 258–64.

Stoll, G., Jung, S., Jander, S., van der, M. P. and Hartung, H. P. (1993a). Tumor necrosis factor-alpha in immune-mediated demyelination and Wallerian degeneration of the rat peripheral nervous system. *J Neuroimmunol* **45**, 175–82.

Stoll, G., Jander, S. *et al.* (1993b). Macrophages and endothelial cells express intercellular adhesion molecule-1 in immune-mediated demyelination but not in Wallerian degeneration of the rat peripheral nervous system. *Lab Invest* **68**, 637–44.

Street, V. A., Bennett, C. L., Goldy, J. D. *et al.* (2003). Mutation of a putative protein degradation gene LITAF/SIMPLE in Charcot–Marie–Tooth disease 1C. *Neurology* **60**, 22–6.

Stumm, R. K., Zhou, C., Ara, T. *et al.* (2003). CXCR4 regulates interneuron migration in the developing neocortex. *J Neurosci* **23**, 5123–30.

Sugimura, K., Haimoto, H., Nagura, H., Kato, K. and Takahashi, A. (1989). Immunohistochemical differential distribution of S-100 alpha and S-100 beta in the peripheral nervous system of the rat. *Muscle Nerve* **12**, 929–35.

Sumner, A., Said, G., Idy, I. and Metral, S. (1982a). Electrophysiological and morphological effects of the injection of Guillain–Barré sera in the sciatic nerve of the rat (author's transl). *Rev Neurol (Paris)* **138**, 17–24.

Sumner, A. J., Saida, K., Saida, T., Silberberg, D. H. and Asbury, A. K. (1982b). Acute conduction block associated with experimental antiserum-mediated demyelination of peripheral nerve. *Ann Neurol* **11**, 469–77.

Sund, M., Vaisanen, T., Kaukinen, S. *et al.* (2001). Distinct expression of type XIII collagen in neuronal structures and other tissues during mouse development. *Matrix Biol* **20**, 215–31.

Suter, U. (2004). PMP22 gene. Lazzarini, R. L. (Ed.) *Myelin Biology and Disorders*. Elsevier, Philadelphia, pp. 547–64.

Suter, U. and Nave, K. A. (1999). Transgenic mouse models of CMT1A and HNPP. *Ann NY Acad Sci* **883**, 247–53.

Suter, U. and Scherer, S. S. (2003). Disease mechanisms in inherited neuropathies. *Nat Neurosci Rev* **4**, 714–26.

Suter, U., Moskow, J. J., Welcher, A. A. *et al.* (1992a). A leucine-to-proline mutation in the putative first transmembrane domain of the 22-kDa peripheral myelin protein in the trembler-J mouse. *Proc Nat Acad Sci USA* **89**, 4382–6.

Suter, U., Welcher, A. A., Ozcelik, T. *et al.* (1992b). Trembler mouse carries a point mutation in a myelin gene. *Nature* **356**, 241–4.

Syroid, D. E., Maycox, P. R., Burrola, P. G. *et al.* (1996). Cell death in the Schwann cell lineage and its regulation by neuregulin. *Proc Nat Acad Sci USA* **93**, 9229–34.

Syroid, D. E., Maycox, P. J., Soilu-Hanninen, M. *et al.* (2000). Induction of postnatal Schwann cell death by the low-affinity neurotrophin receptor in vitro and after axotomy. *J Neurosci* **20**, 5741–7.

Takahashi, M. and Osumi, N. (2005). Identification of a novel type II classical cadherin: rat cadherin19 is expressed in the cranial ganglia and Schwann cell precursors during development. *Develop Dyn* **232**, 200–8.

Takeda, K., Kaisho, T. and Akira, S. (2003). Toll-like receptors. *Ann Rev Immunol* **21**, 335–76.

Tam, S. L. and Gordon, T. (2003). Neuromuscular activity impairs axonal sprouting in partially denervated muscles by inhibiting bridge formation of perisynaptic Schwann cells. *J Neurobiol* **57**, 221–34.

Tamkun, J. W., DeSimone, D. W., Fonda, D. *et al.* (1986). Structure of integrin, a glycoprotein involved in the transmembrane linkage between fibronectin and actin. *Cell* **46**, 271–82.

Taniuchi, M., Clark, H. B. and Johnson, E. M., Jr. (1986). Induction of nerve growth factor receptor in Schwann cells after axotomy. *Proc Nat Acad Sci USA* **83**, 4094–8.

Taniuchi, M., Clark, H. B., Schweitzer, J. B. and Johnson, E. M., Jr. (1988). Expression of nerve growth factor receptors by Schwann cells of axotomized peripheral nerves: ultrastructural location, suppression by axonal contact, and binding properties. *J Neurosci* **8**, 664–81.

Taskinen, H. S. and Roytta, M. (2000). Increased expression of chemokines (MCP-1, MIP-1alpha, RANTES) after peripheral nerve transection. *J Peripher Nerv Syst* **5**, 75–81.

Taveggia, C., Zanazzi, G., Petrylak, A. *et al.* (2005). Neuregulin-1 type III determines the ensheathment fate of axons. *Neuron* **47**, 681–94.

Taylor, J. M. and Pollard, J. D. (2001). Dominance of autoreactive T cell-mediated delayed-type hypersensitivity or antibody-mediated demyelination results in distinct forms of experimental autoimmune neuritis in the Lewis rat. *J Neuropathol Exp Neurol* **60**, 637–46.

Tello, J. F. (1944). Sobre una vaina que envuelve toda la ramificacion del axon en las terminaciones motrices de los musculos estriados. *Trabajos del Laboratorio de Investigaciones Biologicas de la Universidad de Madrid* **36**, 1–59.

Teravainen, H. (1970). Satellite cells of striated muscle after compression injury so slight as not to cause degeneration of the muscle fibres. *Z Zellforsch Mikrosk Anat* **103**, 320–7.

Tham, T. N., Lazarini, F., Franceschini, I. A., Lachapelle, F., Amara, A. and Dubois-Dalcq, M. (2001). Developmental pattern of expression of the alpha chemokine stromal cell-derived factor 1 in the rat central nervous system. *Eur J Neurosci* **13**, 845–56.

Thomas, P. K., Marques, W., Davis, M. B. et al. (1997). The phenotypic manifestations of chromosome 17p11.2 duplication. *Brain* **120**, 465–78.

Tiegs, O. W. (1953). Innervation of voluntary muscle. *Physiol Rev* **33**, 90–144.

Timmerman, V., De Jonghe, P., Ceuterick, C. et al. (1999). Novel missense mutation in the early growth response 2 gene associated with Dejerine–Sottas syndrome phenotype. *Neurology* **52**, 1827–32.

Tobler, A. R., Notterpek, L., Naef, R., Taylor, V., Suter, U. and Shooter, E. M. (1999). Transport of Trembler-J mutant peripheral myelin protein 22 is blocked in the intermediate compartment and affects the transport of the wild-type protein by direct interaction. *J Neurosci* **19**, 2027–36.

Tohyama, K. and Ide, C. (1984). The localization of laminin and fibronectin on the Schwann cell basal lamina. *Arch Histol Jpn* **47**, 519–32.

Tooth, H. (1886). *The Peroneal Type of Progressive Muscular Atrophy*. Lewis, London.

Topilko, P., Schneider-Maunoury, S., Levi, G. et al. (1994). Krox-20 controls myelination in the peripheral nervous system. *Nature* **371**, 796–9.

Topilko, P. and Meijer, D. (2001). Transcription factors that control Schwann cell development and myelination. Jessen, K. R., Richardson, W. D. (Eds.) *Glial Cell Development*. Oxford University Press, Oxford, pp. 223–44.

Trachtenberg, J. T. and Thompson, W. J. (1996). Schwann cell apoptosis at developing neuromuscular junctions is regulated by glial growth factor. *Nature* **379**, 174–7.

Trachtenberg, J. T. and Thompson, W. J. (1997). Nerve terminal withdrawal from rat neuromuscular junctions induced by neuregulin and Schwann cells. *J Neurosci* **17**, 6243–55.

Trapp, B. D. and Kidd, G. J. (2004). Structure of the myelinated axon. Lazzarini, R. L. (Ed.) *Myelin Biology and Disorders*. Elsevier, San Diego, pp. 3–27.

Tricaud, N., Perrin-Tricaud, C., Bruses, J. L. and Rutishauser, U. (2005). Adherens junctions in myelinating Schwann cells stabilize Schmidt–Lanterman incisures via recruitment of p120 catenin to E-cadherin. *J Neurosci* **25**, 3259–69.

Tsai, C. P., Pollard, J. D. and Armati, P. J. (1991). Interferon-gamma inhibition suppresses experimental allergic neuritis: modulation of major histocompatibility complex expression of Schwann cells in vitro. *J Neuroimmunol* **31**, 133–45.

Tsukada, N., Koh, C. S., Inoue, A. and Yanagisawa, N. (1987). Demyelinating neuropathy associated with hepatitis B virus infection. Detection of immune complexes composed of hepatitis B virus surface antigen. *J Neurol Sci* **77**, 203–16.

Tucker, R. P., Hagios, C., Santiago, A. and Chiquet-Ehrismann, R. (2001). Tenascin-Y is concentrated in adult nerve roots and has barrier properties in vitro. *J Neurosci Res* **66**, 439–47.

Tyson, J., Ellis, D., Fairbrother, U. *et al.* (1997). Hereditary demyelinating neuropathy of infancy. A genetically complex syndrome. *Brain* **120**, 47–63.

Uhlenberg, B., Schuelke, M., Ruschendorf, F. *et al.* (2004). Mutations in the gene encoding gap junction protein alpha 12 (connexin 46.6) cause Pelizaeus-Merzbacher-like disease. *Am J Hum Genet* **75**, 251–60.

Ullian, E. M., Sapperstein, S. K., Christopherson, K. S. and Barres, B. A. (2001). Control of synapse number by glia. *Science* **291**, 657–61.

Ullian, E. M., Harris, B. T., Wu, A., Chan, J. R. and Barres, B. A. (2004a). Schwann cells and astrocytes induce synapse formation by spinal motor neurons in culture. *Mol Cell Neurosci* **25**, 241–51.

Ullian, E. M., Christopherson, K. S. and Barres, B. A. (2004b). Role for glia in synaptogenesis. *Glia* **47**, 209–16.

Ulzheimer, J. C., Peles, E., Levinson, S. R. and Martini, R. (2004). Altered expression of ion channel isoforms at the node of Ranvier in P0-deficient myelin mutants. *Mol Cell Neurosci* **25**, 83–94.

Uncini, A., Di, M. A., Di, G. G. *et al.* (1999). Effect of rhTNF-alpha injection into rat sciatic nerve. *J Neuroimmunol* **94**, 88–94.

Unsicker, K., Flanders, K. C., Cissel, D. S., Lafyatis, R. and Sporn, M. B. (1991). Transforming growth factor beta isoforms in the adult rat central and peripheral nervous system. *Neuroscience* **44**, 613–25.

Ushiki, T. and Ide, C. (1988). A modified KOH-collagenase method applied to scanning electron microscopic observations of peripheral nerves. *Arch Histol Cytol* **51**, 223–32.

Uyemura, K., Asou, H. and Takeda, Y. (1995). Structure and function of peripheral nerve myelin proteins. *Prog Brain Res* **105**, 311–18.

Vabnick, I. and Shrager, P. (1998). Ion channel redistribution and function during development of the myelinated axon. *J Neurobiol* **37**, 80–96.

Vabnick, I., Trimmer, J. S., Schwarz, T. L., Levinson, S. R., Risal, D. and Shrager, P. (1999). Dynamic potassium channel distributions during axonal development prevent aberrant firing patterns. *J Neurosci* **19**, 747–58.

Vagnerova, K., Tarumi, Y. S., Proctor, T. M. and Patton, B. L. (2003). A specialized basal lamina at the node of Ranvier. *Soc Neurosci Abs* **29**, 351.18.

Valentijn, L. J., Baas, F., Wolterman, R. A. *et al.* (1992). Identical point mutations of PMP-22 in Trembler-J mouse and Charcot–Marie–Tooth disease type 1A. *Nat Genet* **2**, 288–91.

Vallat, J. M., Sindou, P., Preux, P. M. *et al.* (1996). Ultrastructural PMP22 expression in inherited demyelinating neuropathies. *Ann Neurol* **39**, 813–17.

van der Laan, L. J., Ruuls, S. R., Weber, K. S., Lodder, I. J., Dopp, E. A. and Dijkstra, C. D. (1996). Macrophage phagocytosis of myelin in vitro determined by flow cytometry: phagocytosis is mediated by CR3 and induces production of tumor necrosis factor-alpha and nitric oxide. *J Neuroimmunol* **70**, 145–52.

van der Meche, F. G. and Schmitz, P. I. (1992). A randomized trial comparing intravenous immune globulin and plasma exchange in Guillain–Barré syndrome. Dutch Guillain–Barré Study Group. *N Engl J Med* **326**, 1123–9.

van Doorn, P. A. (2005). Treatment of Guillain–Barré syndrome and CIDP. *J Peripher Nerv Syst* **10**, 113–27.

Van Rhijn, I., Van den Berg, L. H., Bosboom, W. M., Otten, H. G. and Logtenberg, T. (2000a). Expression of accessory molecules for T-cell activation in peripheral nerve of patients with CIDP and vasculitic neuropathy. *Brain* **123** (Part 10), 2020–9.

Van Rhijn, L.W., Jansen, E.J. and Pruijs, H.E. (2000b). Long-term follow-up of conservatively treated popliteal cysts in children. *J Pediatr Orthop B* **9**, 62–4.

Van, K.R., van Doorn, P.A., Schmitz, P.I., Ang, C.W. and van der Meche, F.G. (2000). Mild forms of Guillain-Barré syndrome in an epidemiologic survey in The Netherlands. *Neurology* **54**, 620–5.

Van, K.R., Schmitz, P.I., Meche, F.G., Visser, L.H., Meulstee, J. and van Doorn, P.A. (2004). Effect of methylprednisolone when added to standard treatment with intravenous immunoglobulin for Guillain–Barré syndrome: randomised trial. *Lancet* **363**, 192–6.

Vardhini, D., Suneetha, S., Ahmed, N. et al. (2004). Comparative proteomics of the *Mycobacterium leprae* binding protein myelin P0: its implication in leprosy and other neurodegenerative diseases. *Infect Genet Evol* **4**, 21–8.

Vaughan, R.W., Adam, A.M., Gray, I.A. et al. (1990). Major histocompatibility complex class I and class II polymorphism in chronic idiopathic demyelinating polyradiculoneuropathy. *J Neuroimmunol* **27**, 149–53.

Vega, J.A., Valle-Soto, M.E., Calzada, B. and Alvarez-Mendez, J.C. (1991). Immunohistochemical localization of S-100 protein subunits (alpha and beta) in dorsal root ganglia of the rat. *Cell Mol Biol* **37**, 173–81.

Venstrom, K. and Reichardt, L. (1995). Beta 8 integrins mediate interactions of chick sensory neurons with laminin-1, collagen IV, and fibronectin. *Mol Biol Cell* **6**, 419–31.

Viskochil, D.H. (2003). It takes two to tango: mast cell and Schwann cell interactions in neurofibromas. *J Clin Invest* **112**, 1791–3.

Volterra, A., Magistretti, P.J. and Haydon, P.G. (2002). *The Tripartite Synapse: Glia in Synaptic Transmission*. Oxford University Press, Oxford.

Vriesendorp, F.J., Mishu, B., Blaser, M.J. and Koski, C.L. (1993). Serum antibodies to GM1, GD1b, peripheral nerve myelin, and *Campylobacter jejuni* in patients with Guillain–Barré syndrome and controls: correlation and prognosis. *Ann Neurol* **34**, 130–5.

Vroemen, M. and Weidner, N. (2003). Purification of Schwann cells by selection of p75 low affinity nerve growth factor receptor expressing cells from adult peripheral nerve. *J Neurosci Methods* **124**, 135–43.

Wagner, R. and Myers, R.R. (1996a). Endoneurial injection of TNF-alpha produces neuropathic pain behaviors. *Neuroreport* **7**, 2897–901.

Wagner, R. and Myers, R.R. (1996b). Schwann cells produce tumor necrosis factor alpha: expression in injured and non-injured nerves. *Neuroscience* **73**, 625–9.

Wakamatsu, Y., Maynard, T.M. and Weston, J.A. (2000). Fate determination of neural crest cells by NOTCH-mediated lateral inhibition and asymmetrical cell division during gangliogenesis. *Development* **127**, 2811–21.

Waksman, B.H. and Adams, R.D. (1956). A comparative study of experimental allergic neuritis in the rabbit, guinea pig, and mouse. *J Neuropathol Exp Neurol* **15**, 293–334.

Wallquist, W., Patarroyo, M., Thams, S. et al. (2002). Laminin chains in rat and human peripheral nerve: distribution and regulation during development and after axonal injury. *J Comp Neurol* **454**, 284–93.

Wallquist, W., Plantman, S., Thams, S. et al. (2005). Impeded interaction between Schwann cells and axons in the absence of laminin alpha4. *J Neurosci* **25**, 3692–700.

Walport, M. (1998). Complement. Roitti, I., Brostoff, J., Male, D. (Eds.) *Immunology*. Mosby, Philadelphia, pp. 43–61.

Wanaka, A., Carroll, S.L. and Milbrandt, J. (1993). Developmentally regulated expression of pleiotrophin, a novel heparin binding growth factor, in the nervous system of the rat. *Brain Res Dev Brain Res* **72**, 133–44.

Wang, J.Y., Miller, S.J. and Falls, D.L. (2001). The N-terminal region of neuregulin isoforms determines the accumulation of cell surface and released neuregulin ectodomain. *J Biol Chem* **276**, 2841–51.

Wang, S. and Barres, B.A. (2000). Up a notch: instructing gliogenesis. *Neuron* **27**, 197–200.

Wanner, I., Guerra, N.K., Mahoney, J. *et al.* (2006). Role of N-cadherin in Schwann cell precursors of growing nerves. *Glia* **54**, 439–59.

Warner, L.E., Hilz, M.J., Appel, S.H. *et al.* (1996). Clinical phenotypes of different MPZ (P0) mutations may include Charcot–Marie–Tooth type 1B, Dejerine–Sottas, and congenital hypomyelination. *Neuron* **17**, 451–60.

Warner, L.E., Mancias, P., Butler, I.J. *et al.* (1998). Mutations in the early growth response 2 (EGR2) gene are associated with hereditary myelinopathies. *Nat Genet* **18**, 382–4.

Watson, D.F., Nachtman, F.N., Kuncl, R.W. and Griffin, J.W. (1994). Altered neurofilament phosphorylation and beta tubulin isotypes in Charcot–Marie–Tooth disease type 1. *Neurology* **44**, 2383–7.

Watts, R.J., Schuldiner, O., Perrino, J., Larsen, C. and Luo, L. (2004). Glia engulf degenerating axons during developmental axon pruning. *Curr Biol* **14**, 678–84.

Waxman, S.G. (2005). Cerebellar dysfunction in multiple sclerosis: evidence for an acquired channelopathy. *Prog Brain Res* **148**, 353–65.

Webster, H.D. (1971). The geometry of peripheral myelin sheaths during their formation and growth in rat sciatic nerves. *J Cell Biol* **48**, 348–67.

Webster, H.d.F., Favilla, J.T. (1984). Development of peripheral nerve fibers. Dyck, P.J., Thomas, P.K., Lambert, E.H., Bunge, R.P. (Eds.) *Peripheral Neuropathy*, 2nd Edn. WB Saunders, Philadelphia, pp. 329–59.

Webster, H. (1993). Development of peripheral nerve fibers. Dyck, P.J., Thomas, P.K., Lambert, E.H., Bunge, R.P. (Eds.) *Peripheral Neuropathy*, 3rd Edn. WB Saunders, Philadelphia, pp. 243–66.

Wegner, M. (2000a). Transcriptional control in myelinating glia: the basic recipe. *Glia* **29**, 118–23.

Wegner, M. (2000b). Transcriptional control in myelinating glia: flavors and spices. *Glia* **31**, 1–14.

Wehrle-Haller, B. and Chiquet, M. (1993). Dual function of tenascin: simultaneous promotion of neurite growth and inhibition of glial migration. *J Cell Sci* **106** (Part 2), 597–610.

Weiner, H.L., Friedman, A., Miller, A. *et al.* (1994). Oral tolerance: immunologic mechanisms and treatment of animal and human organ-specific autoimmune diseases by oral administration of autoantigens. *Ann Rev Immunol* **12**, 809–37.

Weiner, J.A. and Chun, J. (1999). Schwann cell survival mediated by the signaling phospholipid lysophosphatidic acid. *Proc Nat Acad Sci USA* **96**, 5233–8.

Weiss, M.D., Luciano, C.A., Semino-Mora, C., Dalakas, M.C. and Quarles, R.H. (1998). Molecular mimicry in chronic inflammatory demyelinating polyneuropathy and melanoma. *Neurology* **51**, 1738–41.

Wekerle, H., Schwab, M., Linington, C. and Meyermann, R. (1986). Antigen presentation in the peripheral nervous system: Schwann cells present endogenous myelin autoantigens to lymphocytes. *Eur J Immunol* **16**, 1551–7.

Wernig, A. and Herrera, A.A. (1986). Sprouting and remodelling at the nerve–muscle junction. *Prog Neurobiol* **27**, 251–91.

Wernig, A., Pecot-Dechavassine, M. and Stover, H. (1980). Sprouting and regression of the nerve at the frog neuromuscular junction in normal conditions and after prolonged paralysis with curare. *J Neurocytol* **9**, 278–303.

Wetmore, C. and Olson, L. (1995). Neuronal and nonneuronal expression of neurotrophins and their receptors in sensory and sympathetic ganglia suggest new intercellular trophic interactions. *J Comp Neurol* **353**, 143–59.

White, P.M., Morrison, S.J., Orimoto, K., Kubu, C.J., Verdi, J.M. and Anderson, D.J. (2001). Neural crest stem cells undergo cell-intrinsic developmental changes in sensitivity to instructive differentiation signals. *Neuron* **29**, 57–71.

Wigston, D.J. (1989). Remodeling of neuromuscular junctions in adult mouse soleus. *J Neurosci* **9**, 639–47.

Wilkinson, R., Leaver, C., Simmons, A. and Pereira, R.A. (1999). Restricted replication of herpes simplex virus in satellite glial cell cultures clonally derived from adult mice. *J Neurovirol* **5**, 384–91.

Williams, L.L., Kissel, J.T., Shannon, B.T., Wright, F.S. and Mendell, J.R. (1992). Expression of Schwann cell and peripheral T-cell activation epitopes in hereditary motor and sensory neuropathy. *J Neuroimmunol* **36**, 147–55.

Willison, H.J. (2005). The immunobiology of Guillain–Barré syndromes. *J Peripher Nerv Syst* **10**, 94–112.

Willison, H.J. and Yuki, N. (2002). Peripheral neuropathies and anti-glycolipid antibodies. *Brain* **125**, 2591–625.

Winer, J.B., Hughes, R.A. and Osmond, C. (1988). A prospective study of acute idiopathic neuropathy. I. Clinical features and their prognostic value. *J Neurol Neurosurg Psychiat* **51**, 605–12.

Winseck, A.K., Caldero, J., Ciutat, D. et al. (2002). In vivo analysis of Schwann cell programmed cell death in the embryonic chick: regulation by axons and glial growth factor. *J Neurosci* **22**, 4509–21.

Wohlleben, G., Hartung, H.P. and Gold, R. (1999). Humoral and cellular immune functions of cytokine-treated Schwann cells. *Adv Exp Med Biol* **468**, 151–6.

Wohlleben, G., Ibrahim, S.M., Schmidt, J., Toyka, K.V., Hartung, H.P. and Gold, R. (2000). Regulation of Fas and FasL expression on rat Schwann cells. *Glia* **30**, 373–81.

Woldeyesus, M.T., Britsch, S., Riethmacher, D. et al. (1999). Peripheral nervous system defects in erbB2 mutants following genetic rescue of heart development. *Genes Dev* **13**, 2538–48.

Wolpowitz, D., Mason, T.B., Dietrich, P., Mendelsohn, M., Talmage, D.A. and Role, L.W. (2000). Cysteine-rich domain isoforms of the neuregulin-1 gene are required for maintenance of peripheral synapses. *Neuron* **25**, 79–91.

Woodhoo, A., Dean, C.H., Droggiti, A., Mirsky, R. and Jessen, K.R. (2004). The trunk neural crest and its early glial derivatives: a study of survival responses, developmental schedules and autocrine mechanisms. *Mol Cell Neurosci* **25**, 30–41.

Woolf, C.J., Reynolds, M.L., Chong, M.S., Emson, P., Irwin, N. and Benowitz, L.I. (1992). Denervation of the motor endplate results in the rapid expression by terminal Schwann cells of the growth-associated protein GAP-43. *J Neurosci* **12**, 3999–4010.

Wrabetz, L., Feltri, M.L., Quattrini, A. et al. (2000). P(0) glycoprotein overexpression causes congenital hypomyelination of peripheral nerves. *J Cell Biol* **148**, 1021–34.

Wrabetz, L., Feltri, M. L., Kleopa, K. A. and Scherer, S. S. (2004). Inherited neuropathies – clinical, genetic, and biological features. Lazzarini, R. L. (Ed.) *Myelin Biology and Disorders.* Elsevier, San Diego, pp. 905–951.

Xiao, Z. C., Revest, J. M., Laeng, P., Rougon, G., Schachner, M. and Montag, D. (1998). Defasciculation of neurites is mediated by tenascin-R and its neuronal receptor F3/11. *J Neurosci Res* **52**, 390–404.

Xiao, Z. C., Ragsdale, D. S., Malhotra, J. D. et al. (1999). Tenascin-R is a functional modulator of sodium channel beta subunits. *J Biol Chem* **274**, 26511–17.

Xu, H., Wu, X. R., Wewer, U. M. and Engvall, E. (1994). Murine muscular dystrophy caused by a mutation in the laminin alpha 2 (Lama2) gene. *Nat Genet* **8**, 297–302.

Xu, W., Manichella, D., Jiang, H. et al. (2000). Absence of P0 leads to the dysregulation of myelin gene expression and myelin morphogenesis. *J Neurosci Res* **60**, 714–24.

Xu, W., Shy, M., Kamholz, J., Elferink, L., Xu, G., Lilien, J. and Balsamo, J. (2001). Mutations in the cytoplasmic domain of P0 reveal a role for PKC- mediated phosphorylation in adhesion and myelination. *J Cell Biol* **155**, 439–46.

Yamada, H., Shimizu, T., Tanaka, T., Campbell, K. P. and Matsumura, K. (1994). Dystroglycan is a binding protein of laminin and merosin in peripheral nerve. *FEBS Lett* **352**, 49–53.

Yamada, H., Chiba, A., Endo, T. et al. (1996a). Characterization of dystroglycan-laminin interaction in peripheral nerve. *J Neurochem* **66**, 1518–24.

Yamada, H., Denzer, A. J., Hori, H. et al. (1996b). Dystroglycan is a dual receptor for agrin and laminin-2 in Schwann cell membrane. *J Biol Chem* **271**, 23418–23.

Yamamoto, M., Fan, L., Wakayama, T., Amano, O. and Iseki, S. (2001). Constitutive expression of the 27-kDa heat-shock protein in neurons and satellite cells in the peripheral nervous system of the rat. *Anat Rec* **262**, 213–20.

Yamashita, N., Sakai, K., Furuya, S. and Watanabe, M. (2003). Selective expression of L-serine synthetic enzyme 3PGDH in Schwann cells, perineuronal glia, and endoneurial fibroblasts along rat sciatic nerves and its upregulation after crush injury. *Arch Histol Cytol* **66**, 429–36.

Yamauchi, J., Chan, J. R. and Shooter, E. M. (2004). Neurotrophins regulate Schwann cell migration by activating divergent signaling pathways dependent on Rho GTPases. *Proc Nat Acad Sci USA* **101**, 8774–9.

Yan, W. X., Taylor, J., Ndrias-Kauba, S. and Pollard, J. D. (2000). Passive transfer of demyelination by serum or IgG from chronic inflammatory demyelinating polyneuropathy patients. *Ann Neurol* **47**, 765–75.

Yan, W. X., Archelos, J. J., Hartung, H. P. and Pollard, J. D. (2001). P0 protein is a target antigen in chronic inflammatory demyelinating polyradiculoneuropathy. *Ann Neurol* **50**, 286–92.

Yanase, H., Shimizu, H., Yamada, K. and Iwanaga, T. (2002). Cellular localization of the diazepam binding inhibitor in glial cells with special reference to its coexistence with brain-type fatty acid binding protein. *Arch Histol Cytol* **65**, 27–36.

Yang, D., Bierman, J., Tarumi, Y. S. et al. (2005). Coordinate control of axon defasciculation and myelination by laminin-2 and -8. *J Cell Biol.* **168**, 655–66.

Yang, F. C., Ingram, D. A., Chen, S. et al. (2003). Neurofibromin-deficient Schwann cells secrete a potent migratory stimulus for Nf1+/- mast cells. *J Clin Invest* **112**, 1851–61.

Yang, H., Xiao, Z. C., Becker, B., Hillenbrand, R., Rougon, G. and Schachner, M. (1999). Role for myelin-associated glycoprotein as a functional tenascin-R receptor. *J Neurosci Res* **55**, 687–701.

Yang, J. F., Cao, G., Koirala, S., Reddy, L. V. and Ko, C. P. (2001). Schwann cells express active agrin and enhance aggregation of acetylcholine receptors on muscle fibers. *J Neurosci* **21**, 9572–84.

Yang, J. T., Rayburn, H. and Hynes, R. O. (1993). Embryonic mesodermal defects in alpha 5 integrin-deficient mice. *Development* **119**, 1093–105.

Yang, Y., LacasGervais, S., Morest, D. K., Solimena, M. and Rasband, M. N. (2004). beta IV spectrins are essential for membrane stability and the molecular organization of nodes of Ranvier. *J Neurosci* **24**, 7230–40.

Yntema, C. L. (1943). Deficient efferent innervation of the extremities following removal of neural crest in Amblystoma. *J Exp Zool* **94**, 319–49.

Yokoi, H., Tsuruo, Y. and Ishimura, K. (1998). Steroid 5alpha-reductase type 1 immunolocalized in the rat peripheral nervous system and paraganglia. *Histochem J* **30**, 731–9.

Yoshihara, T., Kanda, F., Yamamoto, M. *et al.* (2001). A novel missense mutation in the early growth response 2 gene associated with late-onset Charcot–Marie–Tooth disease type 1. *J Neurol Sci* **184**, 149–53.

Young, J. Z. (1938). The functioning of the giant nerve fibres of the squid. *J Exp Biol* **15**, 170–85.

Young, P., Boussadia, O., Berger, P. *et al.* (2002). E-cadherin is required for the correct formation of autotypic adherens junctions of the outer mesaxon but not for the integrity of myelinated fibers of peripheral nerves. *Mol Cell Neurosci* **21**, 341–51.

Young, P., Nie, J., Wang, X., McGlade, C. J., Rich, M. M. and Feng, G. (2005). LNX1 is a perisynaptic Schwann cell specific E3 ubiquitin ligase that interacts with ErbB2. *Mol Cell Neurosci* **30**, 238–48.

Yu, L. T., Rostami, A., Silvers, W. K., Larossa, D. and Hickey, W. F. (1990). Expression of major histocompatibility complex antigens on inflammatory peripheral nerve lesions. *J Neuroimmunol* **30**, 121–8.

Yu, W. M., Feltri, M. L., Wrabetz, L., Strickland, S. and Chen, Z. L. (2005). Schwann cell-specific ablation of laminin gamma1 causes apoptosis and prevents proliferation. *J Neurosci* **25**, 4463–72.

Yuki, N., Taki, T., Takahashi, M. *et al.* (1994). Molecular mimicry between GQ1b ganglioside and lipopolysaccharides of *Campylobacter jejuni* isolated from patients with Fisher's syndrome. *Ann Neurol* **36**, 791–3.

Yuki, N., Tagawa, Y. and Handa, S. (1996). Autoantibodies to peripheral nerve glycosphingolipids SPG, SLPG, and SGPG in Guillain–Barré syndrome and chronic inflammatory demyelinating polyneuropathy. *J Neuroimmunol* **70**, 1–6.

Yuki, N., Yamada, M., Koga, M. *et al.* (2001). Animal model of axonal Guillain–Barré syndrome induced by sensitization with GM1 ganglioside. *Ann Neurol* **49**, 712–20.

Yuki, N., Susuki, K., Koga, M. *et al.* (2004). Carbohydrate mimicry between human ganglioside GM1 and *Campylobacter jejuni* lipooligosaccharide causes Guillain–Barré syndrome. *Proc Nat Acad Sci USA* **101**, 11404–9.

Yurchenco, P. D., Cheng, Y. S. and Colognato, H. (1992). Laminin forms an independent network in basement membranes. *J Cell Biol* **117**, 1119–33.

Zehntner, S. P., Brisebois, M., Tran, E., Owens, T. and Fournier, S. (2003). Constitutive expression of a costimulatory ligand on antigen-presenting cells in the nervous system drives demyelinating disease. *FASEB J* **17**, 1910–12.

Zhou, D. X., Lambert, S., Malen, P. L., Carpenter, S., Boland, L. M. and Bennett, V. (1998). Ankyrin$_G$ is required for clustering of voltage-gated Na channels at axon initial segments and for normal action potential firing. *J Cell Biol* **143**, 1295–304.

Ziskind-Conhaim, L. (1988). Physiological and morphological changes in developing peripheral nerves of rat embryos. *Brain Res* **470**, 15–28.

Zlotnik, A. and Yoshie, O. (2000). Chemokines: a new classification system and their role in immunity. *Immunity* **12**, 121–7.

Zorick, T. S. and Lemke, G. (1996). Schwann cell differentiation. *Curr Opin Cell Biol* **8**, 870–6.

Zorick, T. S., Syroid, D. E., Arroyo, E., Scherer, S. S. and Lemke, G. (1996). The transcription factors SCIP and Krox-20 mark distinct stages and cell fates in Schwann cell differentiation. *Mol Cell Neurosci* **8**, 129–45.

Zou, L. P., Pelidou, S. H., Abbas, N. *et al.* (1999). Dynamics of production of MIP-1alpha, MCP-1 and MIP-2 and potential role of neutralization of these chemokines in the regulation of immune responses during experimental autoimmune neuritis in Lewis rats. *J Neuroimmunol* **98**, 168–75.

Zuo, Y., Lubischer, J. L., Kang, H. *et al.* (2004). Fluorescent proteins expressed in mouse transgenic lines mark subsets of glia, neurons, macrophages, and dendritic cells for vital examination. *J Neurosci* **24**, 10999–1009.

Index

AChRs 73, 74, 81, 83, 85, 87, 98
ADAM proteases 64
adherens 6, 28, 37, 42
adherens
    catenins 6
    junctions 40, 42
adherins
    E-cadherin 6, 40, 42
ankyrins 46
antigen presentation 10, 110, 121, 122, 180
antigen presenting cells 10, 121, 122, 180
antigen recognition 120, 125
AP2a 20, 23
AP2α 18, 30
autoimmune neuropathies 119
axonal fasciculation 57, 58

BAPTA 88
BFABP 15, 20, 23, 25
blood–nerve barrier 104, 116, 118, 164, 178
BMP2 21, 23, 26, 32

cadherin 19 20
Cajal-Bands 70
cAMP 19, 35, 85, 105, 106, 114, 115, 117
Caspr 37, 49, 50, 52, 53, 129, 137, 145
Caspr2 37, 49, 50, 52
CD4 T cells 103, 110, 121, 122, 167, 180
CD4+ T-cells 103
CD8 T cells 110, 111, 121, 178, 180
cell-mediated immunity 104
Charcot–Marie–Tooth
    CMT disease 11, 12
chemokine receptors 101, 102
chemokines 10, 101, 116, 117, 121, 179

chronic inflammatory demyelinating polyneuropathy 12, 102
chronic inflammatory demyelinating polyradiculoneuropathy 122, 159
Claudin 42
CNS
    astrocytes 9, 14, 15, 31, 72, 73, 84, 86, 99, 100, 106, 107, 150
    oligodendrocytes 9, 14, 44, 59, 113, 149, 150
collagen IV 56
collagen V 57
congenital muscular dystrophy 61
co-stimulatory molecules
    BB-1 122, 180
Cre/LoxP 69, 70
Cx29 37, 40, 44, 50, 51
Cx32 40, 50, 130, 145–9
CXCR4 117
cytokine receptor
    TNF-R 111
cytokines 9, 10, 100–5, 107, 109–15, 117, 118, 121, 123–5, 160, 179
    IFN-γ 110–12, 114, 116, 122
    IL-1 104, 105, 109, 111, 112, 114, 123
    IL-1β 111
    IL-6 104, 112, 114, 123
    lymphotoxin 101, 113
    TGF 85, 86, 104, 106, 108, 109, 112, 113, 123, 125
    TGF-β 111
    TGF-β1 111, 114
    TNF 103–6, 110, 113, 117, 121, 123–5, 179
    TNF-α 111, 112, 114–16
    tumor necrosis factor 101, 121, 125, 179

demyelination 12, 41, 46, 53, 54, 103–5, 107, 116, 119, 120, 131, 135, 140, 142, 144, 145, 153, 160, 162, 163, 169–75, 177, 180, 181
dendritic cells 110, 120, 121, 180
desert hedgehog 31
Duchenne muscular dystrophy 94
dystroglycan 37, 48, 55, 58, 63–6, 69, 70, 152
dystrophic mice 69, 70

EAN 103, 104, 116, 119, 160, 169, 172, 178–81
endoneurial fibroblasts 14, 29–31, 55
endostatin 56
endothelins 30
ErbB 84, 85, 95
ErbB2 26, 28, 31, 33
extracellular matrix 9, 22, 23, 29, 48, 55–8, 64, 71, 73, 76, 89, 93, 116, 152
    basal lamina 4, 21, 30, 34, 39, 41, 58–63, 65, 68, 76, 93, 94, 97, 152, 160, 175, 179
    fibrin 59, 60
    metalloproteinases 104, 179
    nidogen 63

F-actin 6, 75
FGF 23, 34, 57
fibronectin 59, 60
functional genomics 6

galactocerebroside 21, 39, 42, 49, 79, 103, 104, 119, 163, 169, 181, 182
galactoplipids 106, 114
gangliosides 38, 52, 53, 76, 91, 120, 155, 167, 168, 172, 181
gap junctions 37, 44, 145, 147–9
GFAP 20, 23, 26, 79, 91, 95, 105
gliogenesis 13, 23, 25, 26
glycoproteins
    tenascins 58
glypican 57
Guillain–Barré Syndrome 12

heparan sulphate proteoglycans 48

ICAM-1 111, 179
IGF2 30
IL-1R 112
inner mesaxon 47, 51, 52, 130, 137
integrins
    α5β1 integrin 60
    LFA-1α 111
    LFA-1β 111
integrin linked kinase 69
internodal region 51
intracellular calcium 87
intraperiod lines 39

JNK 34
juxtaparanode 49, 129

Krox-20 19, 21, 35, 38

L1 21, 23, 111, 115
laminin receptors 39, 56, 61, 63–6, 68, 69, 71
laminin α4-null mice 68
LIF 30
lipopolysaccharide 120, 125
Louis-Antoine Ranvier 76
LPS 120, 121, 125

macrophage chemoattractant protein 116
macrophage inflammatory protein-1 alpha 116
macrophages 103, 119, 144, 180
MAG 38, 40, 51, 52, 59, 66, 79, 111, 129, 181, 183
MCP-1 116
membrane attack complexes 91, 163
metalloprotease 56
MHC class I molecules 110, 111, 121
MHC class II molecules 4, 110, 121, 125, 180
microvilli 37, 48, 58, 64, 66, 68, 70
Miller–Fisher syndrome 76, 91
miniature endplate potentials 78
MIP-1α 116
molecular mimicry 119, 120, 168
myelin lamellae 3, 24, 28, 37–9, 41, 43, 44, 48–50, 53, 66, 70, 119, 125, 129, 137, 144, 147, 153, 160, 161, 163, 170, 174
myelin-related proteins
    MBP 21, 37, 40, 42, 66, 105, 122, 129, 137, 138
    P0 15, 20, 21, 23, 26, 32, 37, 39–42, 79, 103, 105, 111, 119, 144, 169, 170, 172, 180, 181, 183
    P2 103, 119, 122, 169, 172, 180
    PLP 21, 23
    PMP22 21, 23, 37, 39–42, 127–9, 131–6, 138, 140, 147, 148, 151, 169, 180
myelin-forming Schwann cells 5
myelination
    radial sorting 19, 33, 68

Na_v 37, 38, 44, 46, 48, 49, 54
Na_v channels 45–9
N-cadherin 111
N-CAM 21, 57
neural crest cells 13, 15, 17, 20–2, 25–9, 108
neural crest stem cells 14, 15
neuregulin-1 21, 38
neuroglycan C 57

Index 249

neuromuscular junctions 50, 73–5, 77–84, 86, 88–93, 95, 97–9
neuroproteomics 6
neurotrophic factors 9
NFkB 19, 35
neurotrophic factors
    BDNF 9, 11, 19, 33, 36, 85
    NGF 10, 11, 34, 95, 109, 114
    NT-3 9
NGFRp75 105, 114, 115
nodes of Ranvier 44, 55, 58, 59, 61, 63, 66, 68, 70, 129, 144, 163, 167
non-myelin forming Schwann cells 7, 68
Notch 18, 23, 25, 30, 35
Nr-CAM 37, 38, 47, 48
NRG1 18, 21, 23, 25–8, 31, 33–6, 82, 85, 95
NT3 19, 23, 30, 33, 35, 85
NTR 18, 21, 34

p38MAP kinases 33
paranodal loops 48, 130, 147, 149, 153, 182
Pax-3 19, 35
PDGF-BB 30
perisynaptic Schwann cells 4, 8, 14, 72, 74, 79, 168
PINCH 65, 69
presentation of antigens 110, 119
    See also antigen presentation
presynaptic nerve terminal 8, 72, 75
proteoglycans 57
PSC processess 81, 94

radial sorting 33, 61, 68, 128
RANTES 116
Remak bundles 3, 7
remyelination 46, 60, 114, 116, 144, 174
repressor proteins 36

S100 20, 23, 24, 75, 79, 91
saltatory conduction 37, 38, 44, 129, 132

satellite cells 4, 7
Schmidt–Lanterman incisures 6, 149
Schwann cell differentiation 9, 55, 105, 108, 114, 130
Schwann cell microvilli 45, 70
Schwann cell migration 19, 33, 57–60, 106
Schwann cell precursors 13–15, 17, 18, 20–2, 26–32, 66, 79, 85, 105, 108, 128
Schwann cells junctions 88
Schwannopathy 104
septate-like junctions 49
small myelin protein 32
sodium channels 59, 70, 91, 137
Sox-10 18, 20, 23, 25, 31, 35
splotch 80
β1integrin 33
stem cell factor 103
stromal cell derived factor-1 117
sulphatide 21, 39, 42, 49
synapse
    tripartite synapse 8, 73
synaptic growth 81–5, 99
synaptic remodeling 78, 93
synaptogenesis 81, 82, 84–6, 99
syndecan 57, 58

TAG-1 37, 50, 52, 57
TGFβ 34
Theodor Schwann 1, 2, 76
Toll-like receptors 120, 125
    TLR-2 120
    TLR-4 120
Trembler J mice 46, 53
tripartite synapse 72

unmyelinated fibres 6, 177

vitronectin 59
voltage-gated potassium channels 46

Wallerian degeneration 59, 89, 104, 107, 109, 164